# Topics in Applied Physics  Volume 47

# Topics in Applied Physics Founded by Helmut K. V. Lotsch

# Sputtering by Particle Bombardment I

Physical Sputtering of Single-Element Solids

Edited by R. Behrisch

With Contributions by
H. H. Andersen    H. L. Bay    R. Behrisch
M. T. Robinson    H. E. Roosendaal    P. Sigmund

With 117 Figures

Springer-Verlag Berlin Heidelberg GmbH 1981

Dr. *Rainer Behrisch*

Max-Planck-Institut für Plasmaphysik, Euratom Association
D-8046 Garching/München, Fed. Rep. of Germany

ISBN 978-3-662-30888-2          ISBN 978-3-540-38514-1 (eBook)
DOI 10.1007/978-3-540-38514-1

Library of Congress Cataloging in Publication Data. Main entry under title: Sputtering by particle bombardment. (Topics in applied physics; v. 47, ) Bibliography: v. 1, p. Includes index. Contents: 1. Physics and applications. 1. Sputtering (Physics) 2. Solids–Effect of radiation on. 3. Surfaces (Physics)–Effect of radiation on. I. Behrisch, Rainer. II. Andersen, Hans Henrik. III. Series. IV. Title: Particle bombardment. QC176.8.S72S68 530,4'1   81-4313   AACR2

Originally published by Springer-Verlag Berlin Heidelberg New York in 1981
Softcover reprint of the hardcover 1st edition 1981

2153/3130-543210

# Preface

Sputtering phenomena have become of great importance in physics and technology within the last 25 years. This is demonstrated by the inclusion of sputtering in a large number of national and international conferences dealing with topics such as vacuum physics, surface physics, surface analysis, thin films, electron microscopy, atomic collisions, radiation damage, ion implantation and plasma physics. However, there have been very few conferences dealing with sputtering alone and today's knowledge about this process is widely distributed in the scientific and technical literature.

In the three volumes of this Topics in Applied Physics series, an attempt has been made to collect most of today's information about the experimental and theoretical knowledge on sputtering phenomena as well as to show the applications of this process. This task was not possible in a monograph, but only by the contribution of several experts in this field. Every contribution represents the personal view of each author, but an effort was made to get the articles to fit together in a coherent way. Mostly the symbols are used and cross references between the articles are given.

This first volume deals with the physical basis for sputtering of single element solids. After a general overview, two chapters deal with the theoretical basis for understanding sputtering phenomena in amorphous, polycrystalline and single crystal solids followed by two chapters presenting a collection of the experimental results.

Chapter 2 by P. Sigmund starts with a historical survey about the different models developed for the sputtering process. For knockon sputtering caused by a collision cascade, first-order analytical formulas are derived for the sputtering yield in amorphous materials as well as for the angular and energy distributions of the emitted particles. These are valid for the linear cascade regime, i.e., heavy ions at keV energies. Further, corrections for other regimes are presented.

In Chap. 3, Mark T. Robinson discusses the influence of the crystalline structure of the solid as well as the selvage, i.e., the surface layer, and the surface binding energies on the sputtering process. The ideas of crystal transparency and channeling give a basis for understanding the orientation dependence of the sputtering yields. However, for a comprehensive description of the sputtering process, computer models have to be applied and several examples are given.

In chapter 4, H. H. Andersen and H. L. Bay present an overview of all results reported in the literature about measured sputtering yields and their dependence on different parameters. This contribution shows that there are very few measurements of sputtering yields for the systems other than metals bombarded by particles other than noble gas ions.

The last chapter by H. E. Roosendaal deals with sputtering yield measurements for single crystals, especially their dependence on orientation. These effects can only be observed at low fluences or, if the annealing of defects in the crystal is larger than amorphization due to the radiation damage within the range of the incident ions.

In each chapter the references are numbered consecutively as they appear in the text. In order to facilitate search for the work of one author, an alphabetical author index has been provided and is added, together with the subject index at the end of the book.

The second volume will deal with the sputtering of multicomponent targets such as alloys and compounds, chemical sputtering and sputtering by electrons and neutrons. Two chapters will deal with the surface structures which are developed for heavy and light ion bombardment.

In the third volume, today's knowledge about the angular, energy, mass and charge-state distribution of sputtered particles will be presented. Finally the large variety of applications for the sputtering process will be outlined.

It is a great pleasure to thank the Springer Verlag and all authors for the pleasant collaboration and especially Peter Sigmund and Mark T. Robinson for their encouragement and advice in starting this book.


Garching, March 1981                                    *Rainer Behrisch*

# Contents

# Contributors

Hans Henrik Andersen
  Det Fysiske Institut, Aarhus Universitet
  DK-8000 Aarhus C, Denmark

Helge L. Bay
  Institut für Plasmaphysik, Kernforschungsanlage Jülich,
  Euratom Association, D-5170 Jülich, Fed. Rep. of Germany

Rainer Behrisch
  Max-Planck-Institut für Plasmaphysik, Euratom Association
  D-8046 Garching/München, Fed. Rep. of Germany

Mark T. Robinson
  Solid State Division, Oak Ridge National Laboratory
  Oak Ridge, TN 37830, USA

Hans E. Roosendaal
  Universität Bielefeld, Fakultät für Physik, Universitätsstraße 25
  D-4800 Bielefeld 1, Fed. Rep. of Germany

Peter Sigmund
  Fysiske Institut, Odense Universitet
  DK-5230 Odense M, Denmark

# 1. Introduction and Overview

Rainer Behrisch

Sputtering, i.e., the removal of surface atoms due to energetic particle bombardment, is caused by collisions between the incoming particles and the atoms in the selvage, i.e., the near surface layers of a solid. First discovered more than 125 years ago, it took about 50 years until the physical process involved in sputtering was recognised and 100 years until a quantitative description began to be developed. Sputtering is measured by the sputtering yield $Y$ giving the mean number of atoms removed per incident particle. For describing the energy and angular distribution as well as the state of the emitted particles, differential yields are defined. Most sputtering investigations have been performed up to now on monoatomic targets like metals bombarded with noble gas ions. Sputtering measurements on alloys and compounds as well as sputtering by ions which may react chemically with the atoms of the solid have only recently begun. For crystalline targets, the sputtering yields are influenced by the lattice structure especially for particle incidence and atom emergence in close packed crystal directions.

Today sputtering is no longer just an unwanted effect which destroys cathodes and contaminates a plasma. Sputtering is widely applied to surface cleaning and etching, for thin film deposition, for surface and surface layer analysis and for sputter ion sources.

## 1.1 Overview

If a surface is subjected to irradiation by energetic particles it is eroded and surface atoms are removed. This phenomenon is named "Sputtering" in English, "Zerstäubung" in German, "Raspilenie" (Распыление) in Russian and "Pulverisation" in French. Besides sputtering, several other effects are observed with particle bombardment of surfaces. These are in general, backscattering [1.1] as well as trapping and reemission of incident particles [1.2], desorption of surface layers [1.3], the emission of electrons [1.4], the emission of photons [Ref. 1.5, Chap. 3] and a change in surface structure and topography [Ref. 1.6, Chaps. 6 and 7]. In the three volumes of this series the emphasis lies on the description of the sputtering phenomenon. Other bombardment induced effects are included only when they are related to sputtering. Photon emission is treated in connection with the state of sputtered species [Ref. 1.5, Chap. 3], while the change of surface structure and topography is discussed in some detail because of its influence on the sputtering process [Ref. 1.6, Chaps. 6 and 7].

Sputtering can be observed in nature as well as in laboratory experiments. It occurs if matter in two extreme states, such as a hot plasma and a solid, interact with each other, or if a directed beam of energetic particles hits a surface. Such situations are found, for example, at the surface of the moon and other celestrial bodies due to the impact of plasma particles such as the solar wind [1.7], or on earth where energetic particles are produced by radioactive decay [1.8]. In laboratory experiments it occurs, for example, at the cathode in electric gas discharges and is caused by the current of energetic ions produced in the cathode fall [1.9–11]. Sputtering is also observed in the ion source of accelerators as well as at all diaphragms and targets [1.12].

The investigation of electric gas discharges was the place where sputtering was first discovered more than 125 years ago [1.9–11]. Cathode material was observed to become deposited on the surrounding glass walls [1.9], thus, the name "cathode sputtering" can still be found in the literature. It took about 50 years until the physical process causing sputtering became recognised [1.13, 14] and only within the last 25 years has a more clear quantitative description been developed ([1.15], Chaps. 2 and 3). This was possible due to the collection of a large amount of experimental data performed under continuously improved conditions, i.e., with mass analysed ion beams in high vacuums at well characterised materials ([1.16–18], Chaps. 4 and 5). The effort to understand the sputtering process is also stimulated by a broad application in surface analysis, surface etching and thin film deposition [1.5].

## 1.2 The Sputtering Yield

The erosion in sputtering is measured by the sputtering yield $Y$ defined as the mean number of atoms removed from the surface of a solid per incident particle:

$$Y = \frac{\text{atoms removed}}{\text{incident particle}}.$$

The incident particles may be ions, neutral atoms, neutrons, electrons or energetic photons. For bombardment with monoatomic molecular ions, each incident atom is generally counted separately. In sputtering experiments with other molecules it may be appropriate to define the yield per incident molecule. In counting the atoms removed, only those from the solid are included, while incident particles which become reflected or reemitted are not taken into account. This is different for selfsputtering, i.e., for bombardment with the same ions as the solid where this distinction cannot be made. A selfsputtering yield "one" means that on average, one atom is removed or reflected per incident atom.

The definition of the sputtering yield is meaningful only if the number of removed atoms is proportional to the number of incident particles, which has indeed been found in most cases. Deviations are observed for large incident particles, such as molecular ions, at energies $>10\,\text{keV}$ where spike effects (Chaps. 2 and 4) can cause a dependence of the yield on the number of atoms per molecule.

Sputtering has been investigated in most detail for bombardment of monoatomic solids with Hg ions, with noble gas ions and with H ions (Chap. 4). Except for the light ions, sputtering yields are typically 1 to 5 but may lie between zero and about 100 atoms per particle. They depend on the particle energy, particle mass and angle of incidence, the mass of the target atoms, the crystallinity and the crystal orientation of the solid and on the surface binding energies of the target, but they are nearly independent of the temperature. Below a threshold energy which is about 20 to 40 eV for normal incidence, no sputtering takes place. Above this threshold, the yields increase with incident energy and reach a broad maximum in the energy region of 5 to 50 keV. The decrease of the sputtering yield at higher energies is related to the larger penetration of the ions into the solid and a lower energy deposition in the surface layers. Higher mass particles give mostly larger sputtering yields than lighter ions (Chap. 4).

For ion bombardment at an oblique angle of incidence on amorphous solids and on polycrystalline materials with many randomly oriented crystallites in the bombarded area, the sputtering yield increases monotonically with increasing angle of incidence up to a maximum near 70 to 80 deg. The location of the maximum depends on the bombarding particle energy and mass and the surface topography (Chap. 4).

For single crystals the sputtering yield depends additionally on the direction of particle incidence relative to the crystal orientation. For bombardment close to the three major close-packed crystal axes, the yields are a factor of about 2 to 5 lower than for other directions of incidence. These minima are superimposed on the steady increase in the sputtering yield with increasing angle of incidence. Furthermore, for most of the close-packed directions, the maximum in the energy dependence of the sputtering yield is shifted towards lower energies (Chap. 5).

Sputtering of alloys and compounds has only recently been investigated in some detail [Ref. 1.6, Chap. 2]. Generally, one component is initially removed at a larger rate so that the surface layer is enriched in the other component. At low temperatures where diffusion is suppressed, a steady state condition is reached where the atoms are removed stoichiometrically. However, at high enough temperatures, a depletion up to large depths in the solid is possible. These effects can influence the sputter deposition of thin films and the results of depth profiling measurements when using sputtering [Ref. 1.5, Chaps. 4, 8, and 9].

On bombardment with ions which do not react chemically with the atoms of the solid, a surface layer corresponding to the ion range will be damaged and some ions become trapped [Ref. 1.6, Chaps. 6 and 7]. For bombardment with reactive ions, a compound surface layer generally builds up which has a different composition and structure than the material started with [Ref. 1.6, Chap. 5]. Sputtering effects will then correspond to those of the compound formed.

If a volatile compound is formed on a surface (like $CH_4$ or $H_2O$ with hydrogen ion bombardment), sputtering yields are increased at temperatures where the molecules desorb at the surface. If a very stable solid compound is formed on the surface, for example, a carbide or an oxide for carbon or oxygen bombardment, the sputtering yields can be decreased. This is especially the case if the molecules are oriented so that the atoms of one element, such as O or C, represent the top surface layer. As these top atoms are predominantly removed, sputtering of bulk atoms can be drastically reduced ([1.18–20] and Chap. 4). Generally, the effects involved with bombardment by species other than inert gas ions at different target temperatures have not yet been studied in detail, but may become of great technical relevance ([Ref. 1.6, Chap. 6] and [1.20]).

Surface atom removal, trapping of the incident ions, and radiation damage in the surface layer generally cause a facetting of the surface [Ref. 1.6, Chap. 6]. For bombardment with nondiffusing gaseous ions, blistering may also occur at a fluence of $10^{17}$ to $10^{18}$ ions/cm$^2$, but will disappear at fluence about 10 times greater [Ref. 1.6, Chap. 7].

## 1.3 Distributions in Sputtered Particles

The particles removed from a solid surface by sputtering are emitted with a broad distribution in energy $E_1$ [Ref. 1.5, Chap. 2] at different excitation and charge states $q$ [Ref. 1.5, Chap. 3] in all exit angles $\Omega$ [Ref. 1.5, Chap. 2]. This is described by the differential sputtering yields

$$\frac{\partial Y}{\partial E_1}, \frac{\partial^2 Y}{\partial^2 \Omega_1} \quad \text{and} \quad Y_q$$

with $Y = \sum_q Y_q$. In specific experiments, for example Secondary Ion Mass Spectroscopy (SIMS) [Ref. 1.5, Chaps. 3 and 8], only atoms with one given energy and charge state emitted into a given angle may be measured, i.e.,

$$\frac{\partial^3 Y_q}{\partial E_1 \partial^2 \Omega_1} \, \Delta E_1 \Delta \Omega_1$$

is determined. This quantity is generally not directly proportional to the sputtering yield. The fraction of atoms emitted into different states as well as the energy and angular distributions of the emitted particles are not independent of each other and they can have different dependencies on all parameters other than the sputtering yield.

The energy distribution of the sputtered particles generally has a maximum between half and the full surface binding energy. At high emerging energies the number of sputtered particles mostly decreases proportional to $1/E_1^2$. Deviations are observed for an oblique angle of incidence and especially for the situation of spikes [Ref. 1.5, Chap. 2].

The angular distribution of particles sputtered from single crystals shows maxima in close-packed crystal directions [1.17]. For polycrystalline materials, the emission distribution is a superposition of the distributions of the differently oriented crystallites in the bombarded area. For normal incidence the angular distributions may mostly be described in a first approximation by a cosine distribution. For heavy ions and low bombarding energies close to the threshold more atoms are emitted at large angles, while for light ions and higher energies, more atoms leave the surface in the normal direction. For oblique ion incidence the maximum of the emission distribution is shifted away from the incoming ion beam [Ref. 1.5, Chap. 2]. In sputtering of compounds the different constituents may be sputtered with slightly different angular distributions ([Ref. 1.5, Chap. 2] and [Ref. 1.6, Chap. 2]).

The particles emitted are predominantly neutral atoms in the ground state and generally less than 5% are ions. A certain fraction can be emitted as atom clusters [Ref. 1.5, Chaps. 2 and 3].

The investigations of the differential sputtering yields are still very incomplete. However, these measurements, especially for atomically clean single-crystal surfaces, allow to obtain detailed information about the sputtering process and the influence of the surface topography on the collision cascades.

## 1.4 Physical Understanding

Sputtering at a surface is generally caused by a collision cascade in the selvage, i.e., the surface layers of a solid. The processes are in principle the same as those causing radiation damage in the bulk of a solid. They take place far from thermal equilibrium which means that sputtering is different from evaporation.

Generally, an incoming particle will collide with the atoms of the solid, thereby transferring energy to the atomic nuclei. If more energy is transferred than the binding energy at the lattice site a primary recoil atom is created. The energy transfer may take place in a binary collision between an incoming energetic particle and an atom of the solid at rest. It may also occur by a recoil atom produced in radioactive decay. In insulators, another energy transfer is possible, which is also effective for low energy electrons and photons. They can produce a localized excitation which, after decay, creates a repulsive potential strong enough to replace a lattice atom [Ref. 1.6, Chap. 4].

The primary recoil atoms will collide with other target atoms distributing the energy via a collision cascade. A surface atom becomes sputtered if the energy transferred to it has a component normal to the surface which is larger than the surface binding energy. This is generally approximated by the heat of sublimation which is mostly smaller than the displacement energy necessary to create a stable dislocation, i.e., radiation damage in the bulk of a solid (Chap. 3).

Radiation damage effects are mostly concerned with the irradiation of solids by neutrons, electrons or light ions at high energies ($E > 100$ keV). Here the cross sections for energy transfer to target atoms are small and the ranges of the particles in the solid are large. Collision cascades in the surface region are rare and sputtering yields are low [Ref. 1.6, Chap. 3]. However, sputtering becomes the dominant effect for bombardment with high mass particles at lower energies ($E < 100$ keV). Here the ranges are small and most of the energy is transferred to lattice atoms in a small surface layer of a few 100 Å.

In crystalline materials the probability for collisions to create primary recoil atoms, as well as the development of collision cascades, is influenced by the crystal structure due to channeling, blocking or shadowing and focusing effects ([1.14], Chaps. 3 and 5). A possible different structure in the selvage compared to the bulk of the solid, as well as different bindings, will further influence the collisions leading to sputtering (Chap. 3). In general, the decrease in sputtering yield for particle incidence in close-packed crystal directions can be explained by channeling effects, while the role of focusing for explaining a preferential emission in these directions is very doubtful. Here the details of the last collision between the subsurface layer and the surface can play a major role (Chap. 3).

## 1.5 Sputtering Calculation

For a quantitative description of the sputtering process, the parameters involved in the collisions between the atoms (i.e., the potentials or the differential cross sections), the structure and orientation of the solid, as well as the energy loss to electrons have to be known. Generally, a compromise between the best known values of these quantities and those analytically treatable has to be made ([1.15] and Chap. 2 and 3).

For describing the collision cascade in an amorphous solid at energies where the binary collision approximation is still valid, the Boltzmann transport equation can be used. First-order asymptotic solutions achieved for some parameter ranges then give analytical formulas. These are very useful because they show the dependence on the different parameters ([1.15] and Chap. 2). The absolute values are good reference data which may be improved by a more detailed analysis of the collision process and by including other effects like chemical processes. Polycrystalline solids with randomly oriented crystallites can be approximated to a first order by an amorphous solid. Thus the formulas derived for amorphous materials mostly give values in reasonable agreement with the sputtering yields measured for polycrystalline targets (Chap. 2 and 4).

In order to describe details of the collision cascades in amorphous and single crystals the cascades can be followed by a computer simulation. For an amorphous material this can be done by a Monte Carlo Program, while for a single crystal all lattice positions have to be stored in the computer. For energies above ~100 eV the binary collision approximation can be used, while for lower energies, a classical dynamical model should be applied where the trajectories of all atoms in a small crystallite of about 1000 atoms are followed (Chap. 3). In these computer models it is possible to include an ideal or damaged surface and also damage in the solid. Polycrystalline material can be simulated by a superposition of the results of many differently oriented single crystals. Though very promising results have already been achieved, these calculations are time consuming and expensive (Chap. 3). They are, however, especially useful for comparing the details of the calculated distribution of cascades with the assumptions made to achieve analytical solutions.

## 1.6 Applications of Sputtering

Sputtering has long been regarded just as an undesired dirt effect which destroys the cathodes and grids in gas discharge tubes or ion sources and contaminates a plasma and the surrounding walls.

Plasma contamination by metal atoms is still one major problem in fusion research. In this case an attempt is made to heat a hydrogen plasma to temperatures of $\approx 10\,\text{keV}$ at densities $\approx 10^{14}\,\text{cm}^{-3}$ so that in a mixture of D and T ions, fusion to He and neutrons by energy release will take place [Ref. 1.5, Chap. 10]. Sputtering is also one of the causes for the destruction of diaphragms and targets in accelerators and in high-voltage electron microscopes [Ref. 1.6, Chap. 4]. In ion implantation the removal of surface atoms by sputtering mainly limits the achievable concentration in the implantation range [1.21].

However, sputtering is used today for many applications and has become an indispensible process in modern technology. Both the removal of atoms from a surface and the flux of atoms leaving the surface are successfully applied. Sputtering allows a controlled removal even of very tightly bound surface layers on a nearly atomic scale and the possible submicron spacial resolution if a well focused and/or collimated ion beam is used [Ref. 1.5, Chap. 5].

It is applied, for example, in sputter ion sources [Ref. 1.5, Chap. 9], for obtaining atomically clean surfaces [Ref. 1.5, Chap. 4], in micromachining as well as for depth profiling of thin films [Ref. 1.5, Chap. 7]. The atoms removed can be analysed in a mass spectrometer and this gives information about the surface concentration, or a depth profile at continuous bombardment. With rastered primary ion beams or imaging of the sputtered ions, a high spatial resolution can be achieved [Ref. 1.5, Chap. 8]. One of the largest applications of sputtering is, however, the deposition of thin films on a large variety of substrates. These may have large areas of several $\text{m}^2$ or may be extremely small as in microelectronics [Ref. 1.6, Chap. 6].

Because the investigation of sputtering phenomena dates back more than 100 years, the field has been covered by a large number of review articles in many languages. Those which have appeared within the last 30 years [1.13, 22–36] give a vital demonstration of the large advances made in understanding the sputtering process and its applications.

# References

1.1   E.S.Mashkova, V.A.Molchanov. *Scattering of Ions of Medium Energy at Solid Surfaces* (Atomisdat, Moscow 1980); and Rad. Eff. **16**, 143 (1972); and **23**, 215 (1974)
1.2   G.M.McCracken: Rep. Prog. Phys. **38**, 241 (1975)
1.3   E.Taglauer, W.Heiland: J. Nucl. Mater. **76, 77**, 328 (1978)
1.4   K.H.Krebs: Fortschr. Phys. **16**, 419 (1968)
1.5   R.Behrisch (ed.): *Sputtering by Particle Bombardment* III, Topics in Applied Physics (Springer, Berlin, Heidelberg, New York 1982) to be published
1.6   R.Behrisch (ed.): *Sputtering by Particle Bombardment* II, Topics in Applied Physics (Springer, Berlin, Heidelberg, New York 1981) to be published
1.7   G.K. Wehner, C.E. Kenknight, D.L. Rosenberg: Planet. Space Sci. **11**, 885 (1963)
1.8   R.Sizmann: Phys. Verh. **12**, 78 (1961)
1.9   W.R.Grove: Philos. Mag. **5**, 203 (1853)
1.10  J.P.Gassiot: Philos. Trans. R. Soc. London, **148**, 1 (1858)
1.11  J.Plücker: Ann. Phys. Leipzig **103**, 88, 90 (1858)
1.12  E.Goldstein: Verh. Dtsch. Phys. Ges. **4**, 228, 237 (1902)
1.13  V.Kohlschütter: Jahrb. Radioakt. Elektron. **9**, 355 (1912)
1.14  J.Stark, G.Wendt: Ann. Phys. **38**, 921 und 941 (1912)
1.15  P.Sigmund: Phys. Rev. **184**, 383 (1969)
1.16  D.Rosenberg, G.K.Wehner: J. Appl. Phys. **33**, 1842 (1962)
1.17  P.K.Rol: PhD Thesis, Amsterdam (1960)
1.18  O. Almén, G.Bruce: Nucl. Instrum. Methods **11**, 275 and 279 (1961)
1.19  R.Behrisch, J.Roth, J.Bohdansky, A.P.Martinelli, B.Schweer, D.Rusbüldt, E.Hintz: J. Nucl. Mater. **93 & 94**, 645 (1980)
1.20  D.M.Gruen, S.Vepřek, R.B.Wright: "Plasma-Materials Interaction and Impurity Control in Magnetically Confined Thermonuclear Fusion Machines", in *Plasma Chemistry I*, ed. by S.Vepřek, M.Venugopalan, Topics in Current Chemistry, Vol. 89 (Springer, Berlin, Heidelberg, New York 1980) pp. 45–106
1.21  G.Dearnaley, J.H.Freeman, R.S.Nelson, J.Stephen: in *Ion Implantation* (North Holland, Amsterdam 1973)
1.22  A.Güntherschulze: J. Vac. Sci. Technol. **3**, 360 (1953)
1.23  G.K.Wehner: Adv. Electron. Electron Phys. **7**, 239 (1955)
1.24  R.Behrisch: Ergeb. Exakten Naturwiss. **35**, 295 (1964)
1.25  M. Kaminsky: *Atomic and Ionic Impact Phenomena on Metal Surfaces*, Struktur und Eigenschaften der Materie in Einzeldarstellungen (Springer, Berlin, Heidelberg, New York 1965)
1.26  G.Carter, J.S. Colligon: *Ion Bombardment of Solids* (Heinemann, London 1968)
1.27  R.S.Nelson: *The Observation of Atomic Collisions in Crystalline Solids* (North Holland, Amsterdam 1968)
1.28  N.V.Pleshivtsev: *Cathode Sputtering* (Atomisdat, Moscow 1968)
1.29  M.W.Thompson: *Defects and Radiation Damage in Metals* (Cambridge Univ. Press, 1969)
1.30  R.Behrisch, W.Heiland, W.Poschenrieder, P.Staib, H.Verbeek (eds.): *Ion Surface Interaction, Sputtering and Related Phenomena* (Gordon & Breach, London 1973)
1.31  H.Oechsner: Appl. Phys. **8**, 185 (1975)
1.32  A.W.Czanderna (ed.): *Methods in Surface Analysis*, Vol. 1 (Elsevier, Amsterdam 1975) (Contribution by G.K.Wehner)
1.33  *Physics of Ionized Gases*, Proc. Yugoslav Symp. (SPIG): In 7, ed. by V.Vujnovic (Zagreb 1974), H.H.Andersen, p. 361; in 8, ed. by B.Navinšek (Lubljana 1976), G.Carter, p. 281, H.Oechser, p. 461, V.E.Yurasova, p. 493; in 9, ed. by R.K.Janev (Beograd 1978), M.W.Thompson, p. 289, J.L.Whitten, p. 335
1.34  N.H.Tolk, J.C.Tully, W.Heiland, C.W.White (eds.): *Inelastic Ion Surface Collisions* (Academic, New York 1972)
1.35  A.Benninghoven, C.A. Evans,Jr., R. A. Powell, R.Shimizu, H. A.Storms: *Secondary Ion Mass Spectrometry SIMS II*, Springer Ser. Chem. Phys., Vol. 9 (Springer, Berlin, Heidelberg, New York 1979)
1.36  P.Varga, G.Betz, F.P.Viehböck (eds.): Proc. Symp. on Sputtering, Tech. Univ. Wien 1980

# 2. Sputtering by Ion Bombardment: Theoretical Concepts

Peter Sigmund

With 24 Figures

This chapter is an introduction to sputtering theory rather than a review. The first section contains a brief historical survey as well as an attempt to both define and classify sputtering processes. In the second section, some pertinent results from atomic scattering, penetration, and collision cascade theory are summarized. The central section is an updated summary of the linear cascade theory of sputtering, based in part on still unpublished work by the author more than a decade ago. The limitations of linear cascade theory in the isotropic limit are discussed, and comments are made on feasible ways to go beyond this limit. The discussion of spike phenomena in the fourth section was no longer up to date at the time of the page proofs, but a substantial revision was not made because the field is still moving rapidly. On the other hand, an effort was made to clarify and specify the predictions and implications of linear cascade theory with regard to sputtering of multicomponent targets in the last section.

## 2.1 Introductory and Historical Survey

### 2.1.1 Identification of Sputtering Events

Surfaces of solids erode under particle bombardment. This phenomenon, now called sputtering, was first observed in gas discharges in the middle of the last century [2.1, 2] but is now known to be rather universal. All kinds of massive particles can erode all kinds of material surfaces, although with widely different efficiencies, and even light particles like electrons and photons can give rise to considerable erosion effects on certain classes of materials.

Erosion rates are characterized primarily by the *sputtering yield Y* which is defined as the mean number of emitted atoms per incident particle. The sputtering yield depends in general on the type and state of the bombarded material, in particular the detailed structure and composition of the material surface, the characteristics of the incident particle, and the experimental geometry. Reliable experimental values of $Y$ usually lie in the region of $10^{-5} \lesssim Y \lesssim 10^3$ atoms per incident particle (Chap. 4). Smaller values are in general not easy to measure, and larger values can be expected only for rather special bombardment conditions.

Despite the apparent universality of the phenomenon, a considerable variety of erosion mechanisms have been proposed during the past 125 years. *Lively controversies* have been going on between the supporters of various mechanisms over the whole period. At the present time, the general consensus is that no single erosion mechanism explains all observations. On the contrary, it seems that most erosion mechanisms proposed up till now may operate under appropriate conditions, although not all erosion mechanisms should come under the heading of sputtering.

The elementary event in sputter erosion takes place on an *atomic scale*. This is demonstrated most clearly by the observation that *light* is emitted from the region around the sputtered surface [2.3] representing, among other components, line spectra characteristic of *atoms* from the sputtered material.

The numbers quoted above for the sputtering yield give a clear indication of the fact that the elementary sputtering event, i.e., the erosion effect caused by *one* bombarding particle, is a *statistical variable*, as opposed to a situation where, e.g., each incident particle would be responsible for the emission of *one* target atom.

It may be appropriate at this point to mention the traditional concepts of *physical* and *chemical sputtering*. The former category invokes a transfer of kinetic energy from the incident particle to target atoms and subsequent ejection of those atoms through the target surface which have acquired enough kinetic energy to overcome the binding forces exerted by the target. The latter category invokes a chemical reaction induced by the impinging particles which produces an unstable compound at the target surface.

Physical sputtering effects occur at incident-particle energies ranging from the medium and upper eV region into the MeV region, but tend to be weak in the lower eV region, as opposed to chemical sputtering [Ref. 2.4, Chap. 5] effects which may persist down to much smaller energies. There is no sharp border line between physical and chemical sputtering. At energies high up in the GeV region, all sputtering effects are expected to decrease rapidly in importance because of the decreasing strength of interaction.

The interaction of particles with matter in the range of energies indicated above is a topic belonging to atomic collision physics. It is obvious that the physics of the sputtering process must be closely connected with other atomic collision and penetration phenomena. This close connection has proved quite useful in the understanding of sputtering phenomena.

For historical and many practical reasons, sputtering of *metals* by *ion* bombardment is by far the most investigated topic within the field of physical sputtering. The dominant process in that case is believed to be what is often called *knockon sputtering* [2.5, 6]. The elementary event is an *atomic collision cascade* where the incident ion knocks atoms off their equilibrium sites in the target, thus causing these atoms to move through the material, to undergo further collisions, and eventually to cause the ejection of atoms through the target surface. The general consensus is that this sputtering mechanism is the most universal one, being active for *ionic* bombardment of *all* types of solids at appropriate ion energies, as well as for certain other types of incident particles such as neutrons and high-energy electrons. However, knockon sputtering need not be dominating when competing with other mechanisms. Ionizing radiation like x-rays, electrons, etc. may cause quite pronounced erosion effects.

This latter class of phenomena is less well understood but under active investigation [Ref. 2.4, Chap. 4].

It may be useful to conclude these preliminary considerations by stressing that 1) sputter erosion is not the only observable effect of particle bombardment and, conversely, that 2) not all erosion by particle bombardment should be classified as sputter erosion.

With regard to 1), it may be noted that a few electron volts are sufficient to release an atom from a surface; indeed, energy spectra of sputtered atoms clearly demonstrate that the majority of sputtered atoms have energies in the lower eV range [Ref. 2.7, Chap. 2]. Thus, it is most often only a minor fraction of the incident energy that is consumed in sputtering, the rest ending up in heat and bulk radiation damage, i.e., ionization, disordering, etc. This is important for the understanding of sputtering especially in case of prolonged bombardment.

With regard to 2), this may be the right place to propose a definition of sputtering. The author is not aware of any previous attempt to make such a definition, but he believes that the four criteria listed below are consistent with what is considered to belong to sputtering by most of those who are engaged in basic research in the field.

(i) Sputtering is a class of erosion phenomena observed on a material surface as a consequence of (external or internal) particle bombardment;

(ii) sputtering is observable in the limit of small incident-particle current;

(iii) sputtering is observable in the limit of small incident-particle fluence;

[(iv) sputtering is observable on target materials of homogeneous composition].

Some comments may be appropriate with regard to criteria (ii)–(iv). Criterion (ii) makes explicit the fact that *macroscopic heating* and subsequent *evaporation* of a target by a high-intensity beam is not considered a sputtering phenomenon. Criterion (iii) ensures that a *single* incident particle may indeed initiate a sputtering event. This separates blistering [Ref. 2.4, Chap. 7] (which requires a threshold fluence) from sputtering phenomena. Criterion (iv), if included, separates phenomena like collision-induced *desorption* [Ref. 2.7, Chap. 6] from sputtering.

While the above requirements are consistent with the view of individual sputtering events taking place on an atomic scale, the reader should be aware that both in basic research and applications, sputtering is observed almost invariably from target materials with an inhomogeneous composition at fairly high particle currents and fluence. This does not mean that sputtering is not dominating observable erosion phenomena, but it may complicate the analysis of what goes on (Chap. 4).

### 2.1.2 Historical Overview

The occurrence of a metallic deposit on the glass walls of a discharge tube was apparently first discovered by *Grove* in 1853 [2.1] and *Faraday* in 1854, the latter being consulted by *Gassiot* [2.2]. It took about half a century before *Goldstein* [2.8] presented compelling evidence that the sputtering effect was caused by positive ions of the discharge hitting the cathode. Goldstein performed the first ion-beam sputtering experiment by extracting canal rays through a hole in the cathode and demonstrating the disappearance of a gold coating on the glass wall facing the beam. Despite rather slow progress in understanding sputtering, possible uses of the phenomenon were recognized early. *Plücker* [2.3] mentioned the potential in thin-film coating, and *Wright* [2.9] found interference fringes on sputter deposited films of unusual quality for a wide variety of metals. *Crookes* [2.10], *Granquist* [2.11], and *Holborn* and *Austin* [2.12] initiated sputtering as a field of quantitative experimental research. Their way of posing questions is still valid today, although the requirements and possibilities with regard to experimental accuracy have changed drastically over those past 80 years.

Several suggestions had, of course, been made concerning the origin of sputter erosion. Early hypotheses by *Puluj* [2.13] and *Hittorf* [2.14] of sputtering being due to cathode heating in the discharge were discouraged by Granquist's important observation [2.11] that the sputtering rate is insensitive to the temperature of the cathode over a wide range of temperatures. "Chunk emission", such as in spark discharges, was pointed out repeatedly, first in 1882 by *Wächter* [2.15], but has never been considered a universal phenomenon. Early suggestions by *Berliner* [2.16] that the occluded gas from the discharge may cause macroscopic erosion phenomena were clarified by *Stark* and *Wendt* [2.17] and verified much later experimentally. The phenomenon is now called the blistering effect [Ref. 2.4, Chap. 7].

The concept of an individual *sputtering event* on an atomic scale initiated by positive ions was proposed and analyzed extensively by *Stark*. He originated the so-called hot-spot model of sputtering [2.18], and subsequently a collision theory viewing sputtering as a sequence of binary collision events initiated by one bombarding ion at a time [2.5]. *Stark* was also aware of the effect of chemical sputtering [2.5], i.e., the formation of volatile compounds by chemical reactions between incident ions and surface atoms, but unlike several of his contemporaries, especially *Kohlschütter* [2.19], he clearly recognized the less universal nature of that process.

*Stark* applied the conservation laws of the theory of elastic collisions in his analysis of the sputtering event [2.5]. He also had a qualitative understanding of collision cross sections and applied it correctly in the interpretation of the observed energy dependence of the sputtering yield $Y$ of hydrogen ions bombarding a metallic target. At low ion energies, $Y$ increases with increasing energy because an increasing amount of energy is transferred to target atoms near the surface. At higher energies, increased penetration of the incident ion because of its decreasing "size" causes a relatively smaller sputtering effect and thus a plateau in the yield curve. The fact that sputtering yields always *decrease* at high enough ion energies for this same reason, was not demonstrated experimentally until the late 1950's when the use of accelerators became common in this field.

One should note in particular that *Stark* considered his hot-spot model, i.e., evaporation of target material from a microscopically small region with a high local temperature due to individual ion impact, and his collision theory of sputtering, as two different views of one and the same process. This attitude was also taken by *v. Hippel* and *Blechschmidt* [2.20] who made the first of a long series of attempts to formulate a sputtering theory on the basis of local heating, and who expressed the view of this being the only feasible procedure to explicitly treat the statistics of the complex series of collision processes occurring in a sputtering event.

Subsequent investigators took a different stand on this issue. After the successful application of Stark's collision theory in the analysis of ion-induced desorption of monolayers by *Kingdon* and *Langmuir* [2.21], the (erroneous) impression arose that the collision theory would imply that sputtering was a single-collision process resulting in a strongly peaked angular distribution of the flux of sputtered particles. Therefore, the experimental observation of a sputtered flux following the Knudsen cosine law [2.22] was believed to provide unambiguous evidence against the collision theory of sputtering. Only after the demonstration by *Wehner* [2.23] of crystal structure effects in the flux of sputtered particles did it become evident that local evaporation alone could hardly explain the sputtering phenomenon.

It is interesting, however, that *Lamar* and *Compton* [2.24] in 1934, pointed to the dominance of binary collision processes in light-ion sputtering and of local evaporation in heavy-ion sputtering. When rephrased slightly, this view comes close to a modern view of physical sputtering.

The most outstanding obstacle against a quantitative analysis of sputtering phenomena was the virtually complete lack of sound knowledge about the motion of low-energy (1–1000 eV) atomic particles in solids. Present improved understanding of sputtering is not so much due to an (still not very substantial) improvement in the knowledge of low-energy atomic motion, but rather due to the use of high-energy particle beams in sputtering. High-energy sputtering experiments were reported first by *Smith* [2.25], and subsequently by *Almén, Thompson, Rol, Yurasova, Molchanov* [2.26–30] and many others.

As mentioned above, *Wehner*'s observation of spot patterns in single crystal sputtering [2.23] revived the interest in the collision theory of sputtering. Independently, *Keywell* [2.6, 31] made a first attempt to formulate Stark's multiple-collision model in terms familiar from neutron transport theory. His, as well as subsequent calculations by *Harrison* [2.32] were important steps in the sense that *probability concepts* as expressed by *collision cross sections* finally made their entrance into sputtering theory. The theories were less successful in the sense that the number of unknown parameters was prohibitive to allow quantitative conclusions.

Less ambitious but more successful attempts to formulate a collision theory of sputtering were made by *Almén* [2.26], *Rol* [2.28], and notably by *Pease*

[2.33]. These treatments split the sputtering event into two steps: the creation of a primary recoil atom by an incident ion and the subsequent development of a cascade of recoil atoms, of which some are ejected through the surface. This genuine high-energy concept of sputtering, originating in the theory of radiation damage [2.34] has proved quite successful. It allows the general behaviour of the sputtering yield to be predicted as a function of ion type, energy, and angle of incidence *without* explicit reference to the details of the recoil cascade. Those details become important mainly when interest is directed towards predictions of the absolute magnitude of the sputtering yield, and differential quantities like angular and energy spectra. This great simplification is achieved in the following way. If the impinging ion has a high energy, it will create recoil atoms at a rate determined by its nuclear stopping power (i.e., the mean energy lost in elastic atomic collisions per travelled path length [2.35]. These recoil atoms have high enough energy to generate higher-order recoils. The *number* of recoil atoms is roughly proportional to the available energy, and the *fraction* reaching the target surface depends on the energy deposited per unit depth.

This intuitive picture, when applied to swift light ions [2.33] interacting via Rutherford's law with the target atoms, provided a valid description of the observed decrease of the sputtering yield proportional to $E^{-1}$, $E$ being the ion energy. The stopping power of slower (and heavier) ions became known through the theoretical work of *Lindhard* and coworkers [2.36], supported by experimental work on ion ranges by *Davies* and coworkers [2.37, 38]. *Brandt* and *Laubert* [2.39] demonstrated the similarity of measured sputtering yields with calculated stopping powers with regard to the dependence on ion energy, and they provided a rough estimate of the absolute magnitude of the sputtering yield.

Meanwhile, the theory of collision cascades had advanced due to the work of *Leibfried, Lindhard* et al., *Dederichs, Robinson, Sanders, Thompson* [2.40–45] and the present author [2.46, 47], who also made an attempt to collect the available knowledge in a transport theory of sputtering [2.48]. The proportionality of the sputtering yield with the deposited energy was shown to be valid asymptotically in the limit of high ion energy, and the flux of low-energy recoil atoms was found to be isotropic in that same limit. A rather crude extrapolation was found necessary in the treatment of the motion of low-energy atoms, and the target surface was taken into account by way of a postulate [2.45] rather than firm knowledge. Despite obvious shortcomings, this theory became successful as a reference standard for sputtering yield measurements (Chap. 4). It has also been a reasonable starting point for further refinements of the theory which will be discussed below. It is important, however, to notice that both minor and major discrepancies have been pointed out by the comparison of measured with calculated sputtering yields (Chap. 4). More important and less evident is the fact that unambiguous evidence has never been provided to confirm or disprove the validity of the model invoked for the low-energy recoil cascade.

During the rapid development of the theory of collision cascades, the hot-spot (or "spike") theory received little attention, despite repeated claims of a thermal component observed in energy spectra of sputtered atoms [2.49]. Definite evidence for the high density of deposited energy in very heavy-ion bombardment emerged gradually [2.50, 51]; the importance of this fact to sputtering yields was demonstrated by *Andersen* and *Bay* [2.52] who found that the sputtering yield of a diatomic molecule may be substantially higher than twice the sputtering yield of the individual atoms at the same velocity.

In the period from 1955 to 1965, the dominant effort in the study of sputtering on metals was actually spent in the investigation of crystal lattice effects. The discovery [2.23] of spot patterns suggested that much could be learnt about the slowing down of low-energy atoms in crystals from measurements of this kind [2.53]. Since this is an important issue, a special article has been devoted to it, to which the reader is referred (Chap. 3). That contribution also contains an extensive review on the anisotropy of the sputtering yield of crystals with respect to the incident beam, a phenomenon that was found experimentally by *Rol* et al. [2.54] and *Almén* [2.26].

This historical overview concentrated on the sputtering of "pure" metals, and hence on knockon sputtering.

### 2.1.3 Classification of Sputtering Events

In the following, an attempt will be made to classify sputtering events by means of qualitative considerations, mainly with the purpose of giving an indication of what physical input is needed in the formulation of a theoretical description. The situation is simplest in the case of sputtering by elastic collisions (knockon sputtering), while electronic processes, when important, may give rise to a wider variety of events.

### a) Knockon Sputtering

For metallic targets, elastic collision processes are most important. Indeed, a 10 keV argon ion has a velocity of about $10^{-3}$ times the velocity of light. It takes about $\sim 10^{-13}$ s for such an ion to travel 100 Å in vacuum. This time is very long compared with relaxation times of conduction electrons ($\sim 10^{-19}$ s). Thus, even though an incident particle may spend a larger or smaller fraction of its energy in exciting electrons, this energy will be immediately shared by all electrons without providing atoms a chance to escape.

It is convenient to distinguish between three qualitatively different situations (Fig. 2.1), the single-knockon regime, the linear cascade regime, and the spike regime. In the single-knockon regime, the bombarding ion transfers energy to target atoms which, possibly after having undergone a (small) number of further collisions, are ejected through the surface if energetic enough to overcome binding forces. In the two other cases, recoil atoms are energetic

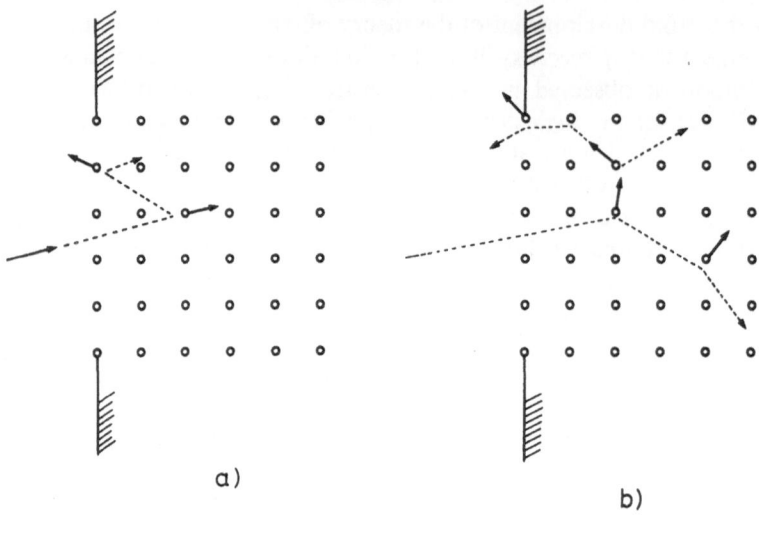

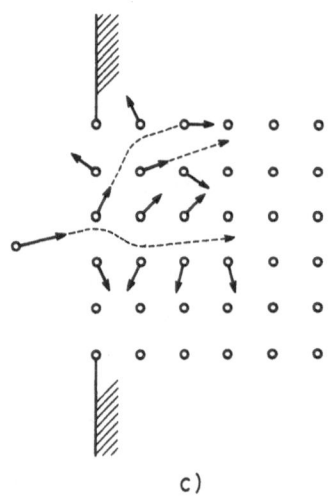

**Fig. 2.1a–c.** Three regimes of sputtering by elastic collisions. (**a**) The single-knockon regime. Recoil atoms from ion-target collisions receive sufficiently high energy to get sputtered, but not enough to generate recoil cascades. (**b**) The linear cascade regime. Recoil atoms from ion-target collisions receive sufficiently high energy to generate recoil cascades. The density of recoil atoms is sufficiently low so that knock-on collisions dominate and collisions between moving atoms are infrequent. (**c**) The spike regime. The density of recoil atoms is so high that the majority of atoms within a certain volume (the spike volume) are in motion

enough to generate secondary and higher-generation recoils, of which some may approach the target surface and overcome the barrier. The linear cascade differs from the spike regime by the spatial density of moving atoms which is small in the former and large in the latter case. The difference is most directly tested by diatomic molecular ion bombardment [2.52]. A molecule hitting a solid will dissociate almost immediately. The statistical nature of the slowing-down process suggests a linear superposition of the two cascades if sufficiently dilute. Thus, the sputtering yield will be approximately twice that for bombardment with an atomic ion. In the spike case, twice the amount of energy is shared

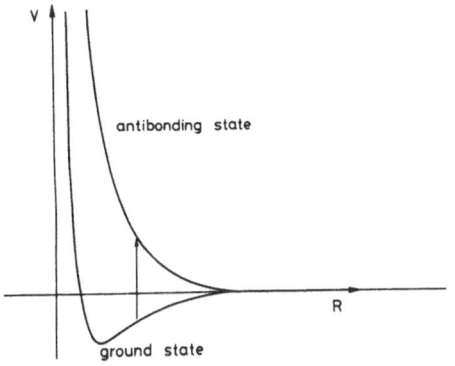

antibonding state

ground state

V

R

**Fig. 2.2.** Bonding and antibonding states of a diatomic molecule (schematically). V is the interaction energy between the atoms and R the internuclear distance

by all atoms within roughly the same volume as for atomic-ion bombardment. Depending on the energy distribution of those atoms, one may expect a drastic increase of the fraction of atoms able to overcome the surface barrier. Therefore, the sputtering yield of a molecule may be more than twice that for atomic-ion bombardment in the spike case.

Asserting, tentatively, that atoms are ejected from a more or less well-defined layer of thickness $\Delta x_0$, one expects the sputtering yield to be proportional to the number of recoil atoms generated in that layer. In the single-knockon regime, that number is essentially given by the pertinent *cross section*; in the linear cascade regime it is proportional to the available *energy*, i.e., the *energy deposited per unit depth*; in the spike regime, one might be inclined to associate a temperature with the *energy deposited per unit volume*, and determine the yield from the vapor pressure.

Qualitatively, the single-knockon regime falls into the lower and medium eV region except for very light ions where, because of inefficient energy transfer, it extends up into the lower keV region. The linear cascade region is characteristic for keV and MeV ions except for the heaviest ions which are stopped rapidly and tend to generate spikes.

### b) Sputtering by Electronic Excitation

In insulators, lifetimes of excited electronic states may be long enough to allow excitation energy to be transferred to atomic motion. There are numerous ways of how this can be accomplished in detail; a prototype is the excitation of the bonding ground state of a molecule into an antibonding excited state (Fig. 2.2). Dissociation occurs if the lifetime of the antibonding state is comparable to or greater than the time it takes the nuclei to separate under their mutual repulsive interaction.

One may again consider three regimes, similar to those shown in Fig. 2.1. In the threshold regime, ionizing or dissociating events are isolated in space. Such events may be generated by uv photons and low-energy ($\lesssim 100 \, \text{eV}$) electrons, and also low energy ions (ion velocity $v \lesssim e^2/\hbar$). The linear ionization cascade regime represents the case where high-energy secondary electrons ($\gtrsim 100\,\mathrm{eV}$) may produce secondary ionizations. This situation is pronounced in the case of high-speed ($v \gg e^2/\hbar$) ions. Finally, heavy ions at intermediate velocity ($v \gtrsim e^2/\hbar$) may create dense *ionization spikes*.

One may note that in the two latter cases, sputtering events are expected to be accompanied by (and correlated with) emission of *electrons* and/or *light*. Unfortunately, neither sputtering by electronic excitation nor its connection with other emission phenomena has been studied in great detail, although observed effects may be dramatic ([Ref. 2.4, Chap. 4] and [2.55, 56]).

It is obvious that within this classification scheme, the (extensively investigated [Ref. 2.4, Chap. 5]) process of chemical sputtering is a special case of sputtering by electronic processes. The traditional concepts of physical and chemical sputtering might be abandoned for this reason.

The following, more explicit discussion will be restricted to sputtering by elastic collisions.

## 2.2 Needs and Tools

This section summarizes some results from the theory of atomic collision and penetration phenomena that are useful prerequisites in the theory of knockon sputtering. The reader who is interested in a more detailed introduction to this topic is referred to previous reviews by the author [2.58, 59].

### 2.2.1 Cross Sections

Let us recall the elementary concept of a *cross section*. An atomic collision process involving two atomic particles 1 (projectile) and 2 (target) can be characterized by a cross section $\sigma$; this means that the average fraction of beam particles of type 1 hitting a target of thickness $x$ and density $N$ [2-atoms/volume] undergoing the considered process is given by [2.35]

$$N x \sigma . \tag{2.2.1}$$

Likewise, (2.2.1) represents the probability to undergo this process for a 1-atom penetrating a (small) path length $x$ in a homogeneous medium of randomly distributed 2-atoms. The statement is limited to cases where $N x \sigma \ll 1$.

We shall deal mainly with cross sections for elastic collisions in which *kinetic* energy is transferred from atom 1 to atom 2, and with cross sections for electronic excitation and/or ionization.

Cross sections for elastic collisions are conveniently specified in terms of the differential cross section,

$$d\sigma(E, T) \equiv \frac{d\sigma}{dT} dT ,$$

where $E$ is the initial and $T$ the transferred energy in a single collision. The total cross section $\sigma_{tot} = \int d\sigma(E, T)$ is most often a divergent quantity in classical collision theory. This divergence is, however, immaterial, since measurable quantities do not depend on the total cross section but on higher moments like the stopping cross section, as will be discussed below.

### 2.2.2 Some Results from Elastic-Collision Theory

The energy transferred in an elastic collision between two atoms is limited by the conservation laws of energy and momentum. An atom 1 with initial energy $E$ can at most transfer an energy [2.35]

$$T_m = \gamma E$$
$$= \frac{4M_1 M_2}{(M_1 + M_2)^2} E \tag{2.2.2}$$

to an atom 2 with zero initial energy, and this requires a central (head-on) collision. $M_1$ and $M_2$ are the respective atomic masses.

The probability distribution for the energy transfer $T$ is determined by (2.2.1) and the cross section $d\sigma(E, T)$. For Rutherford scattering, i.e., for energies high enough so that the scattering is determined by the Coulomb repulsion between the nuclei, one finds [2.35]

$$d\sigma(E, T) = \pi \frac{M_1}{M_2} Z_1^2 Z_2^2 e^4 \frac{dT}{ET^2} ; \quad 0 \leq T \leq T_m , \tag{2.2.3}$$

where $Z_1 e$ and $Z_2 e$ are the nuclear charges. This cross section strongly prefers collisions with small energy transfers $(T \ll T_m)$ and, moreover, decreases in absolute magnitude with increasing $E$.

The cross section (2.2.3) is approximately valid only for [2.57] $\varepsilon \gg 1$, where

$$\varepsilon = \frac{M_2 E}{M_1 + M_2} \cdot \frac{a}{Z_1 Z_2 e^2} , \tag{2.2.4}$$

and

$$a \cong 0.885 a_0 (Z_1^{2/3} + Z_2^{2/3})^{-1/2} ; \quad a_0 = 0.529 \text{ Å} , \tag{2.2.5}$$

i.e., at high enough energies so that the nuclei approach closer to each other than the screening radius $a$. Representative values of this energy are listed in Table 2.1.

For lower energies, $\varepsilon \lesssim 1$, the screening of the Coulomb interaction is essential; in this regime, the cross section can be approximated by the

**Table 2.1.** Values of the Thomas-Fermi energy unit (2.2.4), $E_{TF} = E/\varepsilon = (1 + M_1/M_2) \cdot Z_1 Z_2 e^2/a$ for representative ion-target combinations. $E_{TF}$ in keV, $M_1$ and $M_2$ in a.m.u.

| Ion | | H | He | Ne | Ar | Kr | Xe | U |
|---|---|---|---|---|---|---|---|---|
| Target | $M_1$ | 1 | 4 | 20 | 40 | 84 | 131 | 238 |
| $M_2$ | | | | | | | | |
| H | 1 | 0.0869 | 0.494 | 15.33 | 63.6 | 324 | 850 | 3120 |
| C | 12 | 0.414 | 1.087 | 13.86 | 46.8 | 200 | 498 | 1710 |
| Si | 28 | 1.163 | 2.68 | 23.8 | 68.0 | 254 | 585 | 1910 |
| Cu | 64 | 2.93 | 6.30 | 44.1 | 107.0 | 338 | 722 | 2130 |
| Ag | 108 | 5.46 | 11.45 | 72.0 | 160.0 | 453 | 903 | 2260 |
| Au | 197 | 10.75 | 22.2 | 128.4 | 266 | 676 | 1250 | 3060 |

**Table 2.2.** Values of $\lambda_m$ in (2.2.7)

| $m$ | 1.000 | 0.500 | 0.333 | 0.191 | 0.055 | 0.000 |
|---|---|---|---|---|---|---|
| $\lambda_m$ | 0.500 | 0.327 [2.36] | 1.309 [2.50] | 2.92 [2.162] | 15 [2.77] | 24 [2.98] |

expression [2.50, 57]

$$d\sigma(E, T) \cong C_m E^{-m} T^{-1-m} dT ; \quad 0 \lesssim T \lesssim T_m \qquad (2.2.6)$$

with

$$C_m = \frac{\pi}{2} \lambda_m a^2 \left(\frac{M_1}{M_2}\right)^m \left(\frac{2Z_1 Z_2 e^2}{a}\right)^{2m} , \qquad (2.2.7)$$

$\lambda_m$ being a dimensionless function of the parameter $m$ which varies slowly from $m = 1$ at high energies [where (2.2.6) goes over into (2.2.3) and $\lambda_1 = 1/2$] down to $m \approx 0$ at very low energies (Table 2.2). The initial justification of (2.2.6) stems from classical collision theory applied to an interatomic potential of the form $V(R) \propto R^{-1/m}$, where $R$ is the distance between colliding nuclei [2.57]. While this basis is sound only for moderate deviations from Rutherford scattering ($m \gtrsim 0.2$ [2.57]), (2.2.6) will also be used for smaller values of $m$, thus approaching a constant cross section

$$d\sigma(E, T) \cong \frac{\pi}{2} \lambda_0 a^2 T^{-1} dT ; \quad 0 \leq T \leq T_m \qquad (2.2.8)$$

for $m \cong 0$. Note that even in this case, soft collisions ($T \ll T_m$) dominate; this is not the case for the collision spectrum of colliding billard balls.

An important quantity is the mean energy $\overline{\Delta E}$ spent in elastic collisions over a travelled path length $x$. From (2.2.1, 6) and the definition of a mean value one

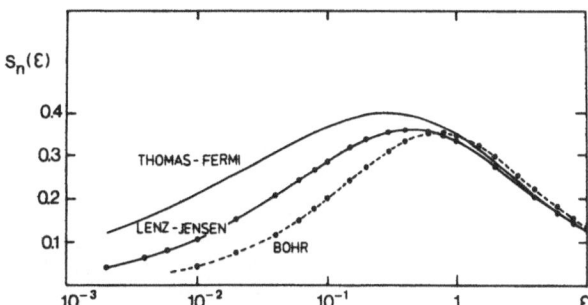

**Fig. 2.3.** Nuclear stopping cross sections in Thomas-Fermi variables. The energy variable $\varepsilon$ is defined in (2.2.4), and $s_n$ is related to the stopping power by (2.2.10, 13). The three curves refer to three different screening functions for the Coulomb interaction between two colliding atoms. The curve labelled "Thomas-Fermi", which is the one used in comparisons of experimental sputtering yields with theory [2.48], is known to be too high, while the curve labelled "Lenz-Jensen" slightly underestimates the stopping cross section. (From [2.57], courtesy of Det Kgl. Danske Videnskabernes Selskab)

finds[1]

$$\overline{\Delta E} = Nx \int d\sigma(E, T) \cdot T \equiv NxS_n(E), \tag{2.2.9}$$

where $S_n(E)$ is called the *nuclear stopping cross section* and

$$S_n(E) = \frac{1}{1-m} C_m \gamma^{1-m} E^{1-2m}. \tag{2.2.10}$$

It is seen that $S_n(E)$ rises approximately proportional to $E$ at low energies ($m \cong 0$), approaches a plateau at intermediate energies ($m \cong 1/2$) and falls off at higher energies ($1/2 < m \lesssim 1$). This behavior can be summarized in a more compact form [2.57]

$$S_n(E) = 4\pi a Z_2 Z_2 e^2 \frac{M_1}{M_1 + M_2} s_n(\varepsilon), \tag{2.2.11}$$

where $s_n(\varepsilon)$ is a universal function depending on the detailed form adopted for the screened Coulomb interaction. Several proposed forms are shown in Fig. 2.3.

The expressions (2.2.6, 7, 10, 11) are based on the so-called Thomas-Fermi model of atomic interaction and have to some extent been tested experimentally. Their accuracy is greatest, in the few-percent region, for weak screening

---

1 In (2.2.9), as well as in many subsequent relationships, no integration limits have been listed. This is justified in view of the fact that a differential cross section occurs in the integrand which in itself defines a range of allowed values for the integration variable $T$ [cf. (2.2.8)].

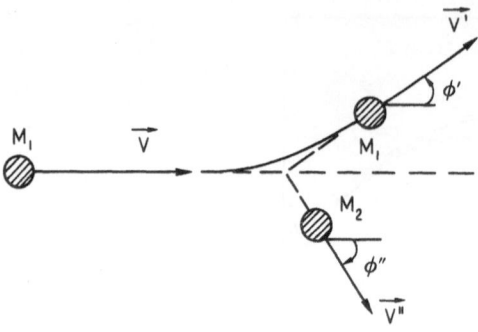

**Fig. 2.4.** Specification of scattering geometry in an elastic collision of particle 1 (mass $M_1$, initial velocity $v_1$) on particle 2 (mass $M_2$, initial velocity zero)

($\varepsilon \gg 1$, $m \cong 1$) and poorest, at best a factor of 2, for heavy screening ($\varepsilon \ll 1$, $m \cong 0$). More details are reviewed in several summaries on collision and penetration theory [2.58–60].

It is noticed that conservation laws provide a unique relationship between the scattering angle and energy $T$ transferred by an atom 1 of energy $E$ hitting an atom 2 *at rest*. With the definition of scattering angles shown in Fig. 2.4, one finds [2.35]

$$\cos\phi' = (1 - T/E)^{1/2} + \tfrac{1}{2}(1 - M_2/M_1)(T/E)(1 - T/E)^{-1/2} \tag{2.2.11a}$$

and

$$\cos\phi'' = (T/\gamma E)^{1/2}. \tag{2.2.11b}$$

Thus, recoil atoms are scattered into the interval $0 \leq \phi'' \leq \pi/2$, while impinging particles are scattered into an interval $0 \leq \phi' \leq \phi_{max}$ where

$$\phi_{max} = \begin{cases} \pi & \text{for} \quad M_1 < M_2, \\ \arcsin\dfrac{M_2}{M_1}(<\pi/2) & \text{for} \quad M_1 > M_2. \end{cases} \tag{2.2.12}$$

It should be stressed that the results of this paragraph, (2.2.2–12), are not applicable to the spike regime where collisions between two moving particles dominate.

### 2.2.3 Some Results from Penetration Theory

The specific energy loss of a particle moving through a random medium is found from (2.2.9) to be given by

$$\frac{dE}{dx} = -NS(E). \tag{2.2.13}$$

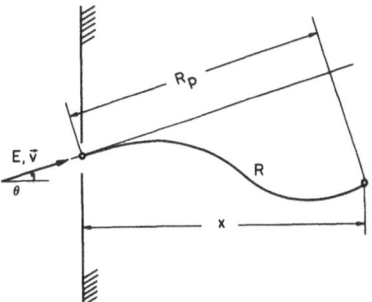

**Fig. 2.5.** Definition of penetration parameters. The initial ion velocity is $v$, its energy $E$ and its angle of incidence to the surface normal is $\theta$. The penetrated path length is $R$, the projected range $R_p$ and the penetration depth $x$

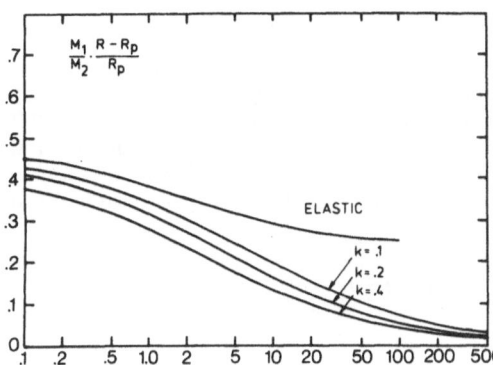

**Fig. 2.6.** Relationship between the average path length $R(E)$ and average projected range $R_p(E)$ (Fig. 2.5) as a function of the mass ratio $M_2/M_1$ and reduced energy $\varepsilon$ (2.2.4) to first-order in $M_2/M_1$. Evaluated for Thomas-Fermi nuclear interaction (Fig. 2.3). The curve labelled "elastic" disregards electronic stopping. The parameter $k$ refers to different strengths of the electronic-stopping function [2.36, 68], (2.2.18). (From [2.36], courtesy of Det Kgl. Danske Videnskabernes Selskab)

The mean penetrated *path length* $R(E)$ of a particle before coming to rest can then be estimated by evaluation of the expression [2.35, 36]

$$R(E) = \int_0^E \frac{dE'}{NS(E')}. \tag{2.2.14}$$

If all energy is spent in elastic collisions, (2.2.10) yields

$$R(E) \cong \frac{1-m}{2m} \gamma^{m-1} \frac{E^{2m}}{NC_m}. \tag{2.2.15}$$

The mean *projected range* $R_p(E)$ is in general smaller than $R(E)$ because of *scattering* of the projectile (Fig. 2.5). The *path length correction* $R_p(E)/R(E)$ is $\ll 1$ for $M_1 \ll M_2$ (for $\varepsilon \lesssim 1$) and close to 1 for $M_1 \gg M_2$, and generally for $\varepsilon \gg 1$ [Fig. 2.6]. Within the validity of (2.2.6), and in the absence of electronic stopping, it depends on the mass ratio $M_2/M_1$ and on $m$, but not *explicitly* on energy.

The *penetration profile* $F_R(x, E, \theta)$ is the statistical distribution in penetration depth $x$ for a beam incident at an angle $\theta$ to the surface normal. It is

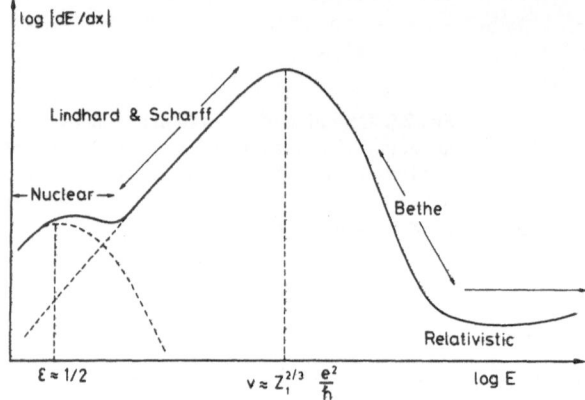

**Fig. 2.7.** Stopping power of an ion as a function of energy (schematically). At the lowest velocities [$\varepsilon \lesssim 1$, (2.2.4)], nuclear stopping dominates for heavy and medium-mass ions, and is competitive for light ions. At higher velocities, for $v \lesssim Z_1^{2/3} e^2/\hbar$, electronic stopping takes over according to (2.2.18). In this regime, the projectile is preferrably neutral. Beyond the stopping-power maximum, the Bethe regime (2.2.17) is approached where the projectile is preferably stripped. When the velocity of light is approached, the stopping power goes through a broad minimum and increases in the extreme relativistic regime

characterized primarily by its mean value [2.36, 50]

$$x(E, \theta) \equiv R_p(E) \cos \theta \tag{2.2.16}$$

and the corresponding standard deviation.

In general, and in particular at high projectile velocities ($\varepsilon \gg 1$), energy loss to electrons dominates the *slowing-down* of the ion, while *scattering* by electrons is a minor effect because of the small mass $m'$ of an electron. Therefore, with regard to electronic parameters, it is the *stopping cross section* that is of interest in sputtering. Figure 2.7 shows a pertinent survey. At high projectile velocities ($v \gg e^2/\hbar$ for protons, and even higher for heavier particles), the description by *Bethe* [2.61] is valid:

$$S_e = \frac{4\pi e_1^2 Z_2 e^2}{mv^2} \left( \log \frac{2mv^2}{I} + \text{correction terms} \right), \tag{2.2.17}$$

where $e_1$ is the projectile charge (if being considered as a point charge) and $-e$ the electron charge. $I$ is the mean ionization potential; for various correction terms cf. [2.62].

The range of validity of (2.2.17) is outside that covering most sputtering experiments. At lower velocities, $v \lesssim Z_1^{2/3} e^2/\hbar$, an estimate by *Lindhard* and *Scharff* [2.63] has proved useful, according to which

$$S_e \cong \xi_e 8\pi e^2 a_0 \frac{Z_1 Z_2}{Z} \frac{v}{e^2/\hbar}, \tag{2.2.18}$$

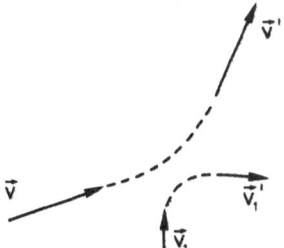

Fig. 2.8. Scattering geometry for two colliding particles

where $Z = (Z_1^{2/3} + Z_2^{2/3})^{1/2}$ and $\xi_e$ is a function of the atomic number $Z_1$. $\xi_e$ has empirically been shown to oscillate around the proposed dependence $\xi_e \sim Z_1^{1/6}$. For details cf. [2.62]. Equation (2.2.18) is less well-established than (2.2.17), and it is also less accurate. Yet it is a satisfactory tool in knockon sputtering where it enters as a correction. Some caution is indicated in the use of (2.2.18) in ionization sputtering, once a quantitative theory of that phenomenon should emerge.

### 2.2.4 Boltzmann's Equation

Multiple collision processes of the type indicated in Figs. 2.1a–2.1c are conveniently treated within the framework of *transport theory*. Although most essential features of that theory are implicit in the original formulation by *Boltzmann* [2.64], it may be worth while to outline the assumptions entering as well as those features which it is essential to include.

Boltzmann's equation determines a function $f(r, v, t)$, where $f(r, v, t) d^3 r d^3 v$ is a statistical average over the number of atoms in a volume element $d^3 r$ at $r$ moving with a velocity $(v, d^3 v)$ at time $t$. In its most common form, for a one-component medium, Boltzmann's equation reads

$$\left( \frac{\partial}{\partial t} + v \cdot \nabla_r + \frac{k}{m} \nabla_v \right) f(r, v, t) = \int K(v, v_1; v', v'') d^3 v_1 d^3 v' d^3 v''$$

$$\cdot |v - v_1| (f'f'' - ff_1) \qquad (2.2.19)$$

where $K(v, v_1; v', v'') d^3 v' d^3 v''$ is the differential cross section for a projectile with initial velocity $v$ and a target atom moving with initial velocity $v_1$ to scatter each other into velocities $(v', d^3 v')$ and $(v'', d^3 v'')$, respectively (cf. Fig. 2.8). Moreover, $f_1 = f(r, v_1, t)$, $f' = f(r, v', t)$, etc., and $k = k(r, v, t)$ is an external force acting on the moving particle.

Extension of (2.2.19) to two- or more-component systems is a matter of adding suitable indices, and results, in general, in a hierarchy of equations which may or may not be easily decoupled.

There is one single, central assumption underlying (2.2.19), the assumption of molecular chaos which implies that after every individual collision, the detailed configuration of particles in real and velocity space randomizes, i.e., the initial conditions for the subsequent collision are determined *only* by the statistical distribution function $f(r, v, t)$ and *not* by the events that happened previously. The degree of validity of this assumption varies substantially over the range of applications of (2.2.19) in penetration phenomena.

Most readers are presumably familiar with some applications of Boltzmann's equation. In the theory of electric conductivity and related phenomena, the external force stems from an electric field, the spatial variable drops out, and the collision term on the right-hand side is often treated in a somewhat rudimentary way. In kinetic gas theory, the general form of the collision term is the basis for a valid derivation of the Maxwell-Boltzmann velocity distribution in the absence of external forces. Also here, the spatial variable is immaterial.

Multiple atomic collisions and ion penetration obey laws formally similar to neutron penetration. External forces can usually be ignored ($k=0$), and the target particles can be assumed at rest initially, for sufficiently fast projectiles, i.e.,

$$f(r, v_1, 0) = N\delta(v_1) \tag{2.2.20}$$

where $\delta$ is the Dirac delta function, and $N$ is the density of target particles in real space.

Two important limiting cases can be specified. If only the fate of the projectile is of interest, the motion of target particles need not be followed, and (2.2.19) reduces to[2]

$$-\left(\frac{\partial}{\partial t} + v \cdot \nabla_r\right) f(r, v, t) = N \int d^3 v' [v f K(v; v') - v' f' K(v'; v)]. \tag{2.2.21}$$

Conversely, when the motion of recoil atoms is important, (2.2.19) can be linearized for a *dilute cascade*, i.e., where only a small fraction of atoms is moving,

$$f(r, v, t) = N\delta(v) + F(r, v, t), \tag{2.2.22}$$

where $F$ is "small", so that (2.2.19) reads:

$$-\left(\frac{\partial}{\partial t} + v \cdot \nabla_r\right) F(r, v, t) = N \int d^3 v' d^3 v'' [v F K(v; v', v'')$$
$$- v' F' K(v'; v, v'') - v'' F'' K(v''; v', v)]. \tag{2.2.23}$$

In (2.2.23), $K(v; v', v'')$ stands for $K(v, 0; v', v'')$, and in (2.2.21),

$$K(v; v') \equiv \int d^3 v'' K(v; v', v'').$$

---

2 The limits of integration in (2.2.21, 23), and similar, subsequent transport equations are determined in one part by the differential cross section [cf. footnote to (2.2.9)] and to the other part by the physical requirement on the distribution functions $f$ and $F$ that the energy of a moving particle cannot exceed the initial energy.

Equation (2.2.21) can be used as a basis for ion penetration theory, while (2.2.23) can serve as a basis for the theory of linear collision cascades. The only formal difference is the importance of recoil effects in the latter.

For reasons of practical convenience, other types of linear transport equations have been used frequently in both penetration and cascade theory. Their physical equivalence has been proved in general terms by *Lindhard* and *Nielsen* [2.65], and the identity of the underlying physical assumptions is evident in actual cases. The present author based his sputtering theory [2.48] on the following equation,

$$-\left(\frac{\partial}{\partial t} + v \cdot \nabla_r\right) G(r, v, v_0, t) = Nv \int K(v; v', v'') d^3 v' d^3 v''(G - G' - G''), \qquad (2.2.24)$$

where $G(r, v, v_0, t) d^3 r d^3 v_0$ is the expected number of atoms moving with velocity $(v_0, d^3 v_0)$ in the volume $(r, d^3 r)$ at time $t$, and $v$ is the initial velocity of the particle initiating a cascade in $r = 0$ at $t = 0$. Thus, (2.2.24) projects events over a finite time interval, unlike (2.2.23) which is differential. After the adoption of initial conditions, the respective solutions reflect identical physical situations. The equivalence of (2.2.23, 24) is shown in Appendix A.

### 2.2.5 Some Results from Linear Cascade Theory

A primary quantity of interest in cascade theory is the expected number of atoms participating in a cascade. The determination of this quantity is a classical problem in the theory of radiation damage [2.34], and it is closely related to that of the number of free electrons generated in ionization cascades initiated by a high-energy electron in a gas [2.66]. In both cases, this number, if much larger than 1, is proportional to the available (i.e., the initial) energy unless energy is consumed for other purposes than setting atoms in motion or ionizing gas atoms, respectively.

More specifically, let $n(E, E_0)$ be the mean number of atoms set in motion with an initial energy greater than some value $E_0$ in a cascade initiated by a primary ion or recoil of initial energy $E(> E_0)$, the average being taken over a large number of cascades. Allowing for elastic collisions only, and ignoring all binding forces between target atoms, one finds the asymptotic expression [2.67]

$$n(E, E_0) \sim \Gamma_m \frac{E}{E_0} \quad \text{for} \quad \gamma E \gg E_0, \qquad (2.2.25)$$

where $\gamma$ is defined by (2.2.2) with $M_1$ the ion mass and $M_2$ the target mass. Furthermore,

$$\Gamma_m = \frac{m}{\psi(1) - \psi(1 - m)} \qquad (2.2.26)$$

**Table 2.3.** Values of $\Gamma_m$ according to (2.2.26)

| $m$ | 0.500 | 0.333 | 0.250 | 0.000 |
|---|---|---|---|---|
| $\Gamma_m$ | 0.361 | 0.452 | 0.491 | 0.608 |

and $\psi(x) = d[\log \Gamma(x)]/dx$, when the cross section (2.2.6) is assumed to be valid. Equation (2.2.25) can be derived as an asymptotic solution of an integral equation that can be traced back to (2.2.23) or (2.2.24). A systematic presentation of several of the various steps involved can be found in [2.58]. The present discussion concentrates on the physical content of (2.2.25) and its various extensions.

One may first note that according to Table 2.3, the parameter $\Gamma_m$ depends only weakly on the exponent $m$ that characterizes the scattering cross section. Evidently, the number of atoms participating in a cascade is rather insensitive to the detailed manner in which energy is shared amongst the atoms (Fig. 2.1b). In particular, only the *energy* of the primary particle enters, but not its mass or atomic number [2.58].

Next, include the effect of electronic stopping. Excitation of electrons acts approximately like a continuous drain of energy, such that only an amount [2.41]

$$v(E) = E - \eta(E) \qquad (2.2.27)$$

is available for the creation of recoil atoms, where $\eta(E)$ is the average amount of energy ending up in electronic excitation during the entire slowing-down process [2.41]. Thus, for nonnegligible electronic stopping, $v(E)$ replaces $E$ in (2.2.25). However, since electronic stopping depends on *velocity* rather than *energy* according to (2.2.17, 18), $v(E)$ depends on the *ionic mass* [2.41]. Tabulations are available [2.68].

From (2.2.25, 27), one readily obtains the mean number of atoms $F(E, E_0)dE_0$ recoiling into a given energy interval $(E_0, dE_0)$ by differentiation:

$$F(E, E_0) \sim \Gamma_m v(E)/E_0^2 \quad \text{for} \quad \gamma E \gg E_0 ; \qquad (2.2.28)$$

$F(E, E_0)$ is called recoil density [2.67]. Since recoils of higher generations dominate, the initial direction of motion of the projectile is immaterial and the distribution in the solid angle $\Omega_0$ is isotropic:

$$F(E, E_0, \Omega_0)dE_0 d^2\Omega_0 \sim \Gamma_m \frac{v(E)}{E_0^2} dE_0 \frac{d^2\Omega_0}{4\pi}, \quad \text{for} \quad \gamma E \gg E_0 . \qquad (2.2.29)$$

Deviations have been shown [2.44] to be smaller by a factor $\propto (M_1 E_0/M_2 E)^{1/2}$ [cf. Sect. 2.3.4, (c)].

Finally, one may ask what is the distribution in space of the recoiling atoms. Since the number of recoil atoms turns out to be proportional to the available energy, cf. (2.2.25, 29), one may be tempted to assert that the spreading of recoil atoms in space, scales according to the spreading of energy. Thus, if we define a density of deposited energy $F_D(E, \Omega, r)$ for an ion starting its slowing down in $r = 0$ such that

$$\int d^3 r F_D(E, \Omega, r) = v(E), \tag{2.2.30}$$

where $\Omega$ is the unit vector in the initial direction of motion, one may suppose that recoil atoms are generated according to the distribution

$$F(E, \Omega, E_0, \Omega_0, r) dE_0 d^2\Omega_0 d^3 r \sim \Gamma_m \frac{F_D(E, \Omega, r)}{E_0^2} dE_0 \frac{d^2\Omega_0}{4\pi} d^3 r. \tag{2.2.31}$$

For elastic collisions, the proof for the validity of this expression has been given explicity by the author in [2.48]. The mathematical argument is rather lengthy and goes over spatial moments of the distribution function. The physical argument is that the gross spatial extension of a cascade is determined by high-energy particles, and that any fine structure must be smeared in a single-particle distribution function. Thus, deviations from the simple decoupled behaviour of high and low-energy atoms indicated in (2.2.31) are of the order of $R(E_0)/R(E) \ll 1$. A similar proof for the case of nonnegligible electronic stopping [2.69] has never been published.

A remark is indicated concerning the value of the exponent $m$ occurring in $\Gamma_m$, (2.2.25–31). Since $m$ depends slowly on energy according to Sect. 2.2.2, the question arises as to *which* energy, $E$ or $E_0$, fixes $m$. The problem is subtle, and has been solved only approximately [2.47]. For $E \gg E_0$, the value $m_0 = m(E_0)$ is most appropriate, the reason being that the majority of collisions takes place at energies closer to $E_0$ than to $E$.

Finally, let us have a look at the deposited-energy density $F_D(E, \Omega, r)$. It may first be appropriate to stress that in the present notion, energy is considered *deposited* when being shared among the atoms down to the very end of linear energy dissipation, i.e., until Fig. 2.1c rather than Fig. 2.1b characterizes the state of motion. This is different from the stopping-power concept that focuses on the energy lost by *one* moving particle, disregarding further relocation of energy by means of recoil atoms.

A full determination of $F_D(E, \Omega, r)$ is quite a complex problem [2.50]. The function can be characterized more easily by its spatial moments such as [2.58]

$$x_D(E, \theta) \equiv \int d^3 r \, x F_D(E, \Omega, r) = X_D(E) \cos\theta, \tag{2.2.32}$$

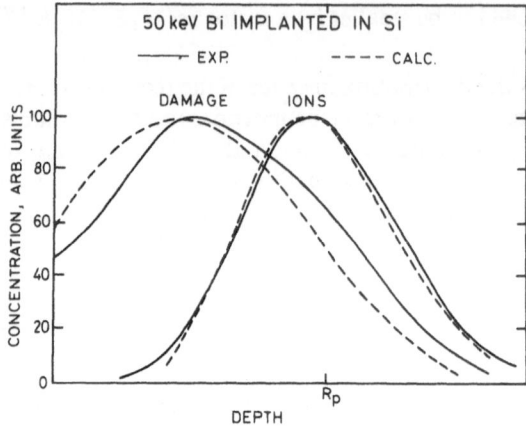

**Fig. 2.9.** Comparison of measured and calculated penetration and damage profiles. Experimental [2.161]: 50 keV bismuth ions implanted into crystalline silicon, analysed by Rutherford backscattering (RBS). Theoretical [2.162]: deposited-energy and penetration profile calculated for Thomas-Fermi interaction including electronic stopping. Both profiles are normalized to peak level. The stopping power in the *RBS analysis* has been adjusted to fit the calculated projected range $R_p$ of Bi

similar to (2.2.16). A major effort has been invested in providing estimates and measurements of the *depth profile* of deposited energy [2.46, 50, 68, 70], i.e.,

$$F_D(E, \theta, x) \equiv \int_{-\infty}^{\infty} dy \int_{-\infty}^{\infty} dz F_D(E, \mathbf{\Omega}, r), \qquad (2.2.33)$$

the x-axis being the target surface normal to which the beam direction $\mathbf{\Omega}$ makes an angle $\theta$. Measurements make use of the proportionality of radiation damage (disorder) with available energy. Thus, the depth profile of disorder follows $F_D(E, \theta, x)$ to the extent that the initial energy $E$ is very large compared to the pertinent threshold, and that noncollisional processes (relaxation, diffusion) do not exert a dominating influence on the profile.

Figure 2.9 shows a comparison between a calculated and a measured profile. The agreement is quite good, and by now, quite typical. Theoretical approaches have been reviewed extensively [2.58, 59, 68, 69]; experimental techniques include Rutherford backscattering spectroscopy, electron microscopy, and others.

For qualitative orientation one may note that the energy deposition profile roughly follows the corresponding penetration profile with regard to its scaling properties, its extension in depth and, to a lesser extent, its shape. The higher the mass ratio $M_2/M_1$, the more both profiles are characterized by successive large-angle scattering events undergone by the projectile [cf. (2.2.11a)], and the less important is recoil motion. Conversely, for a small mass ratio $M_2/M_1$, the heavy ion is stopped rapidly, while recoil atoms, despite lower energy, may have comparable ranges and thus be instrumental in determining the accurate shape and extension of the energy deposition profile.

## 2.3 Sputtering from Linear Collision Cascades

### 2.3.1 Particle Flux in an Infinite Medium

There is no sputtering without a target surface. Despite this, it is convenient initially to disregard the presence of a surface in the evaluation of the flux of recoil atoms initiated by some projectile. Here are some of the reasons. First, conservation laws like (2.2.30) are valid in an infinite medium, and utilizing them turns out to substantially simplify the theoretical procedure [2.36, 46, 50]. Second, even more important is the flexibility with regard to particular experimental geometries, including sputtering of a rough surface [2.71], transmission sputtering [2.48], and the like. A direct attack on the half-space problem is possible, but in practice, is restricted to particularly simple geometries like that involving a plane surface [2.72]. The influence of a target surface will, however, have to be corrected for [cf. Sect. 2.3.4(b)]. Last but not least, the infinite-target geometry may occasionally simulate the experimental situation better than the half-space; this includes such cases where sputtered particles hit a collector from which reflection and subsequent return to the target is possible.

Consider a source supplying $\psi$ projectiles per unit time of initial energy $E$. *This source generates a stationary distribution of moving target atoms.* Denoting $G(E, E_0)dE_0$ as the mean number of atoms moving at any time with energy $(E_0, dE_0)$, one immediately finds that

$$G(E, E_0)dE_0 = \psi n(E, E_0)dt_0 \tag{2.3.1}$$

where

$$dt_0 = \frac{dE_0}{|dE_0/dt|} = \frac{dE_0}{v_0|dE_0/dx|} \tag{2.3.2}$$

is the mean time needed by a recoil atom to slow down from $E_0 + dE_0$ to $E_0$, $v_0$ is the velocity of a target atom of energy $E_0$, and $n(E, E_0)$ is given by (2.2.25) in the simplest case. Indeed, since each recoil atom with energy $> E_0$ passes once and only once through the interval $(E_0, dE_0)$, $G(E, E_0)$ must be proportional to $n(E, E_0)$. Moreover, $\psi dt_0$ is a measure of the "coverage" in time of atoms with energies in that interval. From (2.3.1, 2) follows

$$G(E, E_0) = \frac{\psi n(E, E_0)}{v_0|dE_0/dx|}$$

$$\equiv \psi \frac{E_0}{v_0|dE_0/dx|} \cdot F(E, E_0), \tag{2.3.3}$$

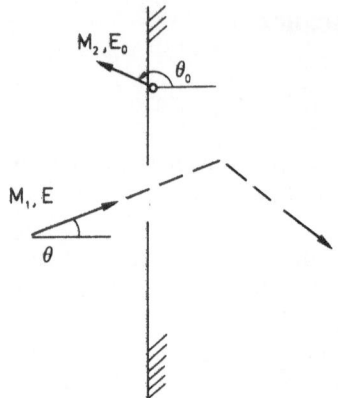

**Fig. 2.10.** Geometry for backsputtering from a semi-infinite target with plane surface

with the recoil density $F(E, E_0)$ given by (2.2.28). The steps leading from (2.2.28) to (2.2.31) also apply to (2.3.3), thus resulting in a mean number

$$\psi \Gamma_m \frac{F_D(E, \Omega, r)}{E_0 v_0 |dE_0/dx|} dE_0 \frac{d^2\Omega_0}{4\pi} d^3r \tag{2.3.4}$$

of atoms moving with energy $(E_0, dE_0)$ in the direction $(\Omega_0, d^2\Omega_0)$ in the volume $(r, d^3r)$ if there is a source at $r=0$ supplying $\psi$ primaries per unit time with energy $E$ and direction $\Omega$. Multiplication of (2.3.4) with $v_0 = v_0 \Omega_0$ yields the current density of target atoms,

$$\psi F_D(E, \Omega, r) \frac{\Gamma_m dE_0}{E_0 |dE_0/dx|} \frac{\Omega_0 d^2\Omega_0}{4\pi} \tag{2.3.5}$$

differential in energy $E_0$ and direction $\Omega_0$. This relationship can serve as a starting point for explicit evaluation of a flux of sputtered particles in the isotropic limit $(E \gg E_0)$ for a wide variety of geometries. The idea is to embed a target surface in the infinite medium and to evaluate the flux through that surface, taking due account of surface binding forces.

### 2.3.2 Backsputtering Yield from a Plane Surface

Consider an ion of energy $E$ impinging at an angle $\theta$ onto a plane surface, and take the $x$-axis along the inward surface normal (Fig. 2.10). Most often, the lateral distribution of the sputtered flux is of secondary interest. Thus, the outward current of target atoms through the surface plane $x=0$ is found by integration similar to (2.2.33), and becomes

$$J(E_0, \Omega_0) dE_0 d^2\Omega_0 = \psi F_D(E, \theta, 0) \cdot \frac{\Gamma_m dE_0}{E_0 |dE_0/dx|} |\cos\theta_0| \frac{d^2\Omega_0}{4\pi}, \tag{2.3.6}$$

still differential in direction and energy. $\theta_0$ is the angle between $\Omega_0$ and the surface normal (Fig. 10).

The expression (2.3.6) has a divergence at $E_0 = 0$ [cf. (2.2.10)]. This is a typical artifact of linear cascade theory. There is no real divergence since the number of moving atoms cannot be infinite in a cascade with finite dimensions. Such a divergence does not occur in a *spike* theory (cf. Chap. 4).

Even in a linear cascade, the sputtered flux becomes finite once surface (or bulk) binding conditions are imposed. Let $P(E_0, \theta_0)$ be the probability for an atom to escape *from the surface*. Then, the sputtering yield $Y$ is found by integrating (2.3.6) over $E_0$ and $\theta_0$ and dividing the sputtered current by the incident current $\psi$,

$$Y = \Lambda F_D(E, \theta, 0) \tag{2.3.7}$$

with

$$\Lambda = \frac{\Gamma_m}{2} \int \frac{dE_0}{E_0 |dE_0/dx|} \int d(\cos\theta_0) |\cos\theta_0| P(E_0, \theta_0). \tag{2.3.8}$$

In (2.3.7), all specific material properties enter into the constant $\Lambda$ which does not contain any parameters characterizing the ion. Conversely, the factor $F_D(E, \theta, 0)$ depends on the type, energy, and direction of the incident ion, and only the target parameters $Z_2$, $M_2$, and $N$ enter, cf. (2.2.4, 11, 18, 24).

Equations (2.3.7, 8) are central results which were first derived in [2.48]; it may be appropriate to recall the main assumptions made in their derivation. First, linear collision cascades were assumed, cf. (2.2.22), i.e., only a small fraction of the target atoms within the cascade volume, in particular the part that extends to the surface, are set in motion. This assumption breaks down for very high ion mass at intermediate energies (cf. Sect. 2.4). Second, large recoil cascades were assumed to ensure the validity of (2.2.25, 28) as well as the isotropy of the low-energy flux (2.2.29). This assumption breaks down, in general, at low energies ($\lesssim 1$ keV) and even at moderately low energies for large differences in ion and target mass, cf. (2.2.25). Third, the presence of a target surface is assumed not to exert an important influence on the development of collision cascades. This assumption breaks down for very light ions hitting heavy targets at low and intermediate energies, and in general for grazing incidence [cf. Sect. 2.3.4(b)]. Fourth, a number of reasonable models concerning elastic and inelastic scattering have entered implicitly in the evaluation. Fifth, bulk binding forces have essentially been ignored [cf. Sect. 2.3.4(a)]. Finally, any directional effects arising from the crystal lattice structure habe been ignored.

The simplest model for surface binding of a metal is based on a planar surface barrier $U_0$. In that case,

$$P(E_o, \theta_o) = \begin{cases} 1 & \text{for} \quad E_0 \cos^2\theta_0 > U_0 \\ 0 & \text{for} \quad E_0 \cos^2\theta_0 < U_0. \end{cases} \tag{2.3.9}$$

After the insertion of (2.3.9) and (2.2.10), (2.3.8) reads

$$\Lambda = \frac{\Gamma_m}{8(1-2m)} \cdot \frac{1}{NC_m U_0^{1-2m}}. \tag{2.3.10}$$

(Note that $\gamma = 1$, since knock-on collisions undergone by recoil atoms are equal-mass events.)

It was mentioned previously (Sect. 2.2.5) that the exponent $m$ occuring in $\Gamma_m$ refers to low-energy collisions. The same is true for the other occurrences of this exponent in (2.3.10), as follows immediately from the observation that the quantity in the dominator is essentially $|dE_0/dx|_{E_0 = U_0}$, i.e., the stopping power of a recoil atom at a very low energy [cf. (2.3.8)]. Since the scattering cross section at energies $E_0 \gtrsim U_0$ is a somewhat doubtful quantity both in practice and in principle, the author proposed [2.48] to insert the value $m = 0$ in (2.3.10) and to remain aware of the considerable inherent uncertainty involved. One then finds the expression

$$\Lambda = \frac{3}{4\pi^2} \frac{1}{NC_0 U_0} \tag{2.3.11}$$

with (2.2.7)

$$C_0 = \frac{\pi}{2} \lambda_0 a_{\text{BM}}^2. \tag{2.3.12}$$

The values of $\lambda$ and $a_{\text{BM}}$ were determined by comparison with previously proposed low-energy interatomic ("Born-Mayer") potentials [2.73],

$$\lambda_0 \cong 24; \, a_{\text{BM}} = 0.219 \, \text{Å}, \tag{2.3.13}$$

the latter, unlike (2.2.5), being chosen independent of $Z_2$. Feasible alternative choices could be given. However, the inherent uncertainty of the binary-collision model at *such* low energies does not encourage an improvement of this isolated feature.

Within the uncertainty of the model of binary random collisions, the assumption that $U_0$ is close to the experimentally measured sublimation energy appears reasonable, although not justified in principle. Empirically, that relationship has long been known [2.26, 74].

The second factor in (2.3.7) can, for dimensional reasons, be written in the form

$$F_D(E, \theta, 0) = \alpha \cdot NS_n(E), \tag{2.3.14}$$

where $\alpha$ is a dimensionless function of the angle of incidence $\theta$, the mass ratio $M_2/M_1$ and the ion energy $E$. For purely elastic collisions and power scattering

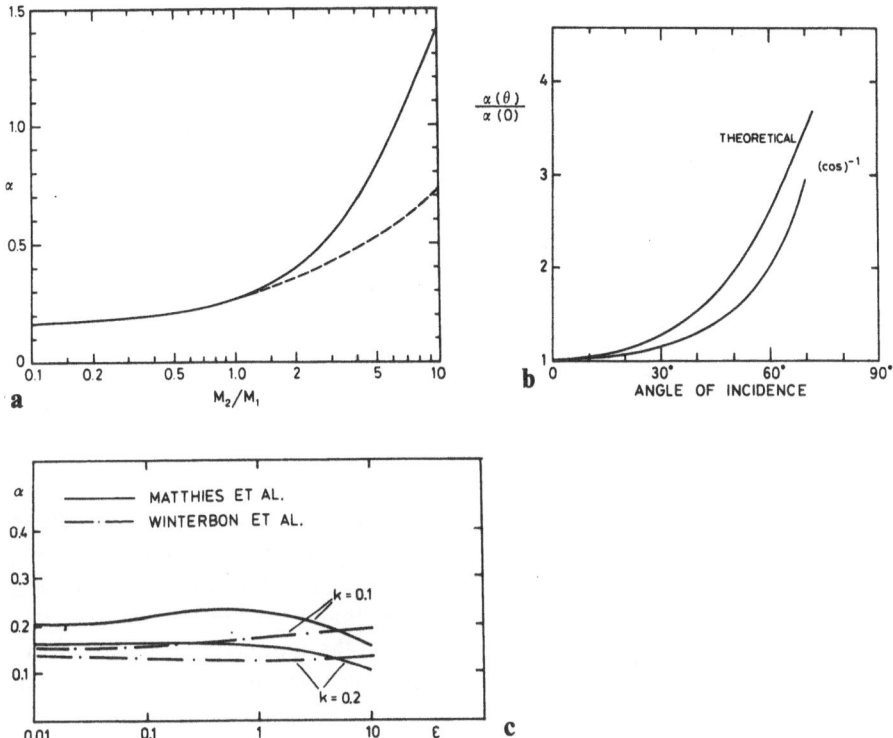

**Fig. 2.11a–c.** Factor $\alpha$ occurring in the backward-sputtering yield formula (2.3.7, 14). (**a**) Dependence on mass ratio $M_2/M_1$. Full-drawn line: theoretical [2.48], evaluated for elastic scattering only, no surface correction applied. Dashed curve: interpolated from experimental sputtering yields for 45 keV ions [2.52] on Si, Cu, Ag, and Au. The difference is mainly due to the neglect of the surface correction at large mass ratios. The increase of $\alpha$ with increasing mass ratio indicates an increasing importance of backscattered ions to the ejection of target atoms. (**b**) Dependence on angle of incidence. Full-drawn curve: theoretical [2.48] for Ar ions on Cu, surface correction applied. Thin curve: $(\cos\theta)^{-1}$ dependence, valid mainly in the high-velocity limit. (**c**) Dependence on energy parameter $\varepsilon$ specified in (2.2.4) for equal masses, $M_1 = M_2$. The parameter $k$ ($\approx 0.15$) determines the strength of electronic stopping [2.36]. Theoretical curves from [2.68, 75]

according to (2.2.6), the only length units entering $F_D$ are $E^{2m}/NC$ [cf. (2.2.15)] and $x$; hence, at $x=0$, $F_D$ must depend on $E$ like $S_n$, (2.2.10). The dimensionless parameter $\alpha$, being independent of $x$, can then not depend on $E$ [2.48, 50]. Electronic stopping, however, produces a slow decrease in $\alpha$ with increasing energy [2.75] which is conveniently tabulated in terms of the parameter $\varepsilon$ [cf. (2.2.4)], which also determines $S_n(E)$ according to (2.2.11).

Figure 2.11 shows several estimates of the factor $\alpha$ as a function of the three pertinent parameters. Obviously, $\alpha$ increases with increasing angle of incidence $\theta$ (except at glancing incidence) because of increasing density of energy deposition near the surface plane $x=0$. The role of the target surface at glancing incidence still needs to be discussed.

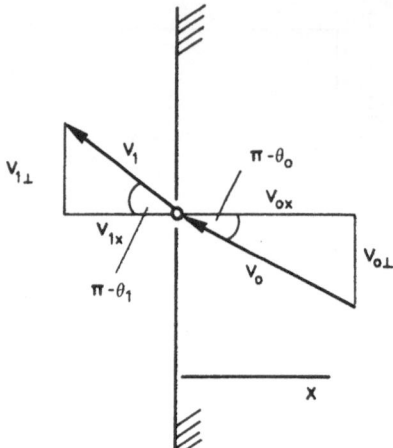

**Fig. 2.12.** Refraction effect during passage through a planar potential barrier (2.3.15a, 15b). Indices 0 and 1 refer to kinematic parameters immediately before and after ejection, respectively

Note also that $\alpha$ increases with increasing mass ratio $M_2/M_1$, an effect which is predominantly caused by the increasing relative importance of large-angle scattering events with decreasing ionic mass. Thus, at equal stopping power, light ions are more efficient sputterers than heavy ones. Except at low energies ($\lesssim 1$ keV), the dominating influence on the sputtering yield stems from the dependence of the stopping power on ion mass which is opposite to that of $\alpha$.

The results presented in this section are all contained in [2.48], although the derivation concentrated on some key points rather than mathematical details.

The results have been utilized rather literally over the past ten years. Numerous modifications are possible and feasible. Before entering into those, let us look into some features of the theory as it stands.

### 2.3.3 Energy and Angular Spectra, Characteristic Depths

The planar surface potential (2.3.9) implies that an atom is refracted upon passing through the surface according to (Fig. 2.12):

$$E_1 \cos^2 \theta_1 = E_0 \cos^2 \theta_0 - U_0 \qquad (2.3.15a)$$

$$E_1 \sin^2 \theta_1 = E_0 \sin^2 \theta_0 . \qquad (2.3.15b)$$

From (2.3.15a), $E_1 \cos\theta_1 d^2\Omega_1 = E_0 \cos\theta_0 d^2\Omega_0$ and hence, (2.3.6) yields

$$d^3 Y = F_D(E, \theta, 0)\frac{\Gamma_m}{4\pi} \frac{E_1 dE_1}{(E_1 + U_0)^2 |dE_0/dx|_{E_0 = E_1 + U_0}} |\cos\theta_1| d^2\Omega_1 \qquad (2.3.16a)$$

for the differential sputtering yield.

One may note that (2.2.4) predicts the *absolute magnitude* of an energy spectrum. Measured spectra have usually been given in arbitrary units.

After insertion of (2.2.10) one obtains

$$\frac{d^3 Y}{dE_1 d^2 \Omega_1} = F_D(E, \theta, 0) \cdot \frac{\Gamma_m}{4\pi} \cdot \frac{1-m}{NC_m} \cdot \frac{E_1}{(E_1 + U_0)^{3-2m}} |\cos\theta_1|. \tag{2.3.16b}$$

Thus, the energy spectrum at a fixed ejection angle $\theta_1$ exhibits a maximum at

$$E_1 = (E_1)_{\text{max}} = \frac{U_0}{2(1-m)}. \tag{2.3.17}$$

The experimental observation of a maximum at roughly this energy was interpreted by *Thompson* [2.45] as evidence in favor of a planar surface potential. This implies that the position of an observed maximum in the energy spectrum of sputtered particles depends on the target only (within the linear cascade regime) and is insensitive to ion type and energy. Existing experimental evidence in favor of this [Ref. 2.7, Chap. 2] is hardly compelling. Alternative models for the ejection probability are available, but also attached to considerable uncertainty [2.76].

In accordance with Sect. 2.2.5, the exponent $m$ in (2.3.16b) varies with energy, i.e.,

$$m = m(E_1),$$

decreasing from a value around $m \sim 0.2$–$0.3$ at $E_1 \sim 1$ keV down to $m \sim 0$ at $E_1 \sim U_0$, somewhat dependent on the target. In the analysis of experiments, caution is indicated in extending (2.3.16a) up to too high energies $E_1$ because of the possible break-down of the isotropy condition $E \gg E_1$.

*Thompson* [2.45], making the assumption that

$$dE_0/dx \cong E_0/D, \tag{2.3.18}$$

where $D$ is the nearest-neighbour distance, arrived at a spectrum similar to (2.3.16b) corresponding to $m = 0$. At energies in the range $\ll 100$ eV, (2.3.18) is presumably as feasible an approximation as the power law (2.2.10). The latter does, however, make it clear that there is a considerable uncertainty, as well as a variation with $E_1$, in the exponent entering into the energy spectrum (2.3.16b). Note in particular that such a variation is not inherent in the recoil density (2.2.28), where $m$ only enters into the numerical factor in the front. [Concerning the influence of binding energies, cf. Sect. 2.3.4(a).]

A cosine law results for the angular distribution of the sputtered flux, according to (2.3.16b). While this dependence is characteristic of an *isotropic* flux *in* the target, observed deviations from such behaviour may have several origins such as surface corrections [cf. Sect. 2.3.4(b)], surface roughness effects, and anisotropy corrections [cf. Sect. 2.3.4(c)]. In addition, texture effects may occur in polycrystalline targets.

Let us consider the pertinent depths characterizing the sputtering process in the present model. In view of the factorization of the sputtered flux (2.3.14), into one part determined by the slowing-down of the ion (and high-energy recoil atoms) and another part representing the spectral properties of sputtered atoms, one deals with at least two types of depth variables that must be suspected to be only loosely related.

First, one may ask from which depth range sputtered atoms originate. It may be useful to differentiate the answer with regard to the different parts of the energy spectrum of sputtered atoms.

Let us look for the mean number of atoms ejected per incident ion with an energy exceeding some arbitrary minimum energy $E_0 (E_0 \gg U_0)$. According to (2.3.6) and (2.2.10), that number is

$$Y(E_0) \cong \frac{\Gamma_m}{4} \frac{1-m}{1-2m} \frac{E_0^{2m}}{NC_m} \frac{F_D(E,\theta,0)}{E_0} \tag{2.3.19}$$

after integration over the hemisphere and insertion of (2.2.10). Now, by comparison with (2.2.25), this can be written as

$$Y(E_0) \cong \frac{1}{4} \cdot \Gamma_m \frac{F_D(E,\theta,0)\Delta x_0}{E_0}, \tag{2.3.20}$$

where $\Gamma_m F_D \Delta x_0 / E_0$ is the number of atoms per ion set in motion with an energy $> E_0$ in the depth interval $(0, \Delta x_0)$, and the factor $1/4$ represents a cosine factor averaged over the hemisphere, i.e., accounts for the fact that not all atoms set in motion move in the direction of the outward surface normal. Obviously, the quantity

$$\Delta x_0 \cong \frac{1-m}{1-2m} \frac{E_0^{2m}}{NC_m} \tag{2.3.21}$$

is a measure of the depth from which those atoms emerge. It appears that *the higher the energy of a sputtered atom, the more likely it comes from deeply inside the target.* This perhaps surprising result is a direct consequence of the strongly peaked energy spectrum of recoil atoms toward small energies $E_0$, cf. (2.2.28). For $E_0$ approaching $U_0$, the above estimate is modified slightly because of surface binding. Regardless of the details, $\Delta x_0$ may become quite small, (less than 10 Å [2.48]), and rather insensitive to energy since here, $m \lesssim 0.2$.

Conversely, the depth variable $x_0$ characterizing the deposited energy profile is of the order of $R \sim E^{2m}/NC$, i.e., dependent on the impinging ion energy. This quantity is of importance mainly with regard to thin-layer or compound-layer sputtering. Indeed, one may inquire about the minimum thickness required for a foil to behave like a semi-infinite target with regard to backsputtering. An accurate prediction depends on the mass ratio in the sense that for high values of $M_2/M_1$, $x_0$ approaches the mean penetration depth, while for $M_2/M_1 \lesssim 1$ it is somewhat smaller [2.48].

### 2.3.4 Corrections to Simple Cascade Theory

#### a) Bulk Binding Forces

In the treatment as sketched up till now, all bulk binding forces were ignored. In this picture, an arbitrarily small amount of energy is sufficient for an atom to be set in motion and to participate in the cascade. This (rather common) picture is less unphysical than it may look. It breaks down, of course, at energies near the lattice binding energy, i.e., a few eV, but atoms of such low energy can only contribute to sputtering if able to overcome the surface potential which is of the same order of magnitude. Thus, the main effect of a lattice binding energy, called $W$ in the following, is to absorb energy from the cascade *during the slowing-down process*. One finds [2.77] that this absorption can be corrected for by a generalization of (2.2.25):

$$n(E, E_0) \sim \Gamma_m \frac{E}{E_0 + (2 - m) W} \tag{2.3.22}$$

for $E_0 \gg W$.

It has often been asserted, in particular in early theories [2.31, 33, 39], that the pertinent bulk binding energy $W$ should be of the order of the threshold energy $E_d$ for permanent lattice displacement. That energy is normally an order of magnitude higher than $U_0$ [2.78]. If this assertion were justified, the surface potential would become immaterial to the sputtering yield. All other parameters remaining fixed, the yield would be expected to become smaller by a factor of $\sim U_0/E_d$. Alternative calculations [2.79] leading to the opposite conclusion also vary other parameters.

However, the displacement threshold energy $E_d$ is consumed mainly *to move an interstitial atom far enough away from a vacant lattice site to prevent recombination* (Fig. 2.13). The critical distance may well exceed the depth range $\Delta x_0$ from which atoms are typically ejected. Thus, an atom may be easily sputtered even though its initial energy is substantially less than $E_d$. Therefore, it appears appropriate to assume that $E_d$ is immaterial in the evaluation of the sputtering yield.

The influence of a binding energy $W$ on the yield and spectrum of sputtered atoms is readily evaluated from (2.3.4, 22). As a result, the energy spectrum in (2.3.16) is modified to

$$\frac{E_1}{(E_1 + U_0)^{3 - 2m}} \rightarrow \frac{E_1}{(E_1 + U_0 + W)(E_1 + U_0)^{2 - 2m}}, \tag{2.3.23}$$

where the peak is shifted toward a higher energy than the one given in (2.3.17). Of course, the total yield corresponding to (2.3.23) is smaller than predicted by (2.3.10).

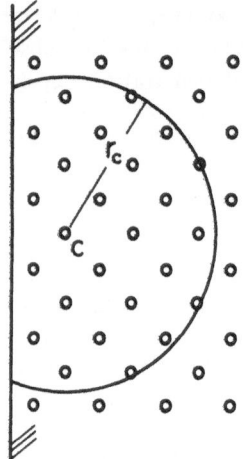

**Fig. 2.13.** Definition of bulk displacement threshold $E_d$. An atom located in point $C$ in the bulk has to receive a minimum recoil energy $E_d$ in order to generate a stable vacancy-interstitial pair. Somewhat oversimplified, this implies that the interstitial forms outside a sphere with radius $r_c$ and the vacancy in $C$. The recombination volume $4\pi r_c^3/3$ contains $\sim 100$–$200$ atomic sites in metals. In actual fact, $E_d$ and the recombination volume are anisotropic [2.78]

## b) Backsputtering from a Semi-Infinite Target

The effect of a target surface on a collision cascade is somewhat similar to that of an absorbing wall on a diffusion profile in unlimited space. An absorbing wall can in turn be simulated by an image source of negative sign to ensure zero particle density at the position of the target surface. Therefore, the flux through a perfectly absorbing wall is twice that through an equivalent reference plane in an infinite medium. The difference between the sputtered flux predicted for an infinite medium and for a half-space is expected to be less pronounced because of the slowing down of moving particles. Unlike in diffusion, particles will not be able to cross a target surface (in an infinite medium) an unlimited number of times because they lose energy while moving.

In this paragraph, three distinct surface effects will be treated, of which the first two tend to compensate each other:

(i) intersection of a uniform, isotropic flux of low-energy atoms with a real target surface, disregarding binding forces;

(ii) nonuniformity of the low-energy flux because of nonuniform deposited-energy distribution near the surface;

(iii) truncation of the collision cascade by the "absorption" of high-energy particles from the medium when reaching the surface.

With regard to (i), let the surface be at $x=0$ and assume a homogeneous isotropic distribution of low-energy atoms, $\varrho(E_0)dE_0d^2\Omega_0/4\pi$ in the region near the target surface and $0\leq x\leq x_1$ with $x_1\gg\Delta x_0$, $\Delta x_0$ being the depth of emergence of sputtered atoms (2.3.21). Because of the linearity of the transport equation (2.2.23), the effect of an absorbing surface can be simulated by the addition of an *image density*

$$\varrho_{im}(E_0)=-\varrho(E_0)\quad\text{for}\quad x\leq0. \tag{2.3.24}$$

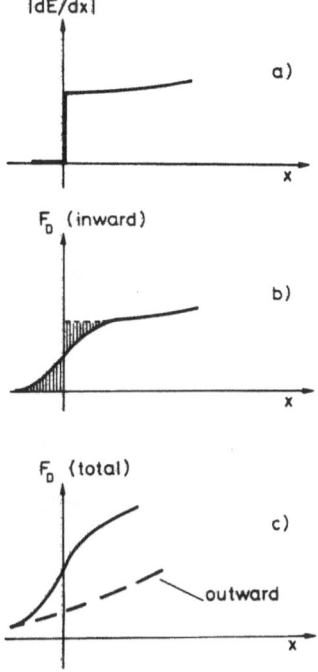

**Fig. 2.14a–c.** Energy deposition near the plane $x=0$ in an infinite medium. (a) Stopping power, i.e. primary energy deposition as a function of penetrated depth for inward motion starting in $x=0$. For initial energy equivalent to $\varepsilon \lesssim 1$, this function increases with $x$ according to Fig. 2.3. (b) Deposited energy due to an inward moving ion. The difference to Fig. 2.14a is caused by energy dissipation of recoil atoms. (c) Total energy deposition including energy dissipation by a backscattered ion moving outward through $x=0$

This causes the total density to be zero at $x=0$ and the flux in the outward direction to increase by a factor of two.

This effect is compensated in part by the fact (ii) that the flux is not actually homogeneous near the surface, even in an infinite medium. This is illustrated in Fig. 2.14. Since impinging ions start dissipating energy at $x=0$, the energy loss as a function of depth has a discontinuity at $x=0$. This discontinuity is smeared by energy transport due to recoiling atoms. A smearing of the order of $\sim \Delta x_0$ is not taken into account in the determination of the deposited-energy distribution $F_D(E, \theta, x)$, since $E_0$ is set equal to zero there. Hence, the calculated distribution $F_D$ normally exhibits a discontinuity at $x=0$ [2.80, 81]. When allowed to be smeared, $F_D(E, \theta, x)$ looks like indicated in Fig. 2.14b, and $F_D(E, \theta, 0)$ reduces to about one half the uncorrected value.

Thus, in case all energy is deposited by incoming ions, the two corrections (i) and (ii) tend to cancel, and the value of the deposited energy determining the sputtering yield is $F_D(E, \theta, x \to 0+)$.

The same argument does not apply to the case where back-scattered ions and high-energy ejected recoil atoms contribute to energy deposition, since such particles deposit energy on both sides of the plane $x=0$ in an infinite medium. This causes the cancellation of the effects mentioned under (i) and (ii) to be only partial.

Consider then (iii), the difference between $F_D$, applying to an infinite medium, and $F_{D,o}$ to a semi-infinite one. An argument used in the analogous problem of ion reflection coefficients can be readily applied [2.82]. Let the flux per ion of ejected recoil atoms from a semi-infinite target be $Y_o(E, \mathbf{\Omega}, E_0, \mathbf{\Omega}_0)dE_0 d^2\mathbf{\Omega}_0$, and of reflected ions $R_o(E, \mathbf{\Omega}, E_0, \mathbf{\Omega}_0)dE_0 d^2\mathbf{\Omega}_0$. Then, $F_D$ can be split into the following contributions for $x \geq 0$:

$$F_D(E, \theta, x) = F_{D,o}(E, \theta, x) + \int Y_o(E, \mathbf{\Omega}, E_0, \mathbf{\Omega}_0)dE_0 d^2\mathbf{\Omega}_0 F_{D,r}(E_0, \theta_0, x)$$
$$+ \int R_o(E, \mathbf{\Omega}, E_0, \mathbf{\Omega}_0)dE_0 d^2\mathbf{\Omega}_0 F_D(E_0, \theta_0, x), \qquad (2.3.25)$$

where $F_{D,r}$ is the energy deposition function in an infinite medium for a recoil atom. Setting $x=0$ one may write

$$F_D(E, \theta, 0) = F_{D,o}(E, \theta, 0) + Y_o \langle F_{D,r} \rangle_o + R_o \langle F_D \rangle_o, \qquad (2.3.26)$$

where $Y_o$ and $R_o$ are the sputtering yield and ion reflection coefficient for a semi-infinite medium, and the averages $\langle \ \rangle_o$ go over the angular and energy spectrum of sputtered atoms and reflected ions for a semi-infinite medium. Multiplying by $\Lambda$ and using (2.3.7), one finds

$$Y_o = \frac{Y - R_o \langle Y \rangle_o}{1 + \langle Y_r \rangle_o}, \qquad (2.3.27)$$

where $Y_o$ and $Y$ are the sputtering yields from a semi-infinite and infinite target, respectively, and $Y_r$ the self-sputtering yield from an infinite target.[3]

This relation has been known for some time [2.69, 83] but has not yet been evaluated quantitatively. The main difficulty involved is the need for full angular and energy dependences of reflected and sputtered particle fluxes at relatively high energies, since the averages $\langle Y \rangle_o$ and $\langle Y_r \rangle_o$ hinge on that spectral part.

The qualitative trends of (2.3.27) are determined mainly by the numerator. Obviously, for a large reflection coefficient $R_o$, in particular if the spectrum of reflected particles is peaked toward the initial energy, the second term tends to cancel the first one, thus causing the sputtering yield $Y_o$ to be small. This is what happens for oblique incidence, $\theta \rightarrow 90°$ (Fig. 2.15). The uncorrected yield $Y$ would approach some finite value which is in general substantially greater than $Y(\theta = 0)$ (Fig. 2.16). Thus, $Y_o$ goes through a maximum and approaches zero with increasing angle of incidence. This behaviour shows up qualitatively in Fig. 2.15. Note however, that the behaviour sketched in (2.3.27) is valid *only* for a planar target surface, and that surface roughness effects may well be important in the analysis of experimental results.

Apart from the crucial importance for the angular dependence of the sputtering yield, the surface correction also affects the sputtering yield at

---

3 Note that the functions $F_{D,r}$ and $F_D$ occurring in the second and third term on the r.h.s. of (2.3.26) are continuous through $x=0$.

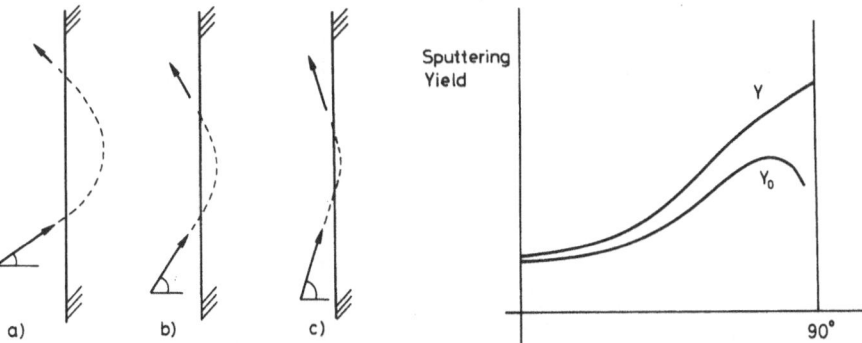

**Fig. 2.15a–c.** Ion reflection by multiple collisions. With increasing angle of incidence $\theta$ [(**a**) through (**c**)], the probability for reflection increases, and the average path length travelled inside the target, and hence the energy loss, decreases

**Fig. 2.16.** Sputtering yield corrected ($Y_o$) and uncorrected ($Y$) for ion reflection, *schematically*. For large mass ratios $M_2/M_1$ and moderately low ion energies, the correction is presumably more pronounced

perpendicular incidence when analyzed as a function of mass ratio $M_2/M_1$ and energy. For example, because of the increase of $R_0$ with increasing $M_2/M_1$, the ratio $Y_o/Y$ decreases. This is in qualitative agreement with the experimental finding [Ref. 2.4, Chap. 5].

The denominator in (2.3.27) has a somewhat different behaviour. Its effect on $Y_o$ is smaller than that of the numerator, causing a slight decrease of $Y_o$ relative to $Y$ with decreasing mass ratio and *increasing* angle of incidence.

Alternative treatments of sputtering from a half-space have recently become available [2.72, 84].

### c) Anisotropy Correction

The most profound simplifying feature in the treatment as sketched up till now is that of an isotropic flux of low-energy recoil atoms as expressed by (2.3.5). Whenever this expression is valid, *the angular and energy distributions of sputtered particles are independent of ion type, energy, and angle of incidence*, and *the variation of the sputtering yield with ion type energy, and angle of incidence is determined by the target parameters $M_2$ and $Z_2$ but not by the surface binding energy and other material parameters.*

It is well documented experimentally [2.85] that for energies in the lower keV region, more or less pronounced deviations from this behaviour may occur, even though $E$ is still quite large compared to $U_0$. Deviations also occur for spike (cf. Sect. 2.4.2) and light-ion sputtering.

The treatment of anisotropy effects was initiated by *Sanders* [2.44], and developed by *Littmark* [2.86, 87]. In these approaches, the anisotropy of the recoil flux is directly related to the momentum of the incoming ions. In the simplest case, i.e., after ignoring all other corrections mentioned in this section,

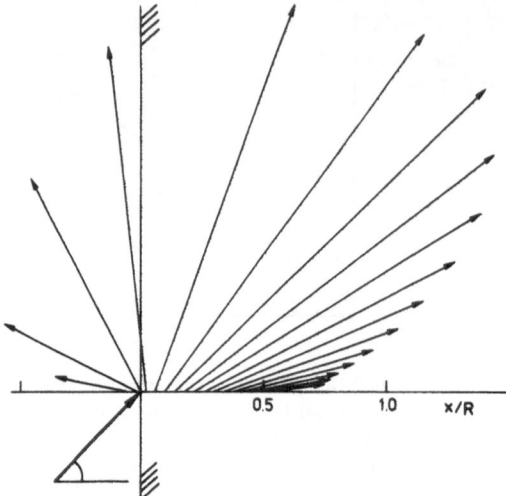

**Fig. 2.17.** Calculated density of deposited momentum $F_p(E, \Omega, x)$, according to [2.86, 87], for $\theta = 45°$, $M_2/M_1 = 1.5$, and elastic scattering according to (2.2.6) with $m = 1/2$. The unit of depth refers to the average path length (2.2.14, 15). The origin of each vector is located in the point $(x, 0)$ [where $x$ is the depth at which $F_p(E, \Omega, x)$ is specified], the direction is that of $F_p$ at $x$, and the length scale is such that $\int_{-\infty}^{\infty} F_p(E, \Omega, x)dx = M_1 v$ [courtesy of U. Littmark]

one obtains the following generalization of (2.3.4) for the density of recoil atoms in phase space [2.69],

$$\psi \frac{\Gamma_m}{4\pi} \frac{dE_0}{v_0|dE_0/dx|} d^2\Omega_0 d^3r \left( \frac{F_D(E, \Omega, r)}{E_0} + 3\Omega_0 \cdot \frac{F_P(E, \Omega, r)}{\sqrt{2M_2E_0}} \right) d^2\Omega_0, \qquad (2.3.28)$$

where $F_P(E, \Omega, r) \cdot d^3r$ is the mean momentum deposited in $(r, d^3r)$ per incoming ion of initial energy $E$ and direction $\Omega$. [Some caution with regard to (2.3.28) is indicated in case of sizeable electronic stopping.]

The momentum distribution is a substantially more complex quantity than the energy deposition profile $F_D$. It is a vector, and at least the component parallel to the incoming ion changes sign in general [2.87] (Fig. 2.17).

Within the range of validity of (2.3.28), one obtains a flux through a plane surface of the form [cf. (2.3.6)]

$$\psi \frac{\Gamma_m}{4\pi} \frac{dE_0}{|dE_0/dx|} |\cos\theta_0| d^2\Omega_0 \left( \frac{F_D(E, \theta, 0)}{E_0} + 3 \frac{\Omega_0 \cdot F_P(E, \Omega, 0)}{\sqrt{2M_2E_0}} \right). \qquad (2.3.29)$$

This makes it clear that the anisotropy correction

(i) causes deviations from the simple cosine dependence of the sputtered flux,

(ii) causes a further complication of the energy spectrum of sputtered particles,

(iii) removes the simple decoupling between the slowing-down characteristics of high and low-energy particles, and

(iv) indicates that deviations from the isotropic limit are of the relative order of $(M_1E_0/M_2E)^{1/2}$.

For a moving projectile, the mean loss of momentum in the direction of motion is

$$\frac{dP}{dx} = \frac{1}{v}\frac{dE}{dx},$$
(2.3.30)

and hence, $F_P(E, \mathbf{\Omega}, 0)$ can be split according to

$$F_P(E, \mathbf{\Omega}, 0) = \frac{1}{v} N S_n(E) \cdot (\alpha_{\parallel} e_x + \alpha_{\perp} e_\varrho),$$
(2.3.31)

where $e_x$ and $e_\varrho$ are unit vectors perpendicular and parallel to the surface plane, and $\alpha_{\parallel}$ and $\alpha_{\perp}$ are dimensionless quantities analogous to $\alpha$ (2.3.14), but with the possibility of taking on *negative* values. For perpendicular incidence, $\alpha_{\parallel}$ is in general, negative [2.87], thus causing a positive correction to the isotropic flux by narrowing the angular distribution and broadening the energy spectrum of sputtered particles.

In the approximation corresponding to (2.3.11), the sputtering yield can be corrected for anisotropy effects according to [2.83]

$$Y(\theta = 0) = N S_n \cdot (\Lambda\alpha + \Lambda'\alpha_{\parallel})$$
(2.3.32)

with

$$\Lambda' = \frac{9}{2\pi^2 N C_0 (\gamma E U_0)^{1/2}}$$
(2.3.33)

and $\gamma$ given by (2.2.2).

These results should be taken as trends rather than quantitative predictions. First, one deals with a slowly convergent series expansion in terms of $(M_1 E_0/M_2 E)^{1/2}$. Second, the corrections treated in the previous paragraphs, in particular surface corrections, are largely unexplored with regard to the momentum profile. Third, even within the present scheme, substantial uncertainties prevail with regard to the detailed behavior of the momentum distribution.

For a number of specific systems, *Betz* et al. [2.88] investigated the deviations from the isotropic limit by means of Monte Carlo simulation of the sputtering process at keV initial ion energies. It was found that sputtered atoms belonging to high generations within the cascade, i.e., as a general rule, those with the lowest energies, followed the predicted cosine law for the emission characteristics, while the lower generations showed a more anisotropic behavior with the beam direction playing the role of a second reference axis in addition to the surface normal; cf. (2.3.29).

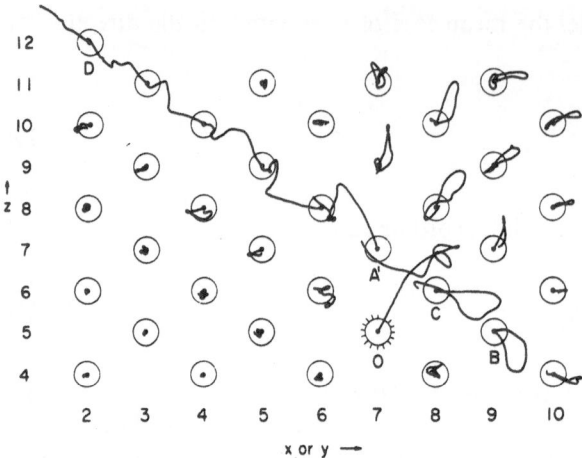

**Fig. 2.18.** Molecular-dynamics simulation of 65 eV Fe recoil in Fe. The event is initiated by atom 0 being knocked off in a direction lying in the (1$\bar{1}$0) crystal plane of an ideal bcc lattice. Thermal displacements are not included. Note the collective motion in the upper right part and the replacement sequence toward the upper left. (From [2.163], courtesy of the American Institute of Physics)

### 2.3.5 Comments on Atomic Motion at Low Energies

The description of atomic motion adopted here does not differ radically from the one developed to describe the motion of MeV alpha particles through gases [2.35], except for the quantitive scale. There are valid reasons to believe that this picture is too naive, and that motion of atoms at low energies differs substantially from what has been assumed here. For example, computer simulation of low-energy events, on the basis of more or less feasible atomic interaction potentials, indicates a dominance of collective types of motion which may be more [2.89] or less [2.90] collimated in specific directions (Fig. 2.18). The claim has been made [2.91] that for this reason, the transport theory is not valid altogether.

While the latter claim disregards the decoupling between high- and low-energy atomic motion, and therefore may be justified only in the regime where asymptotic linear cascade theory becomes inaccurate anyhow, the former objection is serious since it affects the absolute magnitude of the yield as well as the energy spectrum.

While this is not the place to propose an alternative theoretical description of low-energy atomic motion, some comments might be appropriate on the question of *where* cascade theory actually breaks down.

Let us recall the description of low-energy atomic motion adopted above.

(i) Atoms are described as *point particles* associated with a cross section *which determines the probability for the occurrence of collisions.*

(ii) The target is assumed structureless and characterized by a density $N$ of randomly distributed atoms.

(iii) Collisions are assumed binary and elastic.

(iv) Particle motion is described by classical dynamics.

(v) The density of atoms moving with energy in excess of $E_0$ (or $U_0$) is small.

Of those five assumptions, (ii) and (iii) have been extensively questioned in the literature, (iv) has (most often tacitly) been accepted, while (i) and (v) are rarely mentioned as being essential simplifications inherent in the theory.

As far as assumption (iii) is concerned, the present author does not share the opinion of others [2.90] that the very many-body nature of low-energy collisions is essential for the understanding of the sputtering process.[4] Although it is clear that low-energy collisions in solids cannot be strictly binary, there is no obvious way in which many-body effects should *systematically* modify the predictions of a binary-collision picture. Note in particular, that even a binary-collision theory may reveal correlations between different events, although such correlations have only rarely been looked for [2.93, 94]. Moreover, it is usually not appreciated that in penetration theory, many-body effects are fully accounted for to the extent that perturbation theory applies [2.95], in which limit the effects of simultaneous binary collisions are superimposed. Therefore, the limits of the binary-collision model are largely determined by the range of validity of perturbation theory (momentum approximation). This indeed, sets a practical limit to the validity of the binary-collision model, but the discrepancies observed at energies below that limit are not alarming [2.96], taken into account other inherent uncertainties.

The most severe consequence of assumption (ii) is the neglect of lattice structure on the sputtering of crystalline matter. If we restrict our attention to sputtering of polycrystals where effects due to crystal *anisotropy* tend to average out, the question remains as to what *systematic* effects remain in the sputtering of crystalline as opposed to, e.g., amorphous targets. It is the general consensus that with regard to low-energy atomic motion, such an effect would have to be due to focused collision sequences (focusons) ([2.53] and Chap. 3), i.e. connected closely with the breakdown of assumption (i).

Standard arguments based on the size and angular spread of wave packets [2.35] are not suggestive of major quantal corrections to a classical-collision dynamics modelling of sputtering. In case such collision models are analyzed in meticulous detail [2.90], however, questions concerning the validity of classical arguments *with regard to specific processes* should at least be *asked*.

Assumption (v) breaks down by definition under spike conditions. However, in case assumption (i) breaks down it is hard to see how (v) could possibly be valid.

It thus appears that (i) is the key assumption, such that the range of validity of the model adopted for low-energy atomic motion is roughly equivalent to

---

4 Part of the conclusions of these authors are suspected to be spurious because of a rather unusual choice of interatomic potential [2.92]

the energy range within which it is justified to consider atoms as freely moving point particles undergoing (more or less pronounced) collisions at *intervals dependent on the cross section.*

Considerations based on the distance of closest approach between collision partners as well as inspection of computer simulation results [2.89] indicate that in the lower (and possibly medium) eV region, colliding atoms cannot pass each other but instead interchange energy, staying rather far ($\sim 1$ Å) apart from each other. This influences the collision spectrum which then, is no longer determined by the cross section [2.97]; more important, this also has a profound influence on the time scale of slowing down as expressed, e.g., by (2.3.2).

Possible consequences on the cascade theory of sputtering have not been discussed quantitatively. Since it is not essential for many purposes whether it is actually *atoms* that are transported over significant distances or just their *energy*, one might even suppose that as long as energy degrades according to some relation such as (2.3.18), practical consequences on the sputtering effect might be quite limited. To the extent that energy degrades more or less isotropically, such a conclusion may be justified. Energy degradation studies have been made long ago [2.91] and might become useful in that context.

In the case of *directional* transport of energy, such as in the presence of focused collision sequences in crystals, (2.3.3) becomes inapplicable since neither the expression (2.2.25) for $n(E, E_0)$ nor (2.3.2) for $dt_0$ remain valid. One can, however, easily estimate on the basis of an argument like the one leading to (2.3.20), whether or not focused collision sequences contribute substantially to the sputtering yield. Experimental evidence indicates that they do not [2.98].

Indeed, let $\Omega^* = \pi \alpha^{*2}$ be a representative solid angle around a main crystal direction in which focusons propagate, and let $z$ be the total number of such directions in a particular structure. Let $E_f^*$ be a representative initial energy for long-range focusons and $R_f^*$ a representative focuson range. Then, a very rough estimate for the focuson contribution to the sputtering yield is given by

$$Y_f \sim \frac{z \cdot \Omega^*}{4\pi} \cdot \frac{R_f^*}{\Delta x_0} \cdot \frac{U_0}{E_f^*} \, Y, \tag{2.3.34}$$

where $Y$ is the "random" sputtering yield. Equation (2.3.34) follows from (2.3.20), under the assumption of random slowing down until a focuson is generated [2.40], factors of the order of 1 being disregarded. The factor $z\Omega^*/4\pi = z \cdot (\alpha^*/2)^2$ is $\sim 0.2$ for a (generously chosen) acceptance angle of $\alpha^* \cong 15°$. The ratio $U_0/E_f^*$ is of the order of $\lesssim 0.1$ for recommended values of the focusing energy (Chap. 3). Thus, except if $R_f^*$ should be several hundred Ångströms, one finds

$$Y_f \ll Y.$$

Let us assume, for a moment, that $R_f^*$ is indeed large enough so that $Y_f \gtrsim Y$. Then, the depth of origin $\Delta x_0$ would have to be replaced by a quantity approaching $R_f^*$, i.e., several hundred Ångströms. This would have a profound effect on the dependence on ion type and energy of the sputtering yield in that the yield would drop rapidly as soon as the ion range becomes smaller than $R_f^*$. Observed deviations [Ref. 2.4, Chap. 5] from the dependencies predicted by (2.3.7) are hardly consistent with such behavior.

It should be stressed that the above argument applies to the *total* sputtering yield of a *polycrystal*. In a single crystal, the differential yield in a focusing direction would not contain the factor $z \cdot \Omega^*/4\pi$, and if upper portions of the energy spectrum ($E_1 \sim E_f^*$) are analysed, the term $U_0/E_f^*$ also drops out. Thus, under such conditions, focusons, if existent at all, may be isolated [2.99].

It appears fair to draw the conclusion from this paragraph that the expression (2.3.8) for the material constant $\Lambda$ in the sputtering yield (2.3.7) is open to considerable doubt, and alternative evaluations may be desirable.

### 2.3.6 Transmission Sputtering

When an ion beam hits a foil that is thin enough to allow ions to penetrate, erosion may be expected to be observable on both the near and the far side of the foil. The two effects are called backsputtering and transmission sputtering, respectively. Strictly speaking, transmission sputtering might even occur under conditions where no *ions* are transmitted, provided that only sufficient *energy* is transported to the far side of the foil to allow ejection of atoms.

Transmission sputtering experiments may be performed with various purposes in mind. Quite frequently, the purpose has been to provide direct evidence about the existence and range of focused collision sequences [2.27, 100]. More generally, the purpose was a detailed check on the spatial distribution of atomic collision cascades [2.101].

At this point, transmission sputtering is mentioned mainly as a characteristic application of the general description of sputtering to an experimental geometry that differs from the half-space discussed up till now.

When we consider a slab with thickness $t$, the distribution of recoiling atoms in real and velocity space is still given by (2.2.31). The expression for the particle flux follows in a similar manner as in Sect. 2.3.1, and hence the transmission sputtering yield becomes

$$Y_{tr} = \Lambda F_{D,t}(E, \theta, t) \tag{2.3.35}$$

in complete analogy to (2.3.7), where $\theta$ is the angle of incidence and $F_{D,t}(E, \theta, x)$ the distribution of deposited energy as a function of depth $x$ in a foil of thickness $t$.

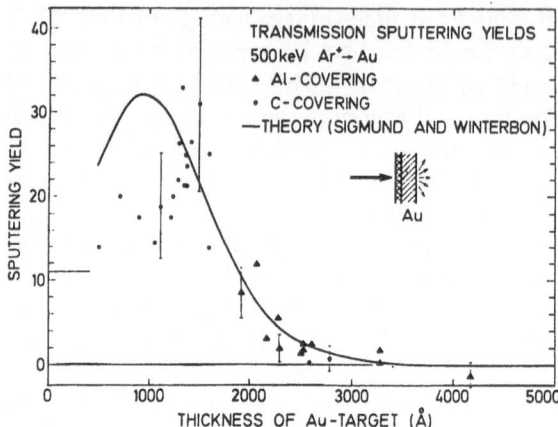

**Fig. 2.19.** Transmission sputtering of gold foils by 500 keV argon bombardment. The surface facing the beam was coated with carbon or aluminium to minimize backsputtering. Experimental results are compared with the theoretical prediction (2.3.35), on the basis of a deposited-energy function uncorrected for surface scattering of the ion beam at the downstream surface. Good agreement is found for target thicknesses exceeding the mean projected range. (From [2.101], courtesy of Gordon and Breach Science Publishers)

For sufficiently large values of $t$ (i.e., well beyond the half-width of $F_D$), the function $F_{D,t}$ will approach the corresponding one for the infinite medium (Fig. 2.19). For somewhat smaller values of $t$, surface corrections of the type considered in Sect. 2.3.4(b) may relate $F_{D,t}$ with $F_D$. For values of $t$ within the half-width of $F_D$ or even less, the determination of $F_D$ should probably utilize the formalism of multiple scattering theory. For very thin targets, the attenuation of the beam becomes small, and primary recoil motion determines the sputtering yield.

## 2.4 Single-Knockon and Spike Regimes

The theoretical framework described in Sect. 2.3.2 has up till now served as a reference standard for sputtering experiments even in cases where the bombardment conditions were clearly outside the region in which linear collision cascades dominate the sputtering process. This attitude is justified to the extent that the theoretical description of sputtering in the single-knockon and spike regime is less developed than in the linear collision cascade regime. Various attempts have been made to provide more qualitative descriptions in the other two regimes, but at present, a consensus about the most promising procedures has not yet been achieved. Therefore, the two topics are treated in a rather more cursory manner here.

### 2.4.1 Single-Knockon Regime

#### a) Light-Ion Sputtering

Sputtering of heavy targets by light ions, especially hydrogen and helium ions, is somewhat peculiar from a theoretical point of view. While there is little doubt that the sputtering process should be described well within linear cascade theory, typical cascades are most often too small to justify the assumption of an isotropic recoil cascade which would otherwise be justified at not too small ion energies. Moreover, surface corrections to the infinite-medium description of sputtering are rarely negligible since reflection coefficients for light ions can be quite high.

According to Fig. 2.11a, the factor $\alpha$ in the expression for the sputtering yield (2.3.7, 14) increases rapidly with decreasing ion mass. This is an indication of an increased contribution of those processes to the sputtering yield, in which the incoming ion is scattered near the target surface and thus generates more than one recoil cascade near the surface. This was pointed out in [2.48], and it was demonstrated experimentally [2.102] that for medium-keV proton bombardment, reflected ions were responsible for a sizable fraction of all processes leading to sputtering. Such behavior must clearly be enhanced at even lower energies because of the monotonic increase in reflection coefficients with decreasing ion energy.

The literature on light-ion sputtering and reflection is sizable, despite rather small sputtering yields $(Y \lesssim 10^{-1})$, because of the occurrence of high fluences in solar-wind bombardment as well as in controlled thermonuclear fusion [2.103]. Apart from numerous computer simulations (for a review cf. [2.104]), a few investigations have been made with the aim of deriving simple scaling laws [2.105–106]. In addition, a more comprehensive attempt has been made to solve the basic transport equations with the physical input equivalent to [2.48] under a more appropriate approximation scheme [2.108].

#### b) Near-Threshold Regime

The following considerations refer to sputtering by an arbitrary ion at low energies, i.e., in the medium and upper eV region.

With the linear cascade theory as a reference standard, two important assumptions break down at low energies.

First, the depth variable $x_0$ characterizing the motion of the ion (which is energy-dependent according to Sect. 2.3.3) decreases and becomes comparable with the depth of origin $\Delta x_0$ (2.3.21) of ejected atoms. Thus, the motion of the primary (high-energy) and ejected (low-energy) particles is no longer decoupled. An indication of this is already obvious from (2.3.29), but in the near-threshold regime, a more radical change must be expected. Despite these reservations, the sputtering-yield formula (2.3.7) has been applied to heavy-ion bombardment in the threshold regime [2.48], with

$$F_D(E, 0, 0) = \alpha N S_n(E) = \alpha N C_0 \gamma E, \tag{2.4.1}$$

$S_n(E)$ being consistent with the low-energy cross section (2.2.8) and $\alpha$ being taken to have the adopted standard dependence, $\alpha = \alpha(M_2/M_1)$, for perpendicular incidence [Fig. 2.11a]. Combining this with (2.3.11), one obtains the surprisingly simple yield formula [2.48]

$$Y \cong \frac{3}{4\pi^2} \alpha \frac{\gamma E}{U_0} \quad \text{for} \quad E \gg U_0. \tag{2.4.2}$$

Empirically, (2.4.2) is not grossly wrong. From a theoretical point of view, however, its basis is unsatisfactory.

A more satisfactory treatment on the basis of transport theory would, in the limit of $E \gg U_0$, still lead to a dependence on $E/U_0$ like (2.4.2), provided that $m = 0$ is adopted for all scattering events, so that the main change would be undergone by the factor $\alpha$. Such a treatment is feasible but has not been done to the author's knowledge.

Second, typical amounts of energy transferred from bombarding ions to recoil atoms may be so small that the importance of recoil cascades may be questioned. It is, then, the knockon cross section

$$\sigma_d(E) = \int_{U_0}^{T_{max}} d\sigma(E, T) \tag{2.4.3}$$

rather than the stopping cross section $S_n(E)$ that governs the sputtering event, and hence, (2.3.7) is inapplicable.

These latter considerations are of primary importance for light-ion sputtering.

### c) Threshold Processes

A major fraction of the early effort in sputtering theory dealt with the prediction of sputtering thresholds (for a review cf. [2.109]). In general, the definition of a threshold energy for sputtering is a delicate task, and usually of little value unless accompanied by an upper limit for the observable yield. Indeed, since the sputtering yield is a statistical variable, fluctuations need to be accounted for once the yield is small. The microstructure and composition of the surface may be a major factor in this context.

There is fairly general agreement that quite different processes are responsible for sputtering near threshold for $M_1 \ll M_2$ and $M_1 \gg M_2$, respectively.

In the first case, ions may be reflected with negligible energy loss and knock out surface atoms for $T_{max} = \gamma E \geqq U_0$ such that an effective threshold should be of the order of $E_{th} \cong U_0/\gamma \cong M_2 U_0/4M_1$.

In the latter case, the velocity of the bombarding ion becomes comparable with the thermal velocities of the target atoms; therefore, simple estimates based on the conservation laws of elastic collisions become inapplicable, and the sputtering process becomes rather similar to an evaporation event [2.110, 111].

### 2.4.2 Spike Regime

The main failure of linear cascade theory under spike conditions (Fig. 2.1c) is the breakdown of the $E_0^{-2}$-law (2.2.28). This relation, which can be found essentially by a dimensional argument, certainly loses its meaning at the point where the calculated number of atoms set in motion exceeds the number of atoms available in the cascade volume, and possibly earlier, dependent on whether or not energy is deposited in clumps (subcascades) or more homogeneously. Once the majority of atoms in a certain volume have been set in motion, the energy must dissipate in a manner quite different from that which is found for linear collision cascades; this means that conventional estimates for the spatial extension of such cascades may no longer be valid under spike conditions. While there is fairly general agreement on the qualitative criteria for the significance of spike effects in collision cascades, there is considerable uncertainty as to the mechanism of energy transport in a spike, as well as to the theoretical evaluation of sputtering yields under spike conditions. The present discussion is based largely on concepts developed by the present author [2.51, 112], but the presentation has been influenced by comments made recently on the validity of the underlying model [2.113][5]

Let us first try to find an estimate of the density of energy in a cascade. Quite roughly, for a cascade volume $\Omega(E)$, $E$ being the initial ion energy, the mean energy $\Theta$ per atom is given by

$$\Theta \sim \frac{v(E)}{N\Omega(E)}, \tag{2.4.4}$$

$v(E)$ being defined in (2.2.27). The volume $\Omega(E)$ is not easily specified both in principle and in practice [2.51]. Disregarding this difficulty, one may set

$$\Omega(E) \sim \mathrm{const}\, [R(E)]^3 \propto \left(\frac{\gamma^{m-1}E^{2m}}{NC_m}\right)^3, \tag{2.4.5}$$

$R(E)$ being the ion range (2.2.15), and therefore, if $E$ is small enough ($\varepsilon \ll 1$) so that $v(E) \cong E$,

$$\Theta \propto E^{1-6m}. \tag{2.4.6}$$

---

5 Substantial progress has been made in the calculation of yields and energy spectra in the spike regime after this article was finished [2.165]

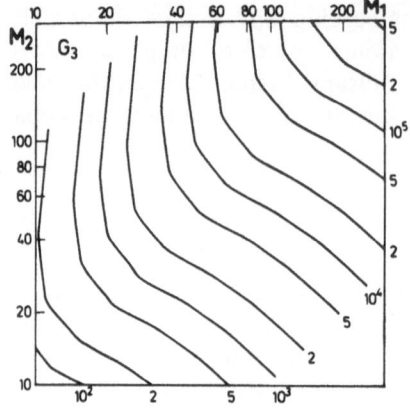

**Fig. 2.20.** Contour plot for parameter $G_3$ determining energy density $\Theta$ in an elastic-collision spike. $\Theta = (G_3 N^2 / E) \cdot (\text{eV\AA}^6 \text{keV})$. (From [2.51], courtesy of the American Institute of Physics)

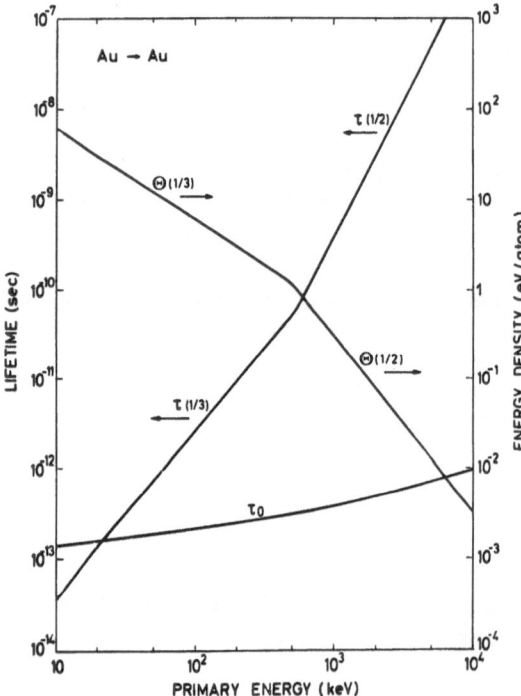

**Fig. 2.21.** Spike parameters for Au ions incident on Au, estimated according to [2.51], versus ion energy $E$. $\Theta$ is the effective maximum energy density in the spike, $\tau$ the time constant for decay, and $\tau_0$ the slowing-down time of the ion. (From [2.164], courtesy of North Holland Publishing Company)

Thus, $\Theta$ decreases with increasing ion energy $E$ and, at constant $E$, increases with increasing ion mass because of decreasing penetration depth. Figures 2.20, 2.21 show (slightly) more quantitative estimates of $\Theta$ [2.51].

One finds that for the heaviest ions, not too high up in the keV region, average energies of several eV per atom are quite common.

From this, it has been concluded [2.112] that down to some recoil energy $E_0^* \gg \Theta$, a collision cascade is sufficiently dilute to develop according to the

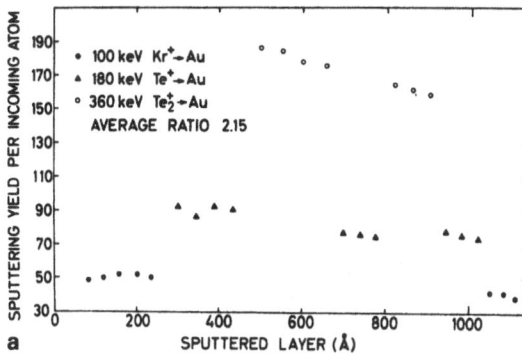

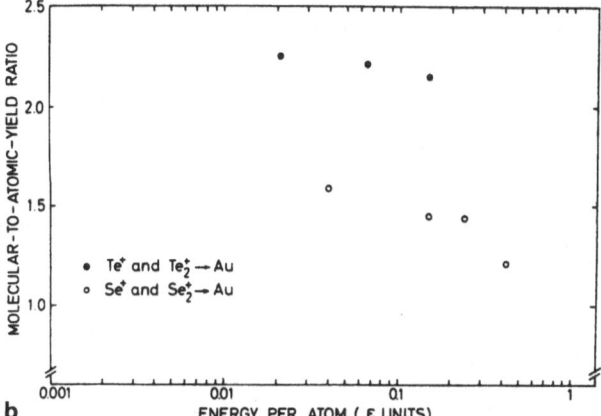

**Fig. 2.22a, b.** Sputtering yields per incident *atomic* particle for gold bombarded by atomic and molecular ions. (From [2.52], courtesy of the American Institute of Physics). (a) Versus sputtered-layer thickness. (b) Versus reduced energy $\varepsilon$, (2.2.4)

scheme discussed in Sect. 2.3, while for $E_0 < E_0^*$, this scheme is no longer valid and we expect spike effects.

Direct evidence in favour of spike formation stems from molecular-ion bombardment [2.52, 113]. Figure 2.22 demonstrates an enhancement of the sputtering yield per incident atomic projectile in case of molecular-ion bombardment as compared to atomic-ion bombardment, as expected according to the discussion in Sect. 2.1.3(a). It is seen that the enhancement increases with increasing ion mass and decreases with increasing ion energy, i.e., in qualitative agreement with the predicted behaviour of the energy density, cf. (2.4.4–6).

Similar evidence in favour of the very existence of spike effects and their significance as a function of ion mass and energy, has emerged from radiation damage measurements [2.114–116]. With regard to quantitative details, several points of uncertainty are obvious. First, the linear dimensions of an individual cascade enter in the third power into (2.4.5) and thus give rise to considerable error. Second, it is not obvious whether $R(E)$ determines the transverse extension of a single cascade at all. Third, because of subcascade formation, even higher energy densities may be expected locally. Fourth, the energy density may drop rapidly from the dense regions inside the cascade toward the surface.

More important, the very nature of the energy transport in a spike is not clear. As a first approximation (based on the observation that at energies of several eV/atom, the structure of the solid is hardly preserved), the author approximated the spike region as an ideal gas at a local temperature $T$ defined by $3kT/2 \cong \Theta$, with $k$ being Boltzmann's constant [2.51]. Because of the high atomic density, the pressure $p$ of such a system is quite high, indeed, $p = NkT = 2v/3\Omega$ may approach $\sim 10^5$ atm. for $\Theta \sim 1$ eV. This presumably gives rise to shock waves at some stage of the slowing-down process [2.117]. It is not clear at the time of writing whether the "thermal" transport of energy or the shock-wave nature of the spike dominates the sputtering yield.

In the case of the former, the sputtering yield can be estimated on the basis of an adopted local Maxwellian velocity distribution of target atoms within a spike. Indeed, a plane surface at temperature $T$ with a planar surface potential $U_0$, cf. (2.3.9), has an evaporation rate $\Phi$ per unit time and surface area given by

$$\Phi = N(kT/2\pi M_2)^{1/2} \exp(-U_0/kT), \tag{2.4.7}$$

$M_2$ being the mass of the target atoms. When the density of energy deposition at the target surface is approximated by

$$F_D(r)/x = 0 = \begin{cases} F_D(0) \ldots \sqrt{y^2 + z^2} \leq \varrho \\ 0 \quad \ldots \sqrt{y^2 + z^2} \geq \varrho, \end{cases} \tag{2.4.8}$$

with $D_D(r)$ being the function specified in (2.2.30) and $\pi\varrho^2$ being the hot surface area, we have a surface temperature $T$ given by

$$\tfrac{3}{2} kT = F_D(0)/N, \tag{2.4.9}$$

and the one-dimensional density (2.2.33)

$$F_D(x = 0) = \pi\varrho^2 F_D(0) = \tfrac{3}{2} NkT \cdot \pi\varrho^2. \tag{2.4.10}$$

For an average containment time $\tau$ per spike, one finds the sputtering yield

$$Y = \tau \cdot \pi\varrho^2 \cdot \Phi \tag{2.4.11}$$

corresponding to a "material" factor $\Lambda$ [cf. (2.3.7)]:

$$\Lambda = \frac{Y}{F_D(x=0)} = \frac{\tau}{\sqrt{3\pi M_2 \cdot \tfrac{3}{2} kT}} \exp(-U_0/kT). \tag{2.4.12}$$

This relationship was mentioned in [2.112]; it is clear, however, from the present argument that the temperature $T$ occuring in (2.4.12) is an effective surface temperature, which may differ substantially from the temperature corresponding to the bulk energy density $\Theta$, cf. (2.4.4) [2.113].

The expression (2.4.12) has been used successfully (Chap. 4 and [2.52]) to qualitatively explain sputtering yields for heavy monatomic ions under spike conditions, by means of containment times $\tau$ estimated on the basis of kinetic gas theory, and by use of bulk cascade temperatures, predicted [2.51] to be of the order of $kT \sim U_0$ in the genuine spike regime (cf. Fig. 2.21). When (2.4.12) is applied to evaluate sputtering yields for molecular-ion bombardment, a strong enhancement of $\Lambda$ requires that $U_0 \gg kT$. This apparent inconsistency has not yet been resolved. More [2.117, 118] or less [2.114, 119] radical alternative proposals are being discussed.

The literature on thermal effects in sputtering is sizable [2.120]. One main reason for favoring the thermal rather than the shock-wave type of spike theories is the repeated observation of "maxwellian" contributions to the energy spectrum of sputtered atoms [2.49, 121–125]. Again, however, both high $(kT \sim U_0$ [2.124]) and low $(kT \ll U_0$ [2.122]) spike temperatures have been reported; whether or not these differences are caused only by the rather different procedures of analyzing experimental data is not clear at the time of writing.[6]

Unfortunately, no experimental results have yet been reported on energy spectra of sputtered atoms under molecular-ion bombardment.

## 2.5  Related Problems

Several of the topics listed below are presently under active investigation. Within the scope of the present introduction, only a few comments will be made and a number of pertinent references will be given.

### 2.5.1  States of Sputtered Particles

A separate chapter has been devoted in this monograph to the charge and excitation state of sputtered particles [Ref. 2.7, Chap. 3]. Within the majority of situations aimed at in the present chapter, the electronic state of a moving particle is secondary in the sense that although it may depend on the characteristics of the collision cascade, it does not influence the development of the cascade to any major extent. Therefore, a theoretical treatment of the state of a sputtered particle can most often be considered quite independent from the particular sputtering model.

For metals, one usually observes an overwhelming fraction of sputtered atoms to emerge as neutrals in their ground states (with regard to sputtered molecules, cf. comments below). This fact, together with the observation [Ref. 2.7, Chap. 3] that ions and excited atoms are emitted under bombardment conditions comprising all regimes sketched in Fig. 2.1, in particular, for both

---

6 Cf. footnote see page 53.

large and small sputtering yields, indicates that the elementary event in the excitation process is the interaction of *one* emerging atom with the target surface [2.126–130].

The state of a sputtered atom is, then, determined by the respective probabilities for excitation (ionization) and deexcitation (neutralization). As was pointed out recently [2.130], the initial state of the emerging atom *in* the target is immaterial, since survival probabilities for excited states are prohibitively small at typical energies of sputtered atoms ($\sim 10$ eV). This is caused by the fact that atomic energy levels are broadened by the interaction with the solid surface within a layer of several Ångströms from the surface. Thus, the excitation process presumably takes place *during* the passage through that layer, and is caused by the time-dependent interaction force exerted by the solid surface on the emerging atom [2.126–130]. The various treatments differ in details, and considerable uncertainty prevails. A simple jellium model of a solid surface yields the probability $P_a$ for an excited state $a$ to be given by [2.129]

$$P_a \cong \tfrac{2}{\pi} \exp[-\pi(\varepsilon_a - \varepsilon_F)/\hbar\gamma_N v_1], \tag{2.5.1}$$

where $\varepsilon_a - \varepsilon_F$ is the energy of the state above the Fermi level $\varepsilon_F$, $v_1$ the velocity of the emerging atom, and $\gamma_N^{-1}$ the distance from the surface over which the level width decreases to $1/2.781$ of its bulk value. Similar dependences on energy level and particle velocity have been obtained numerically [2.127].

There is ample experimental evidence to support the exponential dependence predicted by (2.5.1) of excitation and ionization probabilities on pertinent energy parameters, as well as on particle velocity [Ref. 2.7, Chap. 3]. Qualitative interpolations [2.130] on the basis of (2.5.1) indicate that this model predicts the order of magnitude of ionization probabilities for clean surfaces as well as for targets with adsorbed surface layers. Empirically, the observed exponential dependence of ionization probabilities on electron affinities and ionization potentials had been interpreted in terms of a Boltzmann factor involving a local target temperature [2.131].

Apart from the state of excitation, one may inquire about the fraction of atoms that are sputtered as molecules. It is well documented that among the *ionized* sputtered species, a substantial fraction may be molecules [Ref. 2.7, Chap. 8], ranging from dimers to multimers. There is every reason to believe that the sputtering of neutral molecules is quite a frequent process [2.132]. Theoretical treatments [2.133, 134] have concentrated on the statistical probability of two independently sputtered particles originating in one collision cascade to cluster *after* emerging from the surface. Simultaneous ejection of neighboring atoms appears feasible, especially under spike conditions [2.135]. A thorough theoretical treatment still does not appear very promising in view of the apparent lack of experimental data on energy spectra of sputtered neutral molecules measured under conditions where the sputtering process as such, is sufficiently well described theoretically.

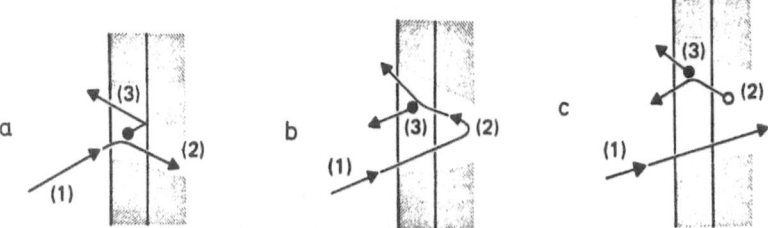

**Fig. 2.23a–c.** Mechanisms of desorption by knock-on collisions [e.g., nitrogen on tungsten]. *(1)*=incident ion, *(2)*=substrate, *(3)*=adsorbate. **(a)** Direct knockon by incoming ion. **(b)** Knockon by reflected ion. **(c)** Knockon by sputtered matrix atom. (From [2.137], courtesy of the American Institute of Physics)

### 2.5.2 Ion-Induced Desorption

It was mentioned in the historical survey that the analysis of *Langmuir's* measurements on collision-induced desorption [2.21] had a considerable impact on the field of sputtering. Similarly, available knowledge on electron-stimulated desorption [2.136] might be useful in the understanding of ion-induced sputtering by electronic processes. The present comments, like the bulk of this article, are restricted to knockon processes, and have been made mainly to caution the reader from making too close a connection between sputtering and desorption phenomena.

It is assumed here that the thickness of an adsorbed layer does not substantially exceed one monolayer; in the opposite case of a much thicker layer, one may instead consider the sputtering of a thick target made up by the adsorbed species.

The process of ejection of an adsorbed atom differs from that of a matrix atom in the following aspects.

i) The emitted particle definitely comes from the upper surface layer, possibly from quite specific positions.

ii) Its binding energy differs in general from that of the matrix atoms.

iii) In the case of widely different masses of adsorbant and adsorbed species, drastic changes in the kinematics of collision processes leading to sputtering may occur.

The third aspect has been investigated theoretically [2.137] for the specific case of a chemisorbed monolayer of nitrogen on tungsten. Figure 2.23 shows three different processes that may lead to desorption of nitrogen, of which only one is related to the sputtering of tungsten atoms. It was shown that the two mechanisms shown in Figs. 2.23a, b lead to surprisingly high desorption yields. Neither process would lead to so large yields if the target were either a massive clean tungsten or a massive solid nitrogen target.

### 2.5.3 Recoil Implantation and Ion-Beam Induced Atomic Mixing

It is obvious that a process leading to sputtering may also give rise to the transport of atoms within the target, e.g., the injection of one species into another through a boundary layer in a sandwiched target, transmission sputtering, or the implantation of an adsorbed species into the bulk. In general such processes may give rise to changes in the depth profile of the composition of a multicomponent target. Progress has been rather slow in this field [2.138–147], but at present, a simple pattern seems to emerge.

One may distinguish between recoil implantation and cascade mixing. The former effect is due to direct ion-target collisions and therefore strongly directional, while the latter process is due to target-target collisions and therefore more or less isotropic. Moreover, recoil implantation involves few atoms with moderate and high energies [cf. (2.2.1, 6)] while cascade mixing strongly peaks toward low recoil energy [cf. (2.2.28)]. In the case of a pronounced difference in atomic mass of the constituents of a target, the cross section for recoil implantation at a given recoil energy is proportional to $Z_2^{2m}M_2^{-m}$, cf. (2.2.6), i.e. largest for the heaviest constituent. However, the range [2.36] at given recoil energy is proportional to $Z_2^{-2m}M_2^{-m}$, and hence, the implantation effect is $\propto M_2^{-2m}$ and thus greatest for the lightest species [2.145].

For cascade mixing, the probability for a given atom to recoil at a given energy of a given species is greatest for the most abundant species [2.77]. For cascade mixing, an atom of a given species will recoil into a given energy interval with a probability that increases with increasing abundance [2.77]. While recoil implantation causes a net displacement of the depth profile of the light species relative to that of the heaviest species in the direction of the beam, cascade mixing is roughly equivalent with an interdiffusion, with a diffusion coefficient essentially independent of target temperature [2.143].

### 2.5.4 Sputtering from Multiple-Component Targets

An understanding of the sputtering behavior of multiple component targets is of primary importance for several reasons.

i) Prolonged bombardment of an elemental target normally leads to an alloy incorporating the bombarding species.

ii) Applications of sputtering in the production and analysis of thin films almost invariably deal with multiple component targets.

iii) Varying composition of the target provides an extra dimension to fundamental sputtering studies that may yield a clue to the understanding of the material dependence of the sputtering yield.

The main complication in the sputtering of multiple-component targets, briefly called alloy sputtering, is the experience that the components need not sputter stoichiometrically; the result of this is that the composition of an alloy can change in a certain depth range beneath the bombarded surface. The depletion of the surface region in the high-yield species tends to counteract the high yield, and eventually a stationary state may be reached where the composition of the flux of sputtered particles reflects that of the bulk alloy.

There is no a priori reason why an alloy should sputter stoichiometrically, i.e., why the flux of sputtered atoms should reflect the composition of the surface layer from which these atoms emerge.

From a theoretical point of view, it is convenient to distinguish between primary and secondary effects. Primary effects comprise the physics of an individual sputtering event on a target with a given composition. Secondary effects include the changes in target composition caused by prolonged bombardment.

### a) Primary Effects

The primary sputtering event can first be looked at in an homogeneous alloy. The sharing of energy in a binary infinite medium has been analysed within the linear cascade regime by *Andersen* and *Sigmund* [2.77]. Straight extension of the formalism sketched in Sects. 2.2.5, 2.3.1, and 2.3.2 yields the following generalization of (2.3.6) for the outward current $J_i dE_0 d^2\Omega_0$ of target atoms of type $i$ ($i = 1, 2$),

$$J_i(E_0, \Omega_0) = \frac{G_i}{(G)_i} \cdot \psi F_D(E, \theta, 0) \cdot \frac{\Gamma_{m_i}}{E_0 |(dE_0/dx)_i|} \cdot \frac{|\cos\theta_0|}{4\pi}, \tag{2.5.2}$$

where $F_D(E, \theta, 0)$ is the energy deposited per unit depth at the alloy surface, $(dE_0/dx)_i$ is the stopping power of an $i$-atom in an elemental $i$-target, and $m_i$ is the cross section exponent $m$ that is most appropriate to characterize the scattering of an $i$-atom at energy $E_0$. The factor $G_i/(G)_i$ is the ratio of the flux functions [as defined according to (2.3.1)] of $i$-atoms in the alloy, $G_i$, and $i$-atoms in an elemental $i$-target, $(G)_i$ respectively. It depends on the atomic concentrations $c_1$ and $c_2$ ($c_1 + c_2 = 1$) and, rather weakly, on the adopted scattering exponents $m_1$ and $m_2$. Figure 2.24 shows that, for roughly equal masses $M_1$, $M_2$ of the constituents, one may approximate

$$\frac{G_i}{(G)_i} \approx c_i, \quad (M_1 \approx M_2) \tag{2.5.3}$$

while for very different masses, the *most abundant* species has a value up to 50% *higher* than (2.5.3) while the least abundant one obeys (2.5.3) more closely. Following the procedure leading to (2.3.7, 10), one obtains the partial sputtering yield $Y_i$ of the $i$-component

$$Y_i = c_i \Lambda_i' F_D(E, \theta, 0), \quad i = 1, 2 \tag{2.5.4}$$

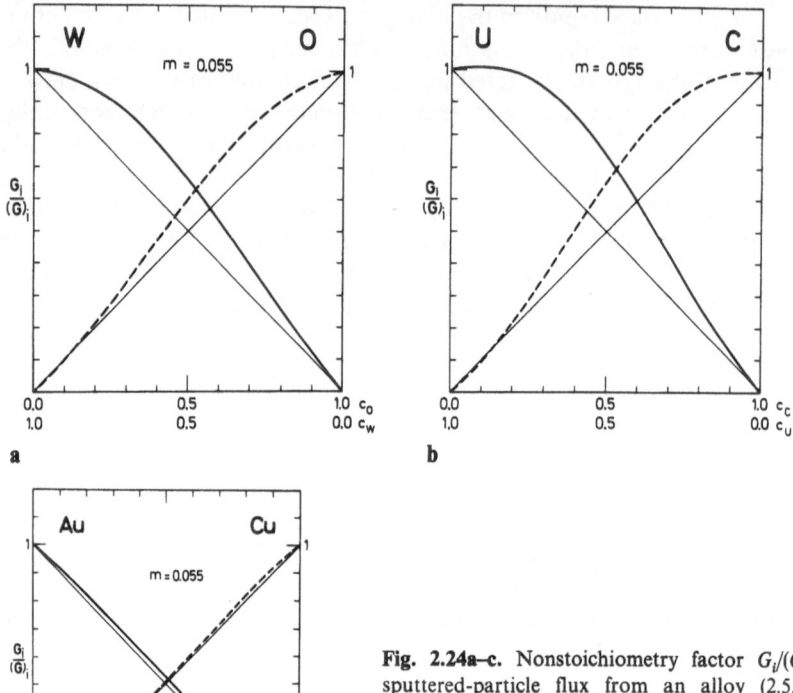

**a**

**b**

**c**

**Fig. 2.24a–c.** Nonstoichiometry factor $G_i/(G)_i$ in a sputtered-particle flux from an alloy (2.5.5). Full drawn line: heavy component, dotted line: light component, thin lines: $G_i/(G)_i = c_i$ (for comparison). (Data from [2.77], courtesy of Det Kgl. Danske Videnskabernes Selskab). (a) Tungsten oxide; (b) uranium carbide; (c) copper-gold

where

$$\Lambda_i' = \Lambda_i \cdot \left(\frac{(U_0)_i}{U_{0i}}\right)^{1-2m_i} \cdot \frac{1}{c_i} \frac{G_i}{(G)_i}. \tag{2.5.5}$$

Here, $\Lambda_i$ is the material constant (2.3.10) or (2.3.11) for an elemental $i$-target, $(U_0)_i$ is the surface potential $U_0$ in an elemental $i$-target, and $U_{0i}$ is the surface potential for an $i$-atom in the alloy.

The situation is particularly simple in the case of practically equal masses of the two constituents where (2.5.4) reduces to

$$Y_i = c_i \left(\frac{(U_0)_i}{U_{0i}}\right)^{1-2m_i} (Y)_i, \tag{2.5.6}$$

and $(Y)_i$ is the sputtering yield of an elemental $i$-target. This shows that the *least bound* species tends to sputter preferentially.

The situation is more complicated in the case of very different masses. Comparison with elemental sputtering yields may not be justified, since the connection of the deposited-energy function $F_D$ with those of the elemental target materials may be rather complex [2.50]. Indeed, the simple intuitive relationship based on a stopping-power argument:

$$F_D(E, \theta, 0) \cong c_1 (F_D(E, \theta, 0))_1 + c_2 (F_D(E, \theta, 0))_2 , \qquad (2.5.7)$$

has a rather limited accuracy, especially when the ion mass is intermediate between two widely different target masses (a case frequently met in applications).

A simple relationship is found for the yield ratio $Y_1/Y_2$ where, according to (2.3.3, 10) and (2.5.4, 5)

$$\frac{Y_1}{Y_2} = \frac{G_1(E_0)(dE_0/dx)_1}{G_2(E_0)(dE_0/dx)_2} \cdot \frac{U_{02}^{1-2m_2}}{U_{01}^{1-2m_1}}, \qquad (2.5.8)$$

where $E_0$ is an arbitrary energy $\ll E$. Equation (2.5.8) can be brought into the form [2.77]

$$\frac{Y_1}{Y_2} = \frac{c_1}{c_2} \cdot \frac{S_{21}(U_{02})}{S_{12}(U_{01})}, \qquad (2.5.9)$$

where $S_{ij}$ is the stopping cross section (2.2.9) for an $i$-atom hitting a $j$-atom. In particular, for $m_1 = m_2$ one finds

$$\frac{Y_1}{Y_2} = \frac{c_1}{c_2} \cdot \left(\frac{M_2}{M_1}\right)^{2m} \cdot \left(\frac{U_{02}}{U_{01}}\right)^{1-2m}, \qquad (2.5.10)$$

which is consistent with (2.5.6) for $M_1 \approx M_2$.

Equation (2.5.10) predicts a tendency for the *lightest* component to sputter preferentially. However, since $m$ is presumably small, $0 \lesssim m \lesssim 0.2$, this non-stochimetric effect is less pronounced than that caused by different surface potentials. Several of these predictions have been mentioned in the literature [2.148–153]. Moreover, from (2.5.2), assuming (2.5.3, 7) to be valid, one obtains the following relationship for the partial sputtering yield,

$$Y_i = c_i \left[ c_1 \left(\frac{(U_0)_1}{U_{0i}}\right)^{1-2m_1} (Y)_1 + c_2 \left(\frac{(U_0)_2}{U_{0i}}\right)^{1-2m_2} (Y)_2 \right] \qquad (2.5.11)$$

which, for $m_1 = m_2 = 0$, has been derived by *Haff* and *Switkowski* [2.154]. According to the above arguments, (2.5.11) should be valid only in the absence of mass differences between the components.

Although the above results have been derived strictly for an homogeneous medium, their validity is not necessarily limited to that situation. In the case of inhomogeneous media, they may be generalized by adopting average compositions over pertinent depth ranges. With regard to the deposited-energy density $F_D(E, \theta, 0)$, the relevant depth range is of the order of $x_0$ as discussed in Sect. 2.3.3, while for $G_i/(G_i)$ in (2.5.2), and hence for $c_1/c_2$ in (2.5.10) and $c_i$ in (2.5.11), the average composition within the sputter depth $\Delta x_0$ [as given by (2.3.21) for an elemental target] should apply. Obviously, the latter average becomes ill-defined in the case of a large concentration gradient.

It is evident from (2.5.2) that within the linear cascade regime, the composition of the sputtered flux is independent of the type, energy, and angle of incidence of the bombarding ion beam. This is equivalent to the fact that within this regime, neither the angular nor the energy distribution of atoms sputtered from an elemental target depend on the beam parameters.

Pursuing this analogy further, one expects such differences to occur in the single-knockon and spike regimes. In the former case, there is clear experimental evidence [2.156–158] as well as a qualitative understanding in similar terms to those applied in desorption studies [2.137], as mentioned briefly in Sect. 2.5.2.

In the spike regime, within the model sketched in Sect. 2.4.2, it appears appropriate to postulate equipartition of energy within the spike volume. Hence, (2.3.7, 12) predict a yield ratio[7]

$$Y_1/Y_2 = (c_1/c_2)(M_2/M_1)^{1/2} \exp(U_{02} - U_{01})/kT \tag{2.5.12}$$

with a spike temperature $T$ dependent on beam and target parameters. As in (2.5.10), the lightest and weakest-bound component is expected to sputter preferentially, but the dependence, in particular, on the surface potential differs from the one in (2.5.10).

An alternative mechanism for particle transport in a spike lead $Haff$ [2.119] to predict an $(M_2/M_1)^{1/4}$ dependence of $Y_1/Y_2$ (for $U_{02} = U_{01}$).

### b) Secondary Effects

As a consequence of preferential sputtering, the composition profile changes near the target surface. Therefore, the contribution of the enriched (i.e., the low-yield) species to the sputtering yield is expected to increase after prolonged bombardment until the composition of the sputtered flux reflects that of the bulk target. Experimental data known to the author, have all been taken at fluences where that stage should have been reached if no competing processes were going on, i.e., after sputtering of a layer a few times $\Delta x_0$ thick. The composition of a surface layer of thickness $\sim \Delta x_0$, should, then read

$$\left(\frac{c_1}{c_2}\right)_{\Delta x_0} \cong \left(\frac{c_1}{c_2}\right)_{bulk} \cdot \left(\frac{c_1 Y_2}{c_2 Y_1}\right), \tag{2.5.13}$$

7   Cf. footnote see page 53.

the second factor on the right-hand side being given by one of the theoretical predictions of the previous paragraph. The ratio $(c_1/c_2)_{\Delta x_0}$ can be determined experimentally by Auger electron spectroscopy [2.151, 152], a technique which presumably samples the composition over a depth somewhat greater than $\Delta x_0$, and by low-energy ion scattering [2.158], a technique that traces the very surface, i.e., a layer that is presumably *thinner* than $\Delta x_0$. Experimental results [2.152] will be summarized in a separate contribution [Ref. 2.4, Chap. 1]. Complications arise due to the fact that ion bombardment causes bulk processes in addition to sputtering, and that, in fact, a much larger fraction of the energy goes into bulk processes than into sputtered atoms. It is well documented experimentally [2.159] that ion-induced composition changes extend well into a layer of the order of the ion range rather than a shallow layer of $\Delta x_0 \cong 5\text{Å}$, although it is not known experimentally whether an *additional* concentration gradient occurs within that shallow layer.

At least four different bulk processes may influence the composition of the penetrated layer:
  (i) segregation,
  (ii) recoil implantation,
  (iii) cascade mixing and
  (iv) diffusion.
In the presence of segregation [2.160], separate phases form at the surface, preventing a stationary state to develop.

Recoil implantation initially causes an enrichment of the heavy species near the surface [2.141, 149], while the lighter species accumulates in a deeper layer; at equilibrium, however, a layer with a thickness of about the ion range is enriched in the *lighter* species [2.145]. To reach this equilibrium, it is necessary to sputter off more than one ion range.

Cascade mixing and (radiation-enhanced) diffusion tend to smear concentration gradients generated by preferential sputtering and/or recoil implantation; thus, even in the absence of recoil implantation, one expects a depleted-layer thickness greater than $\Delta x_0$ (as found experimentally [2.159]) and, consequently, a higher minimum fluence to reach equilibrium.


# Appendix A:
# Connection Between Forward and Backward Boltzmann Equations

It is the purpose of the present appendix to show the connection between the backward and forward forms of the Boltzmann equation. The argument follows the general one presented in [2.65] closely, and will be used to derive (2.2.24) from (2.2.23). Introduce first a propagator function $G$ by way of the definition

$$F(r, v, t) = \int d^3r_0 \int d^3v_0 \, G(r, v, t; r_0, v_0, t_0) F(r_0, v_0, t_0), \tag{A.1}$$

where $t \geq t_0$, such that

$$G(r, v, t_0; r_0, v_0, t_0) = \delta(r - r_0)\delta(v - v_0). \tag{A.2}$$

Obviously, the propagator generates a distribution function at time $t$ on the basis of complete knowledge of the distribution at some (earlier) time $t_0$. It is easily seen that $G$ satisfies (2.2.23) with $G' = G(r, v', t; r_0, v_0, t_0)$ and $G'' = G(r, v'', t; r_0, v_0, t_0)$, i.e., the propagator satisfies the (forward) Boltzmann equation.

Now, take the derivative of (A.1) with respect to $t_0$, and insert (2.2.23) for $\frac{\partial}{\partial t_0} F(r_0, v_0, t_0)$. This yields

$$
\begin{aligned}
0 = &\int d^3 r_0 \int d^3 v_0 F(r_0, v_0, t_0) \frac{\partial}{\partial t_0} G(r, v, t; r_0, v_0, t_0) \\
&- \int d^3 r_0 \int d^3 v_0 G(r, v, t; r_0, v_0, t_0) v_0 \cdot \nabla_{r_0} F(r_0, v_0, t_0) \\
&- \int d^3 r_0 \int d^3 v_0 G(r, v, t; r_0, v_0, t_0) N \int d^3 v' \int d^3 v'' \\
&\quad \cdot [v_0 F(r_0, v_0, t_0) K(v_0; v', v'') - v' F(r_0, v', t_0) K(v'; v_0, v'') \\
&- v'' F(r_0, v'', t_0) K(v''; v', v_0)].
\end{aligned}
\tag{A.3}
$$

Partial integration of the second integral with respect to $r_0$, and replacing integration variables in the third term reduces (A.3) to

$$
\begin{aligned}
0 = &\left(\frac{\partial}{\partial t_0} + v_0 \cdot \nabla_{r_0}\right) G(r, v, t; r_0, v_0, t_0) \\
&- N v_0 \int d^3 v' \int d^3 v'' K(v_0; v', v'')[G(r, v, t; r_0, v_0, t_0) \\
&- G(r, v, t; r_0, v', t_0) - G(r, v, t; r_0, v'', t_0)],
\end{aligned}
\tag{A.4}
$$

since $F(r_0, v_0, t_0)$ is an arbitrary distribution. Because of translational invariance in space and time one has

$$G(r, v, t; r_0, v', t_0) \equiv G(r - r_0, v_0, v, t - t_0), \tag{A.5}$$

so that

$$\left(\frac{\partial}{\partial t_0} + v_0 \cdot \nabla_{r_0}\right) G = -\left(\frac{\partial}{\partial t} + v_0 \cdot \nabla_r\right) G. \tag{A.6}$$

Inserting (A.6) into (A.4), setting $r_0 = 0$, $t_0 = 0$ and interchanging $v$ and $v'$, one finds

$$
\begin{aligned}
-&\left(\frac{\partial}{\partial t} + v \cdot \nabla_r\right) G(r, v, v_0, t) = Nv \int d^3 v' \int d^3 v'' K(v; v', v'')[G(r, v, v_0, t) \\
&- G(r, v', v_0, t) - G(r, v'', v_0, t)]
\end{aligned}
\tag{A.7}
$$

which is identical to (2.2.24), i.e., the propagator $G$ *also* satisfies the backward equation (A.7).

For a collision cascade initiated by one initial particle, one has

$$F(r, v, 0) = \delta(r)\, \delta(v - v_0) \tag{A.8}$$

for $r_0 = 0$, $t_0 = 0$. Comparison of (A.2) and (A.8) shows that, then, the distribution function $F$ becomes identical with the propagator, i.e.,

$$F(r, v, t) \equiv G(r, v, t; 0, v_0, 0)$$
$$\equiv G(r, v_0, v, t) \tag{A.9}$$

in view of (A.5). Thus, it is only a matter of convenience whether the forward or backward form of the Boltzmann equation is used to derive an explicit expression for the distribution function. In [2.48], the backward form was utilized since it was possible from that to derive an equation for the total sputtering yield directly, by integration over the energy and angular distribution of sputtered atoms.

*Acknowledgements.* The author gratefully acknowledges suggestions by Rainer Behrisch, Carsten Claussen, and Alberto Gras-Marti, and the careful assistance of Tove Nyberg, Birgit Sørensen and Søss Terkelsen in the preparation of the manuscript. Part of this work was performed while the author was a visitor at Orsay, where he enjoyed the kind interest of Harry Bernas and Michel Maurette in his work.

# References

2.1   W. R. Grove: Philos. Mag. **5**, 203 (1853)
2.2   J. P. Gassiot: Philos. Trans. Roy. Soc. London **148**, 1 (1858); Proc. R. Soc. London **9**, 146 (1858)
2.3   J. Plücker: Ann. Phys. (Leipzig) **103**, 88, 151 (1858); **104**, 113 (1858); **105**, 67 (1858)
2.4   R. Behrisch (ed.): *Sputtering by Particle Bombardment* II, Topics in Applied Physics (Springer, Berlin, Heidelberg, New York 1982)
2.5   J. Stark: Z. Elektrochem. **14**, 752 (1908); **15**, 509 (1909)
2.6   F. Keywell: Phys. Rev. **97**, 1611 (1955)
2.7   R. Behrisch (ed.): *Sputtering by Particle Bombardment* III, Topics in Applied Physics (Springer, Berlin, Heidelberg, New York 1982)
2.8   E. Goldstein: Verh. Dtsch. Phys. Ges., **4**, 228, 237 (1902)
2.9   A. W. Wright: Am. J. Sci. **13**, 49 (1877); **14**, 169 (1877)
2.10  W. Crookes: Proc. R. Soc. London **50**, 88 (1891)
2.11  G. Granquist: Öfvers. Svenska Vet. Akad. Förh. **54**, 575 (1897); **55**, 709 (1898)
2.12  L. Holborn, L. W. Austin: Abh. d. Physikal. Techn. Reichsanstalt **4**, 101 (1904)
2.13  J. Puluj: Sitzungsber. d. Wiener Akad. Wiss. **81**, 864 (1880)
2.14  W. Hittorf: Ann. Phys. (Leipzig) **20**, 705 (1883); **21**, 90 (1884)
2.15  F. Wächter: Ann. Phys. (Leipzig) **17**, 903 (1882)
2.16  A. Berliner: Ann. Phys. (Leipzig) **33**, 289 (1888)
2.17  J. Stark, G. Wendt: Ann. Phys. (Leipzig) **38**, 921, 941 (1912)
2.18  J. Stark: *Die Elektrizität in Gasen* (Barth, Leipzig 1902)

68     *P. Sigmund*

2.19   V.Kohlschütter: Jahrb. Radioakt. Elektron. **9**, 355 (1912), containing an extensive literature survey including Kohlschütter's own work

2.20   A. v. Hippel, E. Blechschmidt: Ann. Phys. (Leipzig) **80**, 672 (1926); **81**, 999, 1043 (1926); **86**, 1006 (1928)

2.21   K.H. Kingdon, I. Langmuir: Phys. Rev. **20**, 107 (1922); **21**, 210 (1923); **22**, 148 (1923)

2.22   R. Seeliger, K. Sommermeyer: Z. Phys. **93**, 692 (1935); Ann. Phys. (Leipzig) **25**, 481 (1936); Z. Phys. **119**, 482 (1942)

2.23   G.K. Wehner: Phys. Rev. **102**, 690 (1956)

2.24   E.S. Lamar, K.T. Compton: Science **80**, 541 (1934)

2.25   M.L. Smith (ed.): *Electromagnetically Enriched Isotopes and Mass Spectrometry* (Butterworth, London, 1956), p. 53

2.26   O. Almén, G. Bruce: Nucl. Instrum. Methods **11**, 257, 279 (1961)

2.27   M.W. Thompson: Philos. Mag. **4**, 139 (1959)

2.28   P.K. Rol, J.M. Fluit, J. Kistemaker: Physica **26**, 1000 (1960)

2.29   V.E. Yurasova: Zh. Tekh. Fiz. **28**, 1966 (1958) [Engl. transl.: Sov. Phys. Tech. Phys. **3**, 1806 (1959)]

2.30   V.A. Molchanov, V.G. Tel'kovskii: Dokl. Akad. Nauk SSSR **136**, 801 (1961) [Engl. transl.: Sov. Phys. Dokl. **6**, 137 (1961)]

2.31   F. Keywell: Phys. Rev. **87**, 160 (1952)

2.32   D.E. Harrison: Phys. Rev. **102**, 1473 (1956); **105**, 1202 (1957); J. Chem. Phys. **32**, 1336 (1960)

2.33   R.S. Pease: Rendiconti della Scuola Internazionale di Fisica "Enrico Fermi", Corso XIII, 158 (1960)

2.34   G.H. Kinchin, R.S. Pease: Rep. Prog. Phys. **18**, 1 (1955)

2.35   N. Bohr: K. Dan. Vidensk. Selsk. Mat. Fys. Medd. **18**, No. 8 (1948)

2.36   J. Lindhard, M. Scharff, H.E. Schiøtt: K. Dan. Vidensk. Selsk. Mat. Fys. Medd. **33**, No. 14 (1963)

2.37   J.A. Davies, J. Friesen, J.D. McIntyre: Can. J. Chem. **38**, 1526 (1960)

2.38   J.A. Davies, G.C. Ball, F. Brown, B. Domeij: Can. J. Phys. **42**, 1070 (1964)

2.39   W. Brandt, R. Laubert: Nucl. Instrum. Methods **47**, 201 (1967)

2.40   G. Leibfried: J. Appl. Phys. **30**, 1388 (1959)

2.41   J. Lindhard, V. Nielsen, M. Scharff, P.V. Thomsen: K. Dan. Vidensk. Selsk. Mat. Fys. Medd. **33**, No. 10 (1963)

2.42   P.H. Dederichs: Phys. Status Solidi **10**, 303 (1965)

2.43   M.T. Robinson: Philos. Mag. **12**, 145, 741 (1965)

2.44   J.B. Sanders: Thesis, Univ. Leiden (1968)

2.45   M.W. Thompson: Philos. Mag. **18**, 377 (1968)

2.46   P. Sigmund, J.B. Sanders: *Proc. Int. Conf. Applications of Ion Beams to Semiconductor Technology*, ed. by P. Glotin (Editions Ophrys, Grenoble 1967) p. 215

2.47   P. Sigmund: Radiat. Eff. **1**, 15 (1969)

2.48   P. Sigmund: Phys. Rev. **184**, 383 (1969); **187**, 768 (1969)

2.49   M.W. Thompson, R.S. Nelson: Philos. Mag. **7**, 2015 (1962)

2.50   K.B. Winterbon, P. Sigmund, J.B. Sanders: K. Dan. Vidensk. Selsk. Mat. Fys. Medd. **37**, No. 14 (1970)

2.51   P. Sigmund: Appl. Phys. Lett. **25**, 169 (1974); **27**, 52 (1975)

2.52   H.H. Andersen, H.L. Bay: J. Appl. Phys. **45**, 953 (1974); **46**, 2416 (1975)

2.53   R.H. Silsbee: J. Appl. Phys. **28**, 1246 (1957)

2.54   P.K. Rol, J.M. Fluit, F.P. Viehböck, M. de Jong: *Proc. Int. Conf. Ionization Phenomena in Gases*, Vol. 4, (North Holland, Amsterdam 1960) p. 275

2.55   W.L. Brown, L.J. Lanzerotti, J.M. Poate, W.M. Augustyniak: Phys. Rev. Lett. **40**, 1027 (1978)

2.56   J.A. Davies, J. L'Ecuyer, N. Matsunami, R. Ollerhead: Radiat. Eff. **49**, 119 (1980)

2.57   J. Lindhard, V. Nielsen, M. Scharff: K. Dan. Vidensk. Selsk. Mat. Fys. Medd. **36**, No. 10 (1968)

2.58   P. Sigmund: Rev. Roum. Phys. **17**, 823, 969, 1079 (1972)

2.59   P. Sigmund: in *Physics of Ionized Gases* 1972, ed. by M. Kurepa (Inst. of Physics, Univ. Belgrade 1972) p. 137

2.60  P.Sigmund: Ann. Isr. Phys. Soc. **1**, 69 (1977)
2.61  H.A.Bethe: Ann. Phys. (Leipzig) **5**, 325 (1930)
2.62  P.Sigmund: in *Radiation Damage Processes in Materials*, ed. by C. H. S. Dupuy (Noordhoff, Leiden, 1975) p. 3
2.63  J. Lindhard, M.Scharff: Phys. Rev. **124**, 128 (1961)
2.64  L.Boltzmann: Sitzungsber. d. Wiener Akad. Wiss. **66**, 275 (1872)
2.65  J.Lindhard, V.Nielsen: K. Dan. Vidensk. Selsk. Mat. Fys. Medd. **38**, No. 9 (1971)
2.66  R.H.Fowler: Proc. Cambridge Philos. Soc. **21**, 531 (1923)
2.67  P.Sigmund: Appl. Phys. Lett. **14**, 114 (1969)
2.68  K.B.Winterbon: *Ion Implantation Range and Energy Deposition Distributions*, Vol. 2 (Plenum Press, New York and London 1975)
2.69  P.Sigmund: *Theory of Sputtering II* (overdue, unpublished manuscript)
2.70  D.K.Brice: *Ion Implantation Range and Energy Deposition Distributions*, Vol. 1 (Plenum Press, New York and London 1975)
2.71  P.Sigmund: J. Mater. Sci. **8**, 1545 (1973)
2.72  M.M.R.Williams: Philos. Mag. **34**, 669 (1976)
2.73  H.H.Andersen, P.Sigmund: Risø Report No. **103**, (1965)
2.74  D.Rosenberg, G.K.Wehner: J. Appl. Phys. **33**, 1842 (1962)
2.75  P.Sigmund, M.T.Matthies, D.L.Phillips: Radiat. Eff. **11**, 34 (1971)
2.76  D.P.Jackson: Radiat. Eff. **18**, 185 (1973)
2.77  N.Andersen, P.Sigmund: K. Dan. Vidensk. Selsk. Mat. Fys. Medd. **39**, No. 3 (1974)
2.78  P.Vajda: Rev. Mod. Phys. **49**, 481 (1977)
2.79  W.O.Hofer, U.Littmark: Phys. Lett. **71**A, 457 (1979)
2.80  K.B.Winterbon: unpublished note (1972), based on a suggestion made by K. Taulbjerg, cf. also L. V. Spencer, Phys. Rev. **98**, 1597 (1955)
2.81  U.Littmark, G.Maderlechner: in *Physcis of Ionized Gases*, Book of contributed papers, ed. by B.Navinsek, J.Stefan Institute, Ljubljana (Dubrovnik 1976)
2.82  J. Bøttiger, J.A.Davies, P.Sigmund, K.B.Winterbon: Radiat. Eff. **11**, 69 (1971)
2.83  P.Sigmund: *3rd Conf. on Atomic Collisions in Solids*, Kiev; Fl II Monograph 74–07 (H. C. Ørsted Institute, University of Copenhagen 1974)
2.84  M.Imada: J. Phys. Soc. Jpn. **45**, 1957 (1978)
2.85  H.Oechsner: Z. Phys. **238**, 433 (1970)
2.86  U.Littmark: Thesis, Univ. Copenhagen (1974)
2.87  U.Littmark, P.Sigmund: J. Phys. D. **8**, 241 (1975)
2.88  G.Betz, R.Dobrozemsky, F.P.Viehböck: Int. J. Mass Spectrom. Ion Phys. **6**, 451 (1971)
2.89  J.B.Gibson, A.N.Goland, M.Milgram, G.H.Vineyard: Phys. Rev. **120**, 1229 (1960)
2.90  D.E.Harrison,jr., W.L.Moore,jr., H.T.Holcombe: Radiat. Eff. **17**, 167 (1973)
2.91  W.L.Gay, D.E.Harrison,jr.: Phys. Rev. **135**A, 1780 (1964)
2.92  M.T.Robinson: J. Appl. Phys. **40**, 2670 (1969)
2.93  J.E.Westmoreland, P.Sigmund: Radiat. Eff. **6**, 187 (1970)
2.94  P.Sigmund: Phys. Rev. A**14**, 996 (1976)
2.95  N.Bohr: Philos. Mag. **25**, 10 (1913)
2.96  H.H.Andersen, P.Sigmund: K. Dan. Vidensk. Selsk. Mat. Fys. Medd. **34**, No. 15 (1966)
2.97  J.B.Sanders, H.E.Roosendaal: *Proc. Int. Conf. Atomic Collisions in Solids*, Moscow (1977), in press
2.98  C.Lehmann, P.Sigmund: Phys. Status Solidi **16**, 507 (1966)
2.99  M.W.Thompson: *Physics of Ionized Gases*, ed. by R. K. Janev (Inst. Physics, Beograd, 1978) p. 289
2.100  G.Ayrault, R.S.Averback, D.N.Seidman: Scr. Metall. **12**, 119 (1978)
2.101  H.L.Bay, H.H.Andersen, W.O.Hofer: Radiat. Eff. **28**, 87 (1976)
2.102  R.Behrisch, R.Weissmann: Phys. Lett. **30**A, 506 (1970)
2.103  G.M.McCracken: Rep. Prog. Phys. **38**, 241 (1975)
2.104  R.Behrisch, G.Maderlechner, B.M.U.Scherzer, M.T.Robinson: Appl. Phys. **18**, 391 (1979)
2.105  J.Bohdansky, J.Roth, H.L.Bay: J. Appl. Phys. **51**, 2861 (1980)
2.106  J.Vukanic, P.Sigmund: Appl. Phys. **11**, 265 (1976)

2.107  T.Lenskjær, F.Nyholm, S.D.Pedersen, N.B.Petersen: Phys. Lett. **47**A, 63 (1974)

2.108  T.J.Hoffman, H.L.Dodds, M.T.Robinson, D.K.Holmes: Nucl. Sci. Eng. **68**, 204 (1978)

2.109  R.Behrisch: Ergeb. Exakten Naturwiss. **35**, 295 (1964)

2.110  H.M.Windawi: Surf. Sci. **55**, 573 (1976)

2.111  A.S.Dolgov: Fiz. Tverd. Tela **19**, 1263 (1977) [Engl. transl.: Sov. Phys. Solid State **19**, 735 (1977)]

2.112  P.Sigmund: in *Inelastic Ion-Surface Collisions*, ed. by N. H. Tolk, J. C. Tully, W. Heiland, C. W. White (Academic Press, New York, London 1977) p. 121

2.113  D.A.Thompson, S.S.Johar: Appl. Phys. Lett. **34**, 342 (1979)

2.114  D.A.Thompson, R.S.Walker: Radiat. Eff. **36**, 91 (1978)

2.115  M.O.Ruault, J.Chaumont, H.Bernas, P.Sigmund: Phys. Rev. Lett. **36**, 1148 (1976); M.O.Ruault, H.Bernas, J.Chaumont: Philos. Mag. **39**, 757 (1979)

2.116  K.L.Merkle: Radiat. Eff. Lett. **50**, 39 (1980)

2.117  Y.Yamamura, Y.Kitazoe: Radiat. Eff. **39**, 251 (1978)

2.118  G.Carter: Radiat Eff. Lett. **43**, 193 (1979)

2.119  P.Haff: Appl. Phys. Lett. **31**, 259 (1977)

2.120  R.Kelly: Radiat Eff. **32**, 91 (1977)

2.122  G.E.Chapman, B.W.Farmery, M.W.Thompson, I.H.Wilson: Radiat Eff. **13**, 121 (1972)

2.124  M.Szymonski, A.E.de Vries: Phys. Lett. **63**A, 359 (1977)

2.125  A.E.de Vries: Proc. Int. Conf. Atomic Coll. in Solids (Moscow 1977), in press

2.126  J.M.Schroeer, T.N.Rhodin, R.C.Bradley: Surf. Sci. **34**, 571 (1973)

2.127  Z.Sroubek: Surf. Sci. **44**, 47 (1974); Z.Sroubek, J.Zavadil, F.Kubec, K.Zdansky: Surf. Sci. **77**, 603 (1978)

2.128  M.Cini: Surf. Sci. **54**, 71 (1976)

2.129  A.Blandin, A.Nourtier, D.W.Hone: J. Phys. (Paris) **37**, 369 (1976)

2.130  J.K.Nørskov, B.I.Lundqvist: Phys. Rev. B**19**, 5661 (1979)

2.131  C.A.Andersen, J.R.Hinthorne: Anal. Chem. **45**, 1421 (1973)

2.132  H.Oechsner, W.Gerhard: Surf. Sci. **44**, 480 (1974)

2.133  G.P.Können, A.Tip, A.E.de Vries: Radiat. Eff. **21**, 269 (1974); **26**, 23 (1975)

2.134  W.Gerhard: Z. Phys. B**22**, 31 (1975)

2.135  B.J.Garrison, N.Winograd, D.E.Harrison,Jr.: J. Chem. Phys. **69**, 1440 (1978)

2.136  M.J.Drinkwine, D.Lichtman: Prog. Surf. Sci. **8**, 123 (1977)

2.137  H.F.Winters, P.Sigmund: J. Appl. Phys. **45**, 4760 (1974)

2.138  R.S.Nelson: Radiat. Eff. **2**, 47 (1969)

2.139  R.A.Moline, G.W.Reutlinger, J.C.North: in *Atomic Collisions in Solids*, ed. by S. Datz, B. R. Appleton, C. D. Moak (Plenum Press, New York, 1975) p. 159

2.140  J.G.Perkins, P.T.Stroud: Nucl. Instrum. Methods **102**, 109 (1972)

2.141  R.Kelly, J.B.Sanders: Surf. Sci. **57**, 143 (1976); Nucl. Instrum. Methods **132**, 335 (1976)

2.142  K.B.Winterbon: In *Physics of Ionized Gases*, Book of contributed papers, ed. by R.K.Janev, Institute of Physics, Belgrade (Dubrovnik 1978)

2.143  H.H.Andersen: Appl. Phys. **18**, 131 (1979); P.K.Haff, Z.E.Switkowski: J. Appl. Phys. **48**, 3383 (1977); S.A.Schwarz, C.R.Helms: J. Vac. Sci. Technol. **16**, 781 (1979)

2.144  W.O.Hofer, U.Littmark: Phys. Lett. **71**A, 457 (1979)

2.145  P.Sigmund: J. Appl. Phys. **50**, 7261 (1979)

2.146  P.Sigmund, A.Gras-Marti: Nucl. Instrum. Methods **160**, 309 (1980)

2.147  T.Ishitani, R.Shimizu: Appl. Phys. **6**, 241 (1975)

2.148  N.Andersen, P.Sigmund: In *Physics of Ionized Gases*, Book of contributed papers, ed. by V. Vujnovic, University of Zagreb (Rovinj 1974) p. 91

2.149  R.Kelly: Nucl. Instrum. Methods **149**, 553 (1978)

2.150  H.H.Andersen: J. Vac. Sci. Technol. **16**, 770 (1979)

2.151  P.S.Ho, J.E.Lewis, W.K.Chu: Surf. Sci. **85**, 19 (1979)

2.152  G.Betz: Surf. Sci. **92**, 283 (1980)

2.153  M.Szymonski, R.S.Bhattacharya, H.Overeijnder, A.E.de Vries: J. Phys. D**11**, 751 (1978)

2.154  P.K.Haff, Z.E.Switkowski: Appl. Phys. Lett. **29**, 549 (1976)
2.155  N.Saeki, R.Shimizu: Surf. Sci. **71**, 479 (1978)
2.156  G.K.Wehner: Appl. Phys. Lett. **30**, 185 (1977)
2.157  R.R.Olson, M.E.King, G.K.Wehner: J. Appl. Phys. **50**, 3677 (1979)
2.158  E.Taglauer, W.Heiland: Appl. Phys. Lett. **33**, 950 (1978)
2.159  Z.L.Liau, J.W.Mayer, W.L.Brown, J.M.Poate: J. Appl. Phys. **49**, 5295 (1978)
2.160  J.O.Stiegler (ed.): "Proc. of the Workshop on Solute Segregation and Phase Stability during Irradiation", J. Nucl. Mater. **83** (1979)
2.161  E.Bøgh, P.Høggild, I.Stensgaard: Radiat. Eff. **7**, 115 (1971)
2.162  K.B.Winterbon: Radiat. Eff. **13**, 215 (1972)
2.163  C.Erginsoy, G.H.Vineyard, A.Englert: Phys. Rev. **113**, A595 (1964)
2.164  H.L.Bay, H.H.Andersen, W.O.Hofer, O.Nielsen: Nucl. Instrum. Methods **132**, 301 (1976)
2.165  P.Sigmund, C.Claussen: J. Appl. Phys., **52** (1981)

# 3. Theoretical Aspects of Monocrystal Sputtering

Mark T. Robinson

With 16 Figures

The structures of crystalline solids influence the collision cascades which lead to sputtering. The crystal selvage may have a different structure and thermal motion from the interior. The surface is not always flat and may be further modified by the irradiation. As a result atoms in the selvage do not have unique binding energies, but a distribution of values reflecting the structural elements and irregularities present. The cohesive energy of the crystal is only an average value.

Ideas of crystal transparency and channeling supply a conceptual basis for understanding the effects of orientation on the sputtering yields from monocrystals at medium and high ion energies. Embodiments of these ideas in simple models give only qualitative agreement with measured yields. At low energies, crystal effects still persist where such concepts are not applicable. Anisotropies of binding energies alone cannot explain the results.

Linear collision sequences were first proposed to account for crystallographic effects in the ejection of atoms from monocrystals during sputtering. Detailed examination of such sequences shows that their contribution to ordinary sputtering is small. Mechanisms based mainly on the last collisions leading to ejection give only a rough picture of the angular distribution of ejected material. The need for full three-dimensional modelling of the ejection process is indicated.

A comprehensive model of the sputtering process in monocrystals can be based on the computer simulation method, but machine limitations still require somewhat severe approximations to be made for most applications. Recent calculations with unstable dynamical models, as well as those based on the binary collision approximation, have given very promising results. Continued development along these lines should allow quantitative agreement to be achieved between computations and experiments.

## 3.1 Historical Origins

If a single crystal target is irradiated by a well-collimated beam of ions, several interesting features appear in the sputtering behavior which are not present if the target is polycrystalline or amorphous. If the beam is directed parallel to one of the more densely packed rows or planes of atoms in the crystal, the yield of sputtered particles falls well below the corresponding value of polycrystalline material, while for irradiations perpendicular to closely packed planes, the yield may be much higher than the polycrystalline value. Further, the smoothly varying cosine-like angular distribution of ejected material which is characteristic of polycrystalline targets is replaced by an angular distribution containing pronounced maxima in certain directions. Regularities in other aspects of sputtering have also been observed, including differences in the energy distributions of atoms sputtered in different crystallographic directions and angular variations in the relative intensities of sputtered molecular ions. The extensive experimental data are reviewed in detail in Chap. 5 and [3.1a] Chap. 2.

Theoretical treatments of monocrystal sputtering fall into two somewhat distinct groups. One procedure is to develop specialized models to explain particular experiments. Typical examples are the channeling theory of the sputtering yield and the focusing collision theory of the ejection pattern. The other procedure is to construct comprehensive models, capable of dealing in a unified way with the entire sputtering process. This approach is exemplified by the computer simulation models. At present it cannot be said that either procedure has yet been entirely successful. The channeling theory accounts for the angular variations of the yield close to the principal crystal axes and planes, but is not very successful for more general directions. The focusing collision theory gives a qualitative description of the ejection pattern spots observed from cubic metals at high bombarding energies, but cannot successfully interpret low energy irradiations or observations on hexagonal and other low symmetry materials. The computer modelling efforts have made only some promising beginnings so far. This stems mainly from the complex nature of the problem, which requires elaborate models that are very voracious of computer resources.

Although broad interest in monocrystal sputtering developed only recently, it was observed in 1912 that there are differences in the sputtering rates of different surfaces of monocrystal targets exposed to kilovolt ion bombardment [3.1]. The possibilities of two kinds of correlated collision sequence were recognized. In one of these [3.2], a series of collisions, initiated by the incident particle, occurs along a row of atoms in the crystal. Energy is transported away from the irradiated surface, but the idea of momentum focusing was absent. In fact, this mechanism was rejected as implausible in the particular experimental context. The other proposal [3.1–3] was that a series of glancing collisions could steer the incident particle into the region between the atoms and confine it there. The ion would penetrate more deeply in certain crystal directions than in others and the directions of deep penetration would correspond to planes of low sputtering yield.

In spite of this early beginning, there seems to have been no further interest in monocrystal sputtering until *Wehner's* discovery [3.4, 5] that the angular distributions of the atoms ejected from single crystal targets showed maximum intensities in directions corresponding to the more closely spaced atomic rows passing through the target surfaces. The discovery of such ejection patterns or *Wehner* spots inspired *Silsbee's* identification [3.6] of the focusing collision sequence as a means for the long-range transport of momentum in crystals at low energies. Focusing was proposed as an explanation for the Wehner spots. Although more recent work has shown this explanation to be incorrect, focusing is nevertheless a major factor in displacement cascade development in crystals.

At about the same time, experiments from several laboratories [3.7–9] showed that the sputtering yields of monocrystals depend sensitively on the crystallographic direction of the incident beam. This was explained [3.7, 10] in terms of the varying openness seen in stick-and-ball crystal models when viewed from different directions. It was soon realized [3.11] that the experiments of *Almén* and *Bruce* [3.8] implied that ion ranges were greater in some crystal directions than in others. The discovery of ion channeling, first in computer calculations [3.12, 13] and then experimentally [3.14–16], confirmed this suggestion and supplied a secure physical basis for the so-called "transparency" models.

An alternative to the focusing collision theory of the ejection pattern spots was proposed by *Lehmann* and *Sigmund* [3.17]. Their model requires that the target surface have an ordered structure, but does not require long straight rows of atoms intersecting the surface. This model emphasizes the role of surface structure and of the surface binding energy in monocrystal sputtering. The important role of surface properties for sputtering phenomena has also been discussed by *Weijsenfeld* [3.18], who pointed out the effects of surface binding on the directions taken by the ejected atoms.

Before the development of modern high-speed computing machinery, it was not practicable to include the full three-dimensional structures of crystals in theoretical models of sputtering, or, indeed, of other kinds of radiation damage. The investigations of *Vineyard* and his colleagues [3.19] demonstrated the power of computer simulation methods in isolating and understanding the elementary processes of defect production and migration involved in radiation damage. Such techniques were first applied to monocrystal sputtering problems by *Harrison* and his co-workers [3.20]. They proposed ejection mechanisms to account for the Wehner spots which differ from both the focusing collision theory and the model of *Lehmann* and *Sigmund*.

The objectives of this chapter are threefold. First, it supplies a conceptual background for discussing sputtering problems in monocrystal targets. To this end, certain aspects of both surface and bulk crystallography are summarized. Second, a critical account is given of the specialized models which have been used to discuss particular aspects of monocrystal sputtering. Finally, computer simulation techniques are considered in some detail with the view of understanding the current status (and limits) of the art.

## 3.2 Surface Crystallography and Thermodynamics

### 3.2.1 The Structures of Surfaces

A finite monocrystal may be divided into two regions, the *substrate* and *selvage*[1] [3.21]. In the substrate or crystal interior, the structure is periodic in three dimensions. The lattice vibrations have symmetries derived from this structure, as do many other physical properties. In the selvage, however, the structure is at most diperiodic, the translational symmetry perpendicular to the surface being lost.

---

1 This term, also spelled *selvedge*, denotes those edges of a piece of woven fabric which are specially worked to prevent unravelling.

The selvage may consist of one or more layers of atoms. The separations between these layers and their separations from layers in the substrate may be greater or less than the separations between similar layers in the substrate itself. The individual layers of atoms in the selvage may have structures closely resembling those expected from the substrate structure or they may be *reconstructed*. Reconstruction may in effect consist of the introduction of regular arrays of point defects or it may involve drastic changes in structure. The variety of differences between the selvages and substrates of metal crystals, as well as many other aspects of surface crystallography, have recently been reviewed [3.22–27] and will be summarized here only briefly.

Low-energy electron diffraction (LEED) studies [3.28] show the selvage spacings between the {100} and {111} planes of Al, Ni, Cu, and other fcc metals to be within a few percent of the substrate values. The selvage spacings of the {110} planes of Al, Ni, Cu, and Ag, however, appear to be some 5 to 15 % *less* than the substrate values. Contractions have also been reported for Fe {111} and several bcc {001} selvages. In each of these cases, the symmetry of the selvage layer is the same as that of the substrate. The selvages on the {100} surfaces of Ir, Pt, and Au, however, are reordered and may be most easily described as close-packed {111} layers overlying the substrate cube planes. Recent backscattering studies [3.29] have shown the reordered selvage on the Au {100} surface to be just one layer thick.

Theoretical understanding of such structural changes at metal surfaces is very incomplete. Two recent attempts have been made to calculate the relaxations of normally ordered selvage planes for cubic metals [3.30, 31]. Both calculations used the Morse potentials of *Girifalco* and *Weizer* [3.32] to represent the interactions between the atoms of a crystal. The relaxations were found by minimizing the total potential energy of a semi-infinite crystal with the desired surface. The {111}, {100}, and {110} selvages of several fcc metals were all found to relax outward, the first two by rather small amounts, but the last by as much as 15 %, a result in direct conflict with the LEED observations. The outward relaxation comes mainly from the attribute of the potentials used that they are slightly repulsive for nearest neighbor atoms in the crystal and attractive for all more distant neighbors. It seems unlikely that any central potential with this property can successfully account for the contraction of fcc {110} surfaces. Models including both central pair potentials and volume dependent forces can be made free from this criticism, however. A recent calculation [3.33] including only the volume dependent part was able to predict a contraction of about 16 % for Al {110} and much smaller contractions for {100} and {111}. The physical origin of the contraction may be pictured as a surface tension in the electric fluid which represents the conduction electrons. A more elaborate calculation [3.34] has been made of surface relaxations and surface energies in bcc metals using short-range pair potentials and volume-dependent forces. These show much smaller relaxations than were obtained using Morse potentials, although the relaxations of α-Fe and W {100}, {110}, and {111} were all outward at the surface, again in partial conflict with observations [3.28]. Displacements of the second layer were inward, however, in contrast to the Morse potential results [3.31]. A calculation [3.34] for the α-Fe and W {112} showed not only an outward relaxation, but also a tangential one, the outermost plane and the second one moving in opposite directions parallel to ⟨111⟩. Such tangential relaxations could be very important in sputtering, especially at low energies.

The thermal motions of atoms in the selvage are also altered [3.35]. Because the forces acting on the selvage atoms are modified, localized vibrational modes may be produced and some or all of the normal mode frequencies of the substrate crystal may be lowered. The amplitudes of the thermal vibrations are increased and may become anisotropic. The increased vibrational amplitude due to missing pair interactions may be partially offset by electronic effects originating in the "surface tension" of the conduction electrons [3.33].

At least under ideal circumstances, *principal* or *singular* surfaces, such as fcc {100} and {111}, bcc {100} and {110}, and hcp {0001} are believed to be atomically smooth, but this is not true of surfaces making small angles with them. Such nearby or *vicinal* surfaces, on the contrary, consist of atomically smooth terraces separated by monatomic *steps* or *ledges*. The steps occur at regular intervals, their spacings dictated by the average surface orientation. Where possible, the terraces are all of the same width. Otherwise, terraces of two different widths, for instance, may alternate. The regular array of steps causes a characteristic streaking or doubling of the LEED pattern spots from such vicinal surfaces. Photographs of ball models of crystal surfaces [3.36] show this structure very clearly. They show, furthermore, that the edges of the steps need not be smooth, but often contain *kinks*.

Such features of surface crystallography have several possible consequences for sputtering. Since the sputtered atoms themselves originate very largely in the selvage, the structural differences therein must be considered in discussing the geometrical aspects of the ejection process. The positions, widths, and intensities of the Wehner spots must respond to such structural alterations. The trajectories of the incident ions can also be influenced by the surface structure. This would be especially important at low incident ion energies and (probably) on reconstructed or vicinal surfaces.

### 3.2.2 Effects of Radiation Damage on Surface Structures

A fundamental problem in treating single crystal sputtering theoretically is that damage is introduced into the target during the irradiation, especially in the near-surface structure. To what extent is it appropriate to consider the ideal selvage and substrate structures as a basis for sputtering calculations?

Low energy ion bombardment, alternated with thermal annealing is often used to prepare atomically clean monocrystal surfaces [3.37]. The irradiations, for instance with ∼500 eV Ar⁺ ions, are known to cause degradation of LEED patterns from these surfaces. A few systematic studies have been made of the damage introduced into the surfaces of Mo and W [3.38, 39] and of Ni [3.40]. At very low doses ( <0.1 ion per surface atom), the LEED intensities decrease and at slightly higher doses the spots in the patterns become somewhat diffused. Such effects seem to saturate at doses of ~1 ion per surface atom. The electron beams used for LEED observations have coherence lengths parallel to the experimental surface of ~$10^2$ to ~$10^3$ Å [3.41]. Thus, the LEED pattern spots tend to broaden or to develop streaks when the domains of coherently scattering material on the surface become smaller than this (for a fuller discussion, see [3.42]). The LEED results may be interpreted [3.38] as caused by a general roughening of the surface due to the presence of many monatomic steps. There may be regions with dimensions of ~50 Å or more which are largely perfect and of the nominal orientation separated by regions with a high density of small steps (cf. also [3.18]). Similar conclusions were also reached from studies of damaged Ni surfaces by 1 keV $Ar^+$ backscattering [3.40].

In the course of recent studies on noble metal surfaces, *Zehner* [3.43] has recorded many LEED patterns from samples irradiated by 200 to 300 eV $Ar^+$ ions at doses up to ~100 per surface atom. Even on the reconstructed surfaces Au {100} and {110}, the reconstructed selvages persist. Since the doses involved were sufficient for the gravimetric determination of the sputtering yield, these observations support the use of ideal crystal structures in the theoretical modelling of the sputtering of metals.

By contrast, semiconductor surfaces are easily damaged by ion bombardment. The experimental data are summarized in [3.44]. At room temperature, doses as low as 0.1 to 1.0 ion per surface atom may be enough to obliterate a LEED pattern from a Ge crystal. Such damage is sufficient to alter the sputtering behavior of such materials, except at very low ion energies. Both the ejection patterns [3.45] and the sputtering yields [3.46] are characteristic of amorphous media. At temperatures above 600 K, sufficiently rapid annealing of the ion-induced structural damage takes place to permit crystal effects in sputtering to be observed. In materials as structurally sensitive to irradiation as the semiconductors, it is unlikely that models based on the ideal crystal structure would be reliable.

### 3.2.3 Surface Binding Energies

The energy binding the atoms in a crystal surface influences sputtering in two ways. First, it provides a barrier which must be surmounted by escaping atoms and hence plays an important part in determining the yield of sputtered particles. Second, its directional properties are a significant influence on the directions taken by the ejected atoms [3.18]. Both aspects of surface binding are important in structureless media as well as in crystalline ones, but in the latter, there is the additional feature that the surface binding energy depends on crystal orientation.

The *cohesive energy*, $U_0$, is the energy necessary to disperse a solid body into its component atoms. It is evaluated by correcting experimentally measured latent heats of sublimation (or vaporization) to absolute zero. The correction is small

**Table 3.1.** Cohesive energies of selected metals [3.47]

| Element | $U_0$ [eV/atom] | Element | $U_0$ [eV/atom] |
|---------|-----------------|---------|-----------------|
| Mg | 1.51 | Ag | 2.94 |
| Al | 3.39 | Cd | 1.16 |
| Ni | 4.44 | W | 8.79 |
| Cu | 3.48 | Pt | 5.85 |
| Zn | 1.35 | Au | 3.81 |
| Nb | 7.44 | Pb | 2.03 |

**Table 3.2.** Surface binding energies calculated for selected metals [3.49]

| Element | Surface binding energy [eV per atom] | | |
|---------|--------|--------|--------|
|  | {111} | {100} | {110} |
| Al | 3.80 | 3.80 | 3.53 |
| Cu | 4.65 | 4.62 | 4.26 |
| Ag | 4.08 | 3.98 | 3.61 |
| Fe | 4.72 | 5.47 | 5.48 |
| W | 9.75 | 11.52 | 11.86 |

($\sim 0.01$ eV/atom) for most metals. Table 3.1 lists values of $U_0$ for a few metals. The cohesive energy can be regarded as associated with the bonds between pairs of atoms in the crystal. As the solid is vaporized, these bonds are broken, but, since each bond connects two atoms, the number of *distinct* bonds broken is half the number of atoms vaporized (ignoring surfaces for the moment). In contrast, when an atom in the interior of a crystal is removed to the vapor phase, leaving a vacant lattice site behind, all of its bonds to other atoms are broken. Hence, in the absence of relaxations, the binding energy of an atom in the interior of a crystal is $2\,U_0$ [3.18, 48, 49].

On an average, an atom at a crystal surface has half the number of neighbors that an interior atom has. The number for any particular atom will depend upon the orientation of the surface at which it is located and on whether it is in a step or at a kink. In fact, the classical definition of a kink site [3.48] identifies it as a "half-lattice" position. Therefore, the average binding energy of a surface atom is equal to $U_0$, again in the absence of relaxations. This value has often been used in sputtering calculations, especially on structureless media (see Chap. 2 and [3.18, 50]).

There have been two calculations of the binding energies of atoms in ideal surfaces [3.18, 49] based on the Morse potentials of *Girifalco* and *Weizer* [3.32]. Because these potentials were derived to describe the bulk properties of metals, they do not incorporate electron rearrangement effects at metal surfaces. *Weijsenfeld* [3.18] made calculations for Cu, evaluating the binding energies

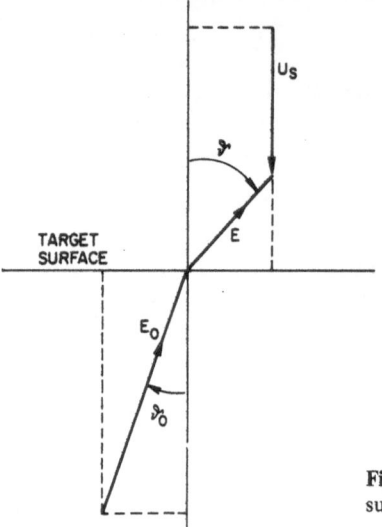

**Fig. 3.1.** The refraction of a sputtered atom by a planar surface binding barrier

without reference to relaxations. *Jackson* [3.49] used a more elaborate model in which a large number of atoms was included and which could take account of relaxations between the layers of the selvage. The effects of these relaxations were small, but additional relaxations of the atoms neighboring the vacant surface site were ignored. A selection from *Jackson*'s results is displayed in Table 3.2. However, calculations of the surface properties of metals must be viewed with some suspicion when they are based on long-range pair potentials derived from bulk materials properties alone [3.26, 34]. The failure of the relaxation calculations for fcc {110} surfaces has been noted.

There have been several recent reviews of theoretical research on the electronic band structures of metals in the vicinity of surfaces [3.51–53]. In the region just outside the surface plane defined by the locations of the atomic centers, the electron density decreases exponentially, while within the selvage it exhibits a series of oscillations. This behavior requires the effective pair potential through which the atoms interact to be different in the selvage than it is in the crystal interior. Unfortunately, no such surface potentials are yet available, so that it is necessary to proceed on the basis of bulk pair potentials, in spite of their limitations in treating surface problems.

On most surfaces at finite temperatures, there will not be a unique binding energy, but a distribution of values reflecting the presence of whatever steps, kinks, and other irregularities are present. The morphologies of real crystal surfaces are of interest in many areas of surface physics and can be studied in part by the methods of statistical thermodynamics [3.54]. Recent stochastic computer simulations indicate the existence of a roughening transition, occurring on solid surfaces at a definite temperature. Below this temperature, the surface is relatively flat and contains a small density of rather isolated clusters of defects (adatoms and surface vacancies). Above the transition, the appearance of the surface is greatly altered. Drawings of such rough surfaces, based on computer simulations [3.54], show the presence of many steps, and of terraces at different elevations. It is tempting to suggest that these are the sorts of surfaces indicated by LEED observations on sputtered materials. If this identification is correct, advances in the statistical mechanical treatment of the roughening transition will prove useful in understanding the topographies of sputtered targets.

The geometrical nature of the surface binding energy can influence the direction taken by a sputtered particle as it leaves the target surface [3.18, 55]. If, as is often assumed [3.18, 20, 50, 55–57], the surface binding energy is a planar barrier parallel to the target surface, only the normal component of the velocity of an ejected atom can surmount it. Consequently, the atom is refracted away from the normal and, if its normal velocity is low enough, may even be reflected back onto the surface. The magnitude of the effect is easily calculated, using Fig. 3.1 as a guide. An ejected particle leaves the surface with energy $E_0$ at an angle $\vartheta_0$ with the normal. After overcoming the surface barrier $U_s$, the energy and direction of the particle are

$$E = E_0 - U_s \tag{3.2.1}$$

$$\mu = \cos\vartheta = [(E_0\mu_0^2 - U_s)/(E_0 - U_s)]^{1/2}, \tag{3.2.2}$$

where $\mu_0 = \cos\vartheta_0$. It may be seen from these equations that particles with energies less than $U_s/\mu_0^2$ will be reflected internally and unable to escape the target. In a structureless medium, approximately one-half of the particles incident on the surface barrier are actually sputtered [3.50], the exact number depending on the angular and energy distributions of the particles approaching the barrier. The refractive effect is closely bound up with the interpretation of ejection patterns.

The plausibility of the *planar binding model* requires some discussion. If the energy binding an atom to a solid surface is regarded as localized in the bonds connecting the atom to its nearest neighbors, this model is applicable only to surfaces having principal orientations, since only in them are the neighbors disposed symmetrically enough to regard their resultant force on the escaping atom as normal to the surface. On vicinal surfaces, the binding of atoms far from steps could be considered as normal to the principal terraces in which they lie, but those at the steps would have a strong component of binding parallel to the terraces. Rough surfaces offer more complicated situations. It is perhaps more reasonable, however, to consider the surface binding as composed of a small local part and of another part coming from the whole crystal. This division is the counterpart of dividing the energy of the crystal into a contribution from pair interactions between the atoms in a crystal and a volume-dependent contribution intended to represent electron gas effects [3.34]. The nonlocal portion can be considered as operating in a direction normal to any surface that is not exceptionally rough. Unfortunately, no calculations seem to have been made which could allow more quantitative assessment of this problem.

Another model of surface binding used in sputtering calculations [3.50, 58–61] has the ejected particles facing an isotropic energy barrier. In this *isotropic binding model*, no refraction of the ejected particles occurs, but the final particle energy is still given by (3.2.1).

Use of the planar binding model is supported, at least for principal orientations, by a computer simulation [3.62] of surface ejection from a {001} surface of W. The atoms of the numerical crystallite interacted with one another through a Morse potential, using the parameters of [3.32], and were placed initially in the relaxed positions that had been found earlier [3.31]. One atom in the surface was given an initial impulse in the desired direction and the equations of motion were integrated until either an atom was ejected from the solid or it was clear that none would be ejected. The final particle direction, energy, and fate (ejected or not) could be described fairly accurately by the planar binding model outlined above when account was taken of the scattering of the initial projectile by its neighbors. Similar behavior of the ejected atoms is shown in other computer simulations [3.63, 64]. Because the model potentials do not give a satisfactory picture of surface relaxations, some doubt must be expressed that they correctly describe the binding geometry. It would be desirable to have further calculations, preferably based on models with more realistic surfaces.

## 3.3 The Sputtering Yields of Monocrystals

The sputtering yields of monocrystalline targets irradiated by heavy (that is, excluding H and He) ions of energies exceeding ~1 keV show a strong dependence on the crystallographic orientation of the incident beam, an effect which increases with the ion energy, at least up to some tens of keV. Such effects have been observed for several different crystal structures and for several different ions, either normally incident on a variety of crystal surfaces or incident in a variety of oblique directions on principal surfaces. The experimental data are reviewed in detail by *Roosendaal* in Chap. 5. It is clear that crystallographic variations in the surface binding energy cannot account for the large effects observed, except perhaps at the lowest energies. Such yield effects are, however, similar, both in their angular behavior and in their energy dependence, to the effects of crystal orientation on ion penetration depths, nuclear reaction rates, and ion backscattering yields, all of which are strongly affected by ion channeling. The general nature of channeling in crystals and its effects on a variety of physical phenomena are reviewed elsewhere [3.65–67]. The essential idea is that a high speed particle moving nearly parallel to a densely packed row or plane of atoms in a crystal experiences a series of correlated glancing collisions which steer it gently away from the row or plane. Such channeled atoms have a probability for close encounters with lattice atoms which is greatly reduced from that for particles on random trajectories.

### 3.3.1 The Concept of Transparency

Orientation effects on the yield were originally discussed in terms of so-called *transparency* models [3.7, 68–73]. Although these models differ in detail, in each case a small sphere of radius $r$ is located at each atomic site in the crystal. The radius is chosen so that $\pi r^2$ is the cross section for a critical interaction between the incident ions and the lattice atoms. Collisions with impact parameter $p \leq r$ contribute to the sputtering in one way, while those with $p > r$ contribute differently [3.68] or not at all [3.69–72]. The distance of the collision point from the surface could also be a factor in the resulting yield [3.71–73]. Thus, from the viewpoint of the incident ion, the target crystal is an assembly of small spheres, whose size can be regard as energy dependent. If the ion beam is incident parallel to a lattice vector, some ions will collide with these spheres, but because of the translational symmetry of the crystal, the others will penetrate deeply without making collisions with $p \leq r$. This property of the lattice of dividing the beam into colliding and noncolliding portions is responsible for the term *transparency*. If the assembly of small spheres is projected onto a plane perpendicular to the incident beam, each row of atoms will appear as a circle of radius $r$. If the beam direction is not parallel to a lattice vector, but is instead perpendicular to a reciprocal lattice vector, the small spheres will project onto the transverse plane as a series of strips of width $2r$, representing the edges of planes of atoms in the crystal. Division of the beam into two portions still occurs. But, if the incident direction is neither parallel to a lattice vector nor perpendicular to a reciprocal lattice vector, the transverse plane will be completely covered and the beam will no longer be divided into distinguishable parts. *Onderdelinden* et al. [3.74–80] have drawn the connection between this description and *Lindhard*'s description of *channeling* [3.81, 82]. Those ions which approach within a critical distance of an atomic row or plane are strongly scattered and lose all memory of their original direction. Because they subsequently encounter the lattice atoms in a more-or-less random manner, they

**Table 3.3.** The most open axes and planes in face-centered cubic crystals (space group Fm3m, $N = 4/a^3$)

| Axis [$uvw$] | Unit Translation $t_{uvw}/a$ | | Plane ($hkl$) | Planar Spacing $d_{hkl}/a$ | |
|---|---|---|---|---|---|
| 011 | $(1/2)^{1/2}$ | 0.7071 | 111 | $(3)^{-1/2}$ | 0.5774 |
| 001 | 1 | 1.0000 | 001 | 1/2 | 0.5000 |
| 112 | $(3/2)^{1/2}$ | 1.2247 | 011 | $(8)^{-1/2}$ | 0.3536 |
| 013 | $(5/2)^{1/2}$ | 1.5811 | 113 | $(11)^{-1/2}$ | 0.3015 |
| 111 | $(3)^{1/2}$ | 1.7321 | 133 | $(19)^{-1/2}$ | 0.2294 |
| 123 | $(7/2)^{1/2}$ | 1.8708 | 012 | $(20)^{-1/2}$ | 0.2236 |
| 114 | $3/(2)^{1/2}$ | 2.1213 | 112 | $(24)^{-1/2}$ | 0.2041 |
| 012 | $(5)^{1/2}$ | 2.2361 | 115 | $(27)^{-1/2}$ | 0.1925 |
| 233 | $(11/2)^{1/2}$ | 2.3452 | 135 | $(35)^{-1/2}$ | 0.1690 |
| 015 } 134 | $(13/2)^{1/2}$ | 2.5495 | 122 | 1/6 | 0.1667 |

are said to have been scattered into the *random beam*. In contrast, those ions which do not approach the atomic rows or planes very closely are steered gently away from them and constitute the axial or planar *aligned beam*. Because channeling effects persist over a significant angular region near each axis or plane, *Onderdelinden*'s model avoids the difficulty of earlier models that the transparency is destroyed by any deviation whatever from the ideal directions. *Lindhard*'s theory of channeling provides a basis for estimating the critical distance of approach to a row or plane of atoms. Estimates can then be given of the effects of crystal orientation, ion energy, and temperature on monocrystal sputtering yields. Before discussing these, the transparency concept will be used to identify those orientations of several common crystal structures for which significant changes in sputtering yield should be expected and to give a rough indication of the magnitudes of these changes.

**Table 3.4.** The most open axes and planes in body-centered cubic crystals (space group Im3m, $N = 2/a^3$)

| Axis $[uvw]$ | Unit Translation $t_{uvw}/a$ | | Plane $(hkl)$ | Planar Spacing $d_{hkl}/a$ | |
|---|---|---|---|---|---|
| 111 | $(3)^{1/2}/2$ | 0.8660 | 011 | $(2)^{-1/2}$ | 0.7071 |
| 001 | 1 | 1.0000 | 001 | 1/2 | 0.5000 |
| 011 | $(2)^{1/2}$ | 1.4142 | 112 | $(6)^{-1/2}$ | 0.4083 |
| 113 | $(11)^{1/2}/2$ | 1.6583 | 013 | $(10)^{-1/2}$ | 0.3162 |
| 133 | $(19)^{1/2}/2$ | 2.1795 | 111 | $(12)^{-1/2}$ | 0.2887 |
| 012 | $(5)^{1/2}$ | 2.2361 | 123 | $(14)^{-1/2}$ | 0.2673 |
| 112 | $(6)^{1/2}$ | 2.4495 | 114 | $(18)^{-1/2}$ | 0.2357 |
| 115 | $3(3)^{1/2}/2$ | 2.5981 | 012 | $(20)^{-1/2}$ | 0.2236 |
| 135 | $(35)^{1/2}/2$ | 2.9580 | 233 | $(22)^{-1/2}$ | 0.2132 |
| 122 | 3 | 3.0000 | 015 ⎱ 134 ⎰ | $(26)^{-1/2}$ | 0.1961 |

**Table 3.5.** The most open axes and planes in crystals with simple cubic (space group Pm3m, $N = 1/a^3$) or NaCl (space group Fm3m, $N = 8/a^3$) structures

| Axis $[uvw]$ | Unit Translation $t_{uvw}/a$ | | Plane $(hkl)$ | Planar Spacing[a] $d_{hkl}/a$ | |
|---|---|---|---|---|---|
| 001 | 1 | 1.000 | 001 | 1 | 1.0000 |
| 011 | $(2)^{1/2}$ | 1.4142 | 011 | $(2)^{-1/2}$ | 0.7071 |
| 111 | $(3)^{1/2}$ | 1.7321 | 111 | $(3)^{-1/2}$ | 0.5774 |
| 012 | $(5)^{1/2}$ | 2.2361 | 012 | $(5)^{-1/2}$ | 0.4472 |
| 112 | $(6)^{1/2}$ | 2.4495 | 112 | $(6)^{-1/2}$ | 0.4083 |
| 122 | 3 | 3.0000 | 122 | 1/3 | 0.3333 |
| 013 | $(10)^{1/2}$ | 3.1623 | 013 | $(10)^{-1/2}$ | 0.3162 |
| 113 | $(11)^{1/2}$ | 3.3166 | 113 | $(11)^{-1/2}$ | 0.3015 |
| 023 | $(13)^{1/2}$ | 3.6056 | 023 | $(13)^{-1/2}$ | 0.2774 |
| 123 | $(14)^{1/2}$ | 3.7417 | 123 | $(14)^{-1/2}$ | 0.2673 |

[a] For crystals with the NaCl structure, these values must be divided by 2.

### 3.3.2 The Open Axes and Planes in Common Crystal Structures

The orientations of the most transparent axes and planes of simple crystal structures can be derived using standard crystallographic tables and formulas [3.83–85]. Examination of crystal models and of photographs and sketches of models [3.36] is also helpful. Some results obtained for several structures are collected in Tables 3.3–3.7, each of which is arranged in order of decreasing transparency. The crystallographic notation used is the following:

**Cubic Crystals:**
Cell edge: $a$
Direction of an axis (a lattice vector): $[uvw]$
Normal to a plane (a reciprocal lattice vector): $(hkl)$

**Hexagonal Crystals:**
Cell edges: $a$, $c$
Axial ratio: $\gamma = c/a$
Direction of an axis: $[uvjw]$ with $u+v+j=0$
Normal to a plane: $(hkil)$ with $h+k+i=0$

**All Crystals:**
Density: $N$ atoms per unit volume
Unit translation along an axis: $t_{uvw}$
Unit translation perpendicular to planes: $d_{hkl}$.

Following common practice, the extra hexagonal (Miller-Bravais) indices $i$ and $j$ will often be omitted to unify the cubic and hexagonal formulas. In general, a lattice vector $[uvw]$ is not perpendicular to the lattice plane $(uvw)$ or, indeed, to any lattice plane at all. The perpendicularity of $[hkl]$ to $(hkl)$ is a special property of

**Table 3.6.** The most open axes and planes in cubic crystals with the diamond (space group Fd3m, $N=8/a^3$) or ZnS (space group F$\bar{4}$3m, $N=8/a^3$) structures

| Axis $[uvw]$ | Unit Translation $t_{uvw}/a$ | | Plane $(hkl)$ | Planar Spacing $d_{hkl}/a$ | |
|---|---|---|---|---|---|
| 011 | $(1/2)^{1/2}$ | 0.7071 | 111[b] | $(3)^{-1/2}$ | 0.5774 |
| 111[a] | $(3)^{1/2}$ | 1.7321 | 011 | $(8)^{-1/2}$ | 0.3536 |
| 001 | 1 | 1.0000 | 113[b] | $(11)^{-1/2}$ | 0.3015 |
| 112 | $(3/2)^{1/2}$ | 1.2247 | 001 | $1/4$ | 0.2500 |
| 013 | $(5/2)^{1/2}$ | 1.5811 | 133[b] | $(19)^{-1/2}$ | 0.2294 |
| 113[a] | $(11)^{1/2}$ | 3.3166 | 112 | $(24)^{-1/2}$ | 0.2041 |
| 123 | $(7/2)^{1/2}$ | 1.8708 | 115[b] | $(270)^{-1/2}$ | 0.1925 |
| 114 | $3/(2)^{1/2}$ | 2.1213 | 135[b] | $(35)^{-1/2}$ | 0.1690 |
| 133[a] | $(19)^{1/2}$ | 4.3589 | 122 | $1/6$ | 0.1667 |
| 012 | $(5)^{1/2}$ | 2.2361 | 013 | $(40)^{-1/2}$ | 0.1581 |

[a] $\kappa=2$ in these rows; the spacings are alternately 1/4 and 3/4 of $t_{uvw}$.
[b] $\kappa=2$ for these planes; the spacings are alternately 1/4 and 3/4 of $d_{hkl}$.

**Table 3.7.** The most open axes and planes in close-packed hexagonal crystals [space group $P6_3/mmc$, $N = (4/3^{1/2}\gamma)/a^3$]

| Axis [$uvjw$] | Unit Translation, $t_{uvjw}/a$ General | Ideal $\gamma$ | Plane[e] ($hkil$) | Planar Spacing, $d_{hkil}/a$ General | Ideal $\gamma$ |
|---|---|---|---|---|---|
| $11\bar{2}0$ | 1 | 1.0000 | $10\bar{1}0$[d] | $(3)^{1/2}/2$ | 0.8660 |
| $0001$ | $\gamma$ | 1.6330 | $0001$[c] | $\gamma/2$ | 0.8165 |
| $10\bar{1}0$ | $(3)^{1/2}$ | 1.7321 | $10\bar{1}1$ | $[3\gamma^2/3 + 4\gamma^2)]^{1/2}$ | 0.7651 |
| $11\bar{2}3$[b] | $(1 + \gamma^2)^{1/2}$ | 1.9149 | $10\bar{1}2$ | $[3\gamma^2/4(1 + \gamma^2)]^{1/2}$ | 0.5941 |
| $10\bar{1}1$[b] | $(3 + \gamma^2)^{1/2}$ | 2.3805 | $11\bar{2}0$[c] | $1/2$ | 0.5000 |
| $22\bar{4}3$[b] | $(4 + \gamma^2)^{1/2}$ | 2.5820 | $10\bar{1}3$ | $[3\gamma^2/(9 + 4\gamma^2)]^{1/2}$ | 0.4609 |
| $41\bar{5}0$ | $(7)^{1/2}$ | 2.6458 | $11\bar{2}2$ | $[\gamma^2/4(1 + \gamma^2)]^{1/2}$ | 0.4264 |
| $20\bar{2}3$[a] | $(12 + 9\gamma^2)^{1/2}$ | 6.0000 | $20\bar{2}1$ | $[3\gamma^2/(3 + 16\gamma^2)]^{1/2}$ | 0.4186 |
| $41\bar{5}3$[b] | $(7 + \gamma^2)^{1/2}$ | 3.1091 | $10\bar{1}1$ | $[3\gamma^2/4(1 + 4\gamma^2)]^{1/2}$ | 0.3693 |
| $11\bar{2}1$[b] | $(9 + \gamma^2)^{1/2}$ | 3.4157 | $20\bar{2}3$ | $[3\gamma^2/(9 + 16\gamma^2)]^{1/2}$ | 0.3389 |
| $11\bar{2}6$[b] | $(1 + 4\gamma^2)^{1/2}$ | | | | |

[a] $\kappa = 2$; the spacings are alternately $1/6$ and $5/6$ of $t_{uvjw}$.
[b] These rows cluster in groups. See the text.
[c] $\kappa = 1$, smooth.
[d] $\kappa = 2$; the spacings are alternately $1/3$ and $2/3$ of $d_{hkil}$.
[e] $\kappa = 1$, rumpled, except as noted.

**Table 3.8.** Some nonuniformly spaced atomic rows in close-packed hexagonal crystals

| Axis [$uvjw$] | Minimum Spacing, General | $t_{uvjw}/6a$ Ideal $\gamma$ | "Equivalent" fcc Axis, [$uvw$] |
|---|---|---|---|
| $20\bar{2}3$ | $(1/3 + \gamma^2/4)^{1/2}$ | 1.0000 | 011 |
| $40\bar{4}3$ | $(4/3 + \gamma^2/4)^{1/2}$ | 1.4142 | 001 |
| $24\bar{6}3$ | $(7/3 + \gamma^2/4)^{1/2}$ | 1.7321 | 112 |
| $26\bar{8}3$ | $(13/3 + \gamma^2/4)^{1/2}$ | 2.2361 | 013 |
| $80\bar{8}3$ | $(16/3 + \gamma^2/4)^{1/2}$ | 2.4495 | 111 |
| $20\bar{2}9$ | $(1/3 + 9\gamma^2/4)^{1/2}$ | 2.5166 | None |

cubic crystals and of special directions in other systems. Throughout this chapter, [$uvw$] and ($hkl$) represent a particular axis and plane, while $\langle uvw \rangle$ and $\{hkl\}$ represent the sets of all crystallographically equivalent axes and planes.

The density of atomic rows in a plane perpendicular to them, whether this is a lattice plane or not, is

$$\varrho_{uvw} = Nt_{uvw}/\kappa, \tag{3.3.1}$$

where $\kappa$ is the number of atoms in the elementary interval along the row. In structures where the atoms are uniformly spaced along the rows, $\kappa = 1$. Nonuniform spacings occur along certain axes in diamond (Table 3.6) and in hcp metals (Tables 3.7 and 3.8). In these cases, $\kappa = 2$. In the transverse plane, the rows

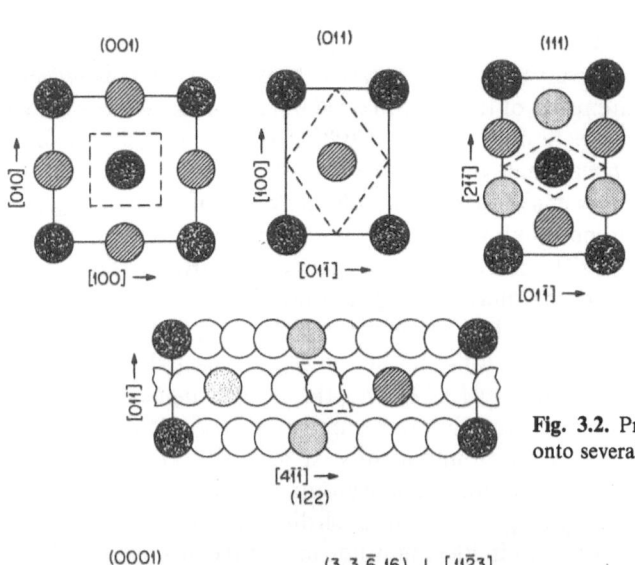

**Fig. 3.2.** Projections of an fcc crystal onto several surfaces

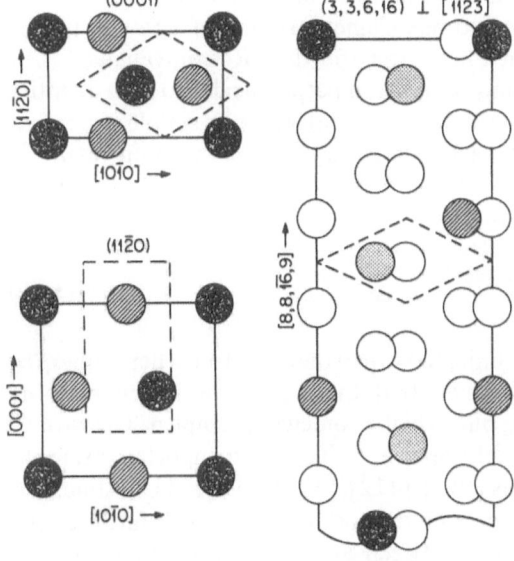

**Fig. 3.3.** Projections of an ideal $[\gamma=(8/3)^{1/2}]$ hcp crystal onto three different surfaces

are arranged in a diperiodic net, of which the elementary area is a *unit mesh* [3.21]. In simple structures, one row is associated with each unit mesh, a situation illustrated by the projections of the fcc structure in Fig. 3.2. Individual rows can always be isolated as long as the critical impact parameter is not too large. Then, as the [122] projection shows, it is more natural to consider the axis as overwhelmed by the plane in which it lies, in this case (01$\bar{1}$). In certain projections of the diamond structure and in all projections of the hcp structure, except that along $\langle 10\bar{1}0\rangle$, however, two rows must be associated with each unit mesh. Figure 3.3 shows the projections of an ideally close-packed $[\gamma = (8/3)^{1/2}]$ hexagonal crystal along three of its most open directions. Only the $\langle 11\bar{2}0 \rangle$ rows can be isolated from each other, and even here the requirements of translational symmetry bring two rows into each unit mesh. The projection of the basal (0001) plane has only trigonal symmetry instead of the hexagonal symmetry of the close-packed (111) projection of the fcc structure. The lower symmetry generates an array of large hexagonal channels bordered by "honeycomb walls" of atomic rows. In the remaining hcp projection, that along $[11\bar{2}3]$, the atomic rows are closely paired, producing a much more open structure than would be thought from (3.3.1). Note that the plane perpendicular to the $[11\bar{2}3]$ axis has the indices $(3, 3, \bar{6}, 16)$ in the ideal hcp structure. Similar clustering occurs for other hcp rows, as indicated in Table 3.7. Since *Lindhard*'s channeling theory [3.82] and the *Onderdelinden* sputtering model both assume that there is only one row per unit mesh, their applicability to hcp structures is limited. Very nonuniformly spaced ($\kappa = 2$) rows are a characteristic feature of the hcp structure. These alternate a short spacing, $t_{uvjw}/6$, with a long one. Consequently, $\varrho_{uvjw}$ is large and such directions are not open ones. The short spacing, however, both in its magnitude and in the disposition of the atoms surrounding it, is equivalent to one elementary unit of a row in the fcc structure. Table 3.8 lists a few such hcp rows and their fcc equivalents.

The planar spacing $d_{hkl}$ is the unit translation perpendicular to a set of lattice planes, or, equivalently, along a row in the reciprocal lattice. In simple structures, one plane can be associated with each elementary interval. Certain planes in diamond and most planes in hcp metals occur in pairs, however. The density of atoms in the individual lattice planes is

$$\varrho_{hkl} = N d_{hkl}/\kappa, \tag{3.3.2}$$

where $\kappa$ is the number of planes in a unit interval. In the case of most hcp planes, the smaller interplanar spacing is so small that these planes are perhaps more naturally regarded as having $\kappa = 1$, but as being somewhat rumpled. The various open planes contain one or more of the open axes. To cite an important example, the fcc (111) plane contains the axes $[01\bar{1}]$, $[11\bar{2}]$, $[12\bar{3}]$, and $[13\bar{4}]$. Distinctions between axial and planar channeling effects may be difficult and will, furthermore, be energy dependent. The transition between an axial channel and a planar channel in which it lies will depend on the magnitudes of the critical impact parameters involved in dechanneling. Limits on these quantities may be deduced from crystallographic data alone. The maximum critical approach distance, $p_c$, at which individual atomic rows may be isolated from one another, can be defined as the radius at which projections of the rows onto the transverse plane just touch. Values may be deduced by inspection of projections such as Figs. 3.3 and 3.4 (see also [3.36]). The corresponding limiting value of the critical angle for axial channeling may be defined by

$$\tan \psi_c = p_c/t_{uvw} \approx \psi_c. \tag{3.3.3}$$

**Table 3.9.** Limiting critical approach distances and critical channeling angles for *fcc* crystals

| Axis [uvw] | Maximum Critical Distance | | Critical Angles (Degrees) | | Most Open Plane |
|---|---|---|---|---|---|
| | $p_c/a$ | | $\psi_c$ | $\psi_\perp$ | |
| 011 | $(3/2)^{1/2}/4$ | 0.3062 | 23.4 | 22.2 | $(11\bar{1})$ |
| 001 | 1/4 | 0.2500 | 14.0 | 14.0 | (100) |
| 112 | $1/4(2)^{1/2}$ | 0.1768 | 8.2 | 13.2 | $(11\bar{1})$ |
| 013 | $1/2(10)^{1/2}$ | 0.1581 | 5.7 | 9.0 | (100) |
| 111 | $1/2(6)^{1/2}$ | 0.2041 | 6.8 | 5.9 | $(01\bar{1})$ |
| 123 | $(3/14)^{1/2}/2$ | 0.2315 | 7.0 | 8.8 | $(11\bar{1})$ |
| 114 | 1/6 | 0.1667 | 4.5 | 4.8 | $(1\bar{1}0)$ |
| 012 | $1/4(5)^{1/2}$ | 0.1118 | 2.9 | 6.4 | (100) |
| 233 | $1/2(11)^{1/2}$ | 0.1508 | 3.7 | 4.3 | $(01\bar{1})$ |
| 015 | $1/2(26)^{1/2}$ | 0.0981 | 2.2 | 5.6 | (100) |
| 134 | $(3/26)^{1/2}/4$ | 0.0849 | 1.9 | 6.5 | $(11\bar{1})$ |

Values of $p_c$ and $\psi_c$ for some fcc axes are collected in Table 3.9. In addition, the table lists the orientation of the most open plane in which each axis is contained and a limiting critical angle with respect to this plane defined by

$$\tan\psi_\perp = d_{hkl}/2t_{uvw} \approx \psi_\perp . \tag{3.3.4}$$

Except for the three principal orientations, $\psi_\perp > \psi_c$, especially for axes lying in {111} and {100}. Thus, these axes will be engulfed (as *Lindhard* puts it [3.82]) in the containing planes when conditions are such that the isolated row critical angle exceeds $\psi_c$.

### 3.3.3 Channeling and Dechanneling

If an incident ion moves swiftly enough, its scattering in encounters with single lattice atoms can be described by the classical impulse approximation [3.86]. The target atoms may be regarded as fixed. Furthermore, because the deflections of the projectile are individually very small, encounters with many successive target atoms are required to turn it away from the atomic row or plane. In this circumstance, the projectile may be regarded as moving in the average potential of the whole row or plane. In the axial case [3.87], the average potential is

$$U_{uvw}(\varrho) = (\kappa/t_{uvw}) \int_{-\infty}^{+\infty} V([z^2 + \varrho^2]^{1/2}) dz , \tag{3.3.5}$$

where $\varrho$ is the distance of the projectile from the row, $z$ is measured along the row, and $V(r)$ is the two-body interaction potential between the swift projectiles and the lattice atoms. In the planar case [3.82], the average potential is

$$U_{hkl}(x) = (2\pi n d_{hkl}/\kappa) \int_x^\infty r V(r) dr,$$    (3.3.6)

where $x$ is the distance of the projectile from the plane. *Lindhard* [3.82] has analyzed the behavior of axially channeled ions using an approximate Thomas-Fermi potential for $V(r)$ in (3.3.5). If the velocity vector of an ion makes an angle $\psi$ with the direction of an atomic row, its distance of closest approach to the row must obey

$$p_m > \psi t_{uvw}$$    (3.3.7)

for it to remain channeled. This condition will be fulfilled as long as $\psi$ does not exceed a critical value. At high energies, the critical angle is

$$\psi_1 = (2Z_1 Z_2 e^2 / E t_{uvw})^{1/2}, \quad E > E_1,$$    (3.3.8)

where

$$E_1 = 2Z_1 Z_2 e^2 t_{uvw} / a_{12}^2.$$    (3.3.9)

In these expressions, $Z_1 e$ and $Z_2 e$ are the nuclear charges on the projectile and target atoms, respectively. The screening length of the Thomas-Fermi interaction between the atoms is [3.88]

$$a_{12} = (9\pi^2/128)^{1/3} a_0 (Z_1^{2/3} + Z_2^{2/3})^{-1/2},$$    (3.3.10)

where $a_0 = \hbar/me^2$ is the Bohr radius (0.529177 Å). At energies greater than $E_1$, the scattering is mainly a result of the internuclear Coulomb repulsion, with electron screening playing a minor role. Consequently, $\psi_1$ is insensitive to the details of the interaction potential. Unfortunately, this energy region is not of much interest in sputtering: for $Ar^+$ ions irradiating Cu monocrystals, $E_1 = 2.8$ MeV for $\langle 011 \rangle$, 4.0 MeV for $\langle 001 \rangle$, and is larger for all other orientations; even for $H^+$ irradiating Al, $E_1 = 48.8$ keV for $\langle 011 \rangle$, 69.0 keV for $\langle 001 \rangle$, and more for other orientations. Since the yields at such energies are small, experiments are likely to be confined to the lower energy region and, until the present, all seem to have been so. In the low energy region, *Lindhard* finds [3.82]

$$\psi_2 = (C a_{12} \psi_1 / 2^{1/2} t_{uvw})^{1/2}, \quad E < E_1$$    (3.3.11)

where $C$ is a constant for which a value near $3^{1/2}$ is suggested. Note that $\psi_1(E_1) = \psi_2(E_1)$ if $C = 2^{1/2}$; in the context of Lindhard's theory, the difference between the two values of $C$ is unimportant, as is also the difference between (3.3.10) and *Firsov's* estimate [3.89]

$$a_{12} = (9\pi^2/128)^{1/3} a_0 (Z_1^{1/2} + Z_2^{1/2})^{-2/3}$$    (3.3.12)

**Table 3.10.** The critical channeling angle $\psi_2$ for $Ar^+$ ions incident upon Cu monocrystals ($C=3^{1/2}$, $a_{12}=0.1160$ Å, $a=3.615$ Å)

| Axis [uvw] | $\psi_2$ [degrees] at E [keV] | | | | | | |
|---|---|---|---|---|---|---|---|
| | 1 | 2 | 5 | 10 | 15 | 20 | 50 |
| 011[b] | 21.03 | 17.69 | 14.07 | 11.83 | 10.69 | 9.95 | 7.91 |
| 001[a] | 16.22 | 13.64 | 10.85 | 9.12 | 8.24 | 7.67 | 6.10 |
| 112[a] | 13.93 | 11.72 | 9.32 | 7.84 | 7.08 | 6.59 | 5.24 |
| 013[b] | 11.50 | 9.67 | 7.69 | 6.47 | 5.85 | 5.44 | 4.33 |
| 111[a] | 10.74 | 9.03 | 7.18 | 6.04 | 5.46 | 5.08 | 4.04 |
| 123[a] | 10.14 | 8.53 | 6.78 | 5.70 | 5.15 | 4.80 | 3.81 |
| 114[a] | 9.23 | 7.76 | 6.17 | 5.19 | 4.69 | 4.36 | 3.47 |
| 012[a] | 8.87 | 7.46 | 5.93 | 4.99 | 4.51 | 4.19 | 3.34 |
| 233[c] | 8.56 | 7.20 | 5.72 | 4.81 | 4.35 | 4.05 | 3.22 |
| 015[b] } 134 | 8.04 | 6.76 | 5.38 | 4.52 | 4.09 | 3.80 | 3.02 |

[a] Observed at 20 keV [3.80].
[b] Observed at 20 keV [3.69].
[c] Possible observed for 45 keV $Kr^+$ ions [3.8].

which is preferred by some authors. An essential point is that (3.3.11) is sensitive to the interaction potential, since the channeled ions do not approach the atomic rows very closely and are thus influenced more strongly by the surrounding distribution of electrons. The dependence of $\psi_2$ on the potential is shown by the appearance of $a_{12}$ in (3.3.11). Table 3.10 presents values of $\psi_2$ for $Ar^+$ ions incident upon Cu monocrystals of the ten most open axial orientations. Comparison of this table with Table 3.9 permits a number of conclusions to be drawn. First, at no energy up to 50 keV can all ten rows be sufficiently isolated to make application of the Lindhard theory entirely plausible. For six directions, indicated in Table 3.10, precise experimental data are available [3.80] using a well-defined (0.3° divergence) beam of 20 keV $Ar^+$ ions. For $\langle 012 \rangle$, $\psi_c < \psi_2 < \psi_\perp$ and for $\langle 114 \rangle$ $\psi_c \approx \psi_2 \approx \psi_\perp$. Planar channeling effects must be very important in these circumstances. Second, those axes lying in {111} and, to a lesser degree, {100}, will show pronounced planar channeling effects, that is, $\psi_c < \psi_2 < \psi_\perp$, down to energies below 5 keV, while axes lying in {110} will not. The relative roles of axial and planar channeling could be shown experimentally by scanning through appropriate directions along different planes. Thus, for instance, the [112] axial channel should be substantially wider when scanned in the ($1\bar{1}0$) plane – the usual way [3.80] – than when scanned in the (111) plane. Third, the order of sputtering yields at energies below $\sim 10$ keV will be rather different to what it is at considerably higher energies. Axial channeling along $\langle 111 \rangle$ will be largely eliminated below 10 keV and no important planes are suitably located to reduce the yield. Hence, apparently less open orientations such as $\langle 112 \rangle$ and $\langle 012 \rangle$, because of significant planar channeling, can have lower yields than does $\langle 111 \rangle$. This conclusion is supported by experiments at and below 5 keV [3.68]. Finally, experimental evidence has been reported [3.7–9, 69, 74–80] for minima in the sputtering yield for all but two of the directions listed in Table 3.10. There is an indication of a yield reduction near $\langle 233 \rangle$ in the only experiment performed in its vicinity. No experimental data are available near $\langle 134 \rangle$. In fact, no attempt seems to have been made to scan an incident beam along a $\{111\}$ plane, for instance by turning a [011] crystal about a [11$\bar{1}$] axis. It seems likely that the minima observed at the $\langle 112 \rangle$ and $\langle 123 \rangle$ axes reflect the effects of $\{111\}$ planar channeling and that their reasonably successful interpretation in axial terms is coincidental.

From the foregoing discussion it can be concluded that a model for sputtering by heavy ions based on Lindhard's theory of axial channeling can have only a rough quantitative significance at energies below 50 keV or so. The sputtering yield in such a model takes the place of the yields of other processes (nuclear reactions, large angle Rutherford scattering – see [3.65–67] for details) that have been discussed in channeling. Two groups of particles contribute to this yield: one is scattered into the random beam immediately by close encounters with near surface atoms; the other is dechanneled by random motions of the ions in the transverse plane and occurs at considerable depths within the crystal. Not all rows terminate in the surface plane of the crystal [3.79]. This may be seen in Figs. 3.2 and 3.3, where the lighter shadings indicate rows that do not reach the surface. Dechanneling is further increased by the thermal motions of the atoms which reduce the translational symmetry of the crystal and lead to close encounters which would otherwise not occur. *Lindhard* [3.82] estimates the minimum yield of a close-encounter process, normalized to the value for a randomly directed beam, to be

$$\chi_m = \begin{cases} \pi N t_{uvw}(2u_1^2 + a_{12}^2) & E > E_1 \\ \pi N t_{uvw}(2u_1^2 + \psi_2^2 t_{uvw}^2) & E < E_1, \end{cases}$$

$$\quad (3.3.13)$$
$$\quad (3.3.14)$$

where $u_1^2$ is the mean square amplitude of one-dimensional thermal vibrations. *Barrett* [3.90] has made a detailed study of minimum yields in channeling using a computer model. He finds that (3.3.13) underestimates thermal dechanneling rather seriously (a factor of about 3), but his calculations do not extend to low enough energy to be applied here.

At low projectile kinetic energies, the impulse approximation breaks down and more dechanneling occurs than was estimated above. In slow collisions, it is not possible to neglect the motion of the target particle and the deflection of the projectile during the collision, even at moderately large impact parameters. Recoiling of the targets permits the projectiles to penetrate more deeply into the rows and planes bordering a channel, lowers the critical angle, and decreases the channeling probability, as compared with predictions using the impulse approximation. Such effects will be most important for heavy projectiles, such as $Xe^+$ or $Hg^+$, at relatively low velocities. Computer simulations [3.91] of the backscattering of various projectiles from Cu $\langle 001 \rangle$ rows, for example, have produced critical angles in reasonable agreement with the Lindhard theory for 0.5 to 5 keV H, N, and Ne atoms, somewhat smaller values for Ar and Cu atoms, and values smaller by a factor two for Xe and Au.

### 3.3.4 The Onderdelinden Model of Monocrystal Sputtering Yields

Within the rough approximation of the *Onderdelinden* model [3.78], the sputtering yield of a single crystal target may be written as

$$Y(\vartheta, E, \psi) = \eta \hat{Y}(\vartheta, E) \chi(\psi, E), \tag{3.3.15}$$

where $E$ is the energy of the incident beam, $\vartheta$ is the polar angle of incidence of the beam, $\psi$ is the deviation angle of the beam from the axis of a nearby channel, and $\hat{Y}(\vartheta, E)$ is the yield of a structureless medium under the same conditions of incidence. The relative dechanneling yield $\chi$ is assumed to be independent of the azimuth in which the beam is incident. Deviations of the "efficiency" $\eta$ from unity measure the degree of approximation in (3.3.15). The dechanneling yield is estimated as follows. The incident ion, entering the crystal with impact parameter $p$ with respect to a channel axis, has the initial potential energy $U(p)$ in the continuum potential of the row. Its initial kinetic energy, projected onto the plane transverse to the row, is $E\psi^2$. Part of this energy is directed radially into the nearest row and may contribute to immediate dechanneling and part is so directed as to carry the ion away from this row. The latter portion may contribute to dechanneling by encounters with other rows at some depth within the crystal. On the average, a transverse kinetic energy $f'E\psi^2$ contributes to dechanneling at shallow enough depths to affect sputtering. Adding this to the potential energy, the condition for remaining channeled may be deduced:

$$p > p_m, \quad \text{where} \quad U(p_m) + f'E\psi^2 = E\psi_2^2. \tag{3.3.16}$$

Since $U(p)$ may be approximated by an inverse quadratic [3.82], (3.3.16) may be solved for $p_m$ and the dechanneling yield written as

$$\chi(\psi) = \pi N t_{uvw} p_m^2 = \chi_m [1 - f'(\psi/\psi_2)^2]^{-1}, \tag{3.3.17}$$

where use has been made of the fact that $\chi(0) = \chi_m$. If it is further assumed that all particles are dechanneled when $\psi = f\psi_2$, where $1 < f < 2$ is to be expected from *Lindhard*'s analysis [3.82], (3.3.17) may be written as

$$\chi(\psi) = \chi_m [1 - (1 - \chi_m)(\psi/f\psi_2)^2]^{-1}. \tag{3.3.18}$$

Equations (3.3.15, 18) can be used to fit experimental sputtering data. From experiments where the angle of incidence of the beam on the target was varied, values of the coefficient $f$ can be inferred. Values of the coefficient $\eta$ can be derived from such experiments also, as well as from investigations where the beam was normally incident on various crystal surfaces.

The number of available experiments is rather small. In a series of investigations [3.77, 78] in which the angle of incidence was varied for 5- to 35-keV $Ar^+$ ions on Cu monocrystals and for 20 keV $Ar^+$ on Al and Au crystals, reasonable agreement with (3.3.18) was found using the value $f = 1.2$. From a more precise investigation [3.80] using 20 keV $Ar^+$ and $Ne^+$ ions on Cu crystals, the value $f = 1.3$ was inferred. A similar value can be extracted from some experiments [3.92] with 27 keV $Ar^+$ ions on a Zn crystal. In the latter instance, however, the width of the minimum at the $\langle 0001 \rangle$ axis is greater than might be inferred from the value of $\psi_2$, probably showing the influence of the arrangement of rows in that case. The magnitude obtained for $f$ is in the range expected and its independence of orientation and ion energy is encouraging.

Analysis of the same sets of $Ar^+$-Cu experiments [3.77, 78, 80] gives values of $\eta$ ranging from 1.3 to 1.6, depending on the orientation. Normal incidence experiments using 1 to 5 keV $Ar^+$ ions on Cu and Au crystals [3.93] indicate values of $\eta$ from 1.1 to 1.4. A further investigation of the sputtering of the principal orientations of Cu monocrystals by 0.1 to 200 keV $Ar^+$ ions [3.94] shows that (3.3.15) can give only a crude accounting of the energy dependence of the monocrystal yields, or, what is the same thing, that $\eta$ is energy dependent. From (3.3.8, 11, 14, 15), the ratio of the yields for two differently oriented monocrystals is

$$(\eta/\eta')(t_{uvw}/t_{u'v'w'}) < Y_{uvw}/Y_{u'v'w'} < (\eta/\eta')(t_{uvw}/t_{u'v'w'})^{3/2}$$

depending on the amount of thermal dechanneling, that is, on the magnitude of the first term in (3.3.14). This inequality predicts, for example, that

$$2.45(\eta_{111}/\eta_{110}) < Y_{111}/Y_{110} < 3.83(\eta_{111}/\eta_{110}),$$

but from the data, the ratio ranges from about 3.5 at 10 keV to about 1.3 at 200 keV and there is no sign of saturation at the higher energy. No choice of constant values of $\eta$ can account for this result. Relative values of $\eta$ can also be inferred from experiments using 1 to 5 keV $Ar^+$ ions on several hexagonal metals [3.95]. The ratios $(\eta_{11\bar{2}0}/\eta_{0001})$ and $(\eta_{10\bar{1}0}/\eta_{0001})$ range from 0.8 to 1.7, depending on the orientation and the metal.

If the thermal vibration amplitude is introduced into the model by using (3.3.14) in (3.3.18), a prediction can be made concerning the temperature dependence of the sputtering yield. The result is not in agreement with the available experimental data [3.79, 80]. The discrepancy could result from a greater amount of thermal dechanneling than the Lindhard theory suggests, or it could reflect a fundamental failure of the model.

In summary, the Onderdelinden model satisfactorily accounts for the angular widths of single crystal axial yield minima at energies above about 10 keV, but does not give a good picture of the magnitude of the yield or of its dependence on incident ion energy or on target temperature. The following are among the factors which may contribute to the limitations of the model.

i) Ions making close encounters with atomic rows are regarded in the model as cast immediately into the random beam. In fact, correlated motion in the transverse plane is probably important. This may take the form of injecting scattered ions into other axial or planar channels than the channel of incidence, or it may take the form of an increased probability of close encounters with neighboring rows. This lattice effect, termed focusing in the transverse plane, has been identified and explored in computer simulations of axial channeling phenomena [3.96].

ii) The model contains no treatment of the way in which the energy of the incident ion reaches the target surface to produce sputtering. The role of focusing collisions, to be discussed in the following section, or of other correlated collision processes involving low energy particles, is ignored.

In view of these limitations, it is unlikely that improvements in the theory along the lines of the transparency models can be made to improve predictions of monocrystal sputtering yields in a fully satisfactory manner.

### 3.3.5 Monocrystal Effects at Low Energies

At incident ion energies below $\sim 10$ keV, interferences between atomic rows prevent channeled particle motion from developing and the relative sputtering yields are no longer correctly predicted by transparency arguments alone. This was already shown for 1 to 5 keV $Ar^+$ ions on Cu crystals [3.68], where the $\langle 111 \rangle$ axial channels are completely closed and the yield from this orientation exceeded that of nominally less open orientations such as $\langle 123 \rangle$ and $\langle 012 \rangle$. Throughout the energy range above 1 keV, however, the three principal orientations display the order

$$Y_{111} > Y_{001} > Y_{011},$$

but below about 500 eV, $Y_{001} > Y_{111}$ [3.94, 97, 98]. The yields from these two orientations differ by about 20% in this energy region, probably too much to be ascribed solely to the orientation dependence of the surface binding energy (compare Table 3.2). No significant theoretical work seems to have been done in the low-energy region, except by computer simulation.

## 3.4 The Angular Distribution of Atoms Ejected from Monocrystals

The lattice correlation effects displayed in the angular distributions of atoms sputtered from crystal surfaces have been studied widely, but their origins are still somewhat controversial. The experimental data are discussed by *Hofer* [Ref. 3.1a, Chap. 2]. The objective of this section is to discuss ejection pattern formation from a theoretical viewpoint, particularly with regard to the role, if any, played by sequences of focusing or replacement collisions. That such *linear collision sequences* occur is not in question. They were observed in the first computer simulation studies of atomic collision phenomena in crystals [3.19] and there is experimental evidence for them as well (see [3.99] for a review). The quantitative interpretation of experiments in this area contains many uncertainties, but there is

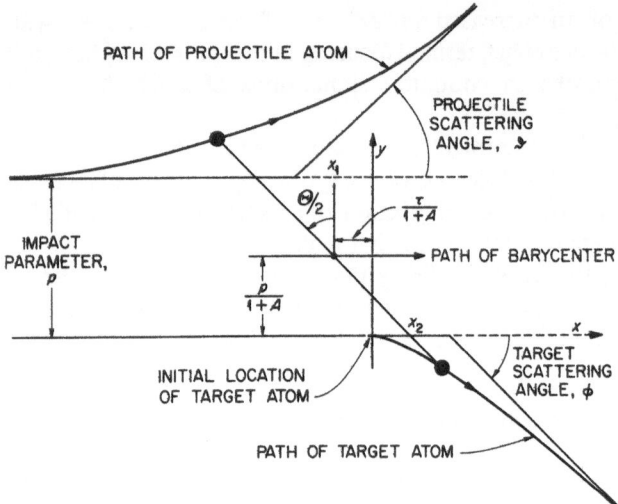

**Fig. 3.4.** The classical scattering of an energetic projectile by an initially stationary target atom

convincing evidence [3.100, 101] that linear collision sequences do occur and that they are an important factor in the development of displacement cascades. There is equally convincing evidence from recent low energy transmission sputtering experiments in gold [3.102, 103] that such sequences are not extremely long ones as has sometimes been suggested. Further experimental and theoretical investigations of these points are needed to resolve the many uncertainties remaining.

### 3.4.1 Classical Scattering Theory

Before embarking upon a detailed discussion of linear collision sequences, it is necessary to summarize some aspects of classical scattering theory [3.104]. Figure 3.4 depicts a single two-body collision in laboratory coordinates and defines several quantities. Solution of the classical equations of motion allows the *barycentric scattering angle* $\Theta$ and the so-called *time integral* $\tau$ to be expressed by the integrals

$$\Theta = \pi - 2p \int_R^\infty dr \, [r^2 g(r)]^{-1}, \tag{3.4.1}$$

$$\tau = (R^2 - p^2)^{1/2} - \int_R^\infty dr \, \{[g(r)]^{-1} - (1 - p^2/r^2)^{-1/2}\}, \tag{3.4.2}$$

with

$$g(r) = [1 - p^2/r^2 - V(r)/E_r]^{1/2}, \tag{3.4.3}$$

where $r$ is the (variable) interatomic separation, $V(r)$ is the interatomic potential energy function, $R$ is the *apsis* of the collision, defined by

$$g(R)=0, \tag{3.4.4}$$

and $E_r$ is the relative kinetic energy of the two atoms,

$$E_r=AE/(1+A), \tag{3.4.5}$$

where $A$ is the ratio of the mass of the target atom to that of the projectile. The energy transferred to the target atom is

$$T=T_m\sin^2\Theta/2, \tag{3.4.6}$$

where

$$T_m=4AE/(1+A)^2. \tag{3.4.7}$$

The laboratory scattering angles of the two particles are given by

$$\tan\vartheta=A\sin\Theta/(1+A\cos\Theta) \tag{3.4.8}$$

$$\phi=(\pi-\Theta)/2. \tag{3.4.9}$$

The intersections of the asymptotic laboratory trajectories of the two particles are located at

$$x_1=[2\tau+(A-1)p\tan\Theta/2]/(1+A) \tag{3.4.10}$$

$$x_2=p\tan\Theta/2-x_1. \tag{3.4.11}$$

In writing (3.4.1–11), it has been assumed that the collision is strictly elastic in the barycentric system. This assumption is appropriate at the particle energies involved in linear collision sequences, but requires modification in high energy collisions.

The details of an atomic collision are controlled by the potential energy function $V(r)$ in (3.4.3). In the following discussion, attention will be confined to two commonly used functions. These are the *Born-Mayer* potential [3.105]:

$$V_{\mathrm{BM}}(r)=C_{\mathrm{BM}}\exp(-r/a_{\mathrm{BM}}); \tag{3.4.12}$$

and the *Molière* approximation [3.106] to the Thomas-Fermi potential [3.88, 89, 107]:

$$V_{\mathrm{M}}(r)=C_{12}(a_{12}/r)(0.35y+0.55y^4+0.10y^{20}), \tag{3.4.13}$$

where

$$C_{12} = Z_1 Z_2 e^2 / a_{12} \tag{3.4.14}$$

and

$$y = \exp(-0.3r/a_{12}). \tag{3.4.15}$$

In these equations, $a_{BM}$, $C_{BM}$, and $a_{12}$ are parameters depending on the atoms involved in the collision. For the Born-Mayer potential, two sets of parameters will be used. One of these is intended to represent interactions between Cu atoms [3.19]:

$$C_{BM} = 22.5 \text{ keV}, a_{BM} = t_{110}/13 = 0.19661 \text{ Å}. \tag{3.4.16}$$

The other is a "universal" set proposed by *Andersen* and *Sigmund* [3.108, 109]:

$$C_{BM} = 52(Z_1 Z_2)^{3/4} \text{ eV}, a_{BM} = 0.219 \text{ Å}. \tag{3.4.17}$$

For the Molière potential, two parameters will also be used. To represent Cu, the value

$$a_{12} = 0.0738 \text{ Å} \tag{3.4.18}$$

is chosen because it makes the Molière potential equal in value to the Born-Mayer potential with parameters (3.4.16) at the nearest neighbor distance in Cu [3.110]. Similar matching between Molière and Born-Mayer potentials for other atoms suggests using the approximate "universal" value

$$a_{12} = 0.0750 \text{ Å}. \tag{3.4.19}$$

See also [3.111] for further comments on potential parameters. Extensive tables of the scattering integrals, (3.4.1, 2) are available [3.112] for the potentials (3.4.12, 13). These tables are the basis of the exact calculations reported below.

In spite of the availability of tabulated scattering integrals, it is often convenient to use analytical approximations. A useful procedure is to replace the potential function $V(r)$ by another function for which the integrals may be evaluated explicitly. The *hard-core approximation* [3.113] is a method of this kind. In the most common version of this procedure, the potential $V(r)$ is replaced by a hard sphere whose radius $R_0$ is the apsis of a head-on ($p=0$) collision. From (3.4.3, 4), this radius is

$$V(R_0) = E_r. \tag{3.4.20}$$

A series of related approximations, based on the use of *truncated power potentials* matched to $V(r)$ in value and in slope at the minimum apsis [3.113–115], can be

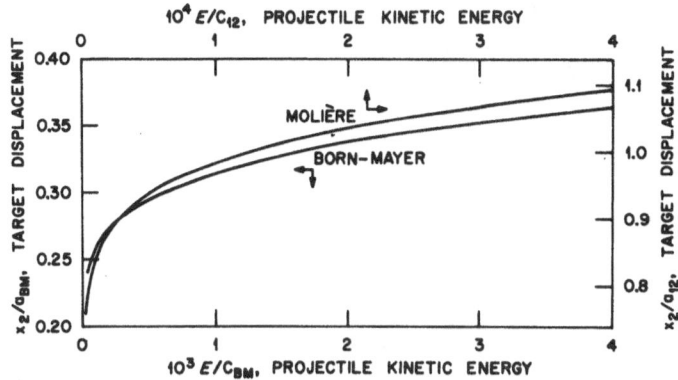

**Fig. 3.5.** The target displacement $x_2$ for head-on collisions between particles according to two interatomic potentials

quite accurate. The approximations mentioned here involve energy-dependent parameters. For instance, the hard-core radius $R_0$ varies with the collision energy, the atoms becoming "larger" as the energy decreases.

A few words must be said about the quantities $x_1$ and $x_2$ in Fig. 3.4, since they are somewhat unfamiliar. For collisions between hard spheres of equal mass, it is easily verified that

$$x_1 = (R_0^2 - p^2)^{1/2}, \quad x_2 = 0,$$

but this is not the case for more realistic treatments of the scattering. Figure 3.5 shows the values of $x_2$ for head-on collisions in the Born-Mayer and Molière potentials. In both cases, $x_2$ is a weak function of impact parameter, decreasing slowly as $p$ increases. In the energy range in which linear collision sequences are important, $x_2$ is much smaller than typical interatomic distances in crystals and its variation with energy is modest. Thus, it is possible to neglect it without significant error. The quantity $x_1$, on the other hand, is much larger and depends more strongly on both impact parameter and energy. In head-on collisions in the energy range of Fig. 3.5, $x_1$ is 1/3 to 1/2 the nearest neighbor distance in Cu.

### 3.4.2 Collision Sequences Along Isolated Rows

Collision sequences along straight rows of uniformly spaced atoms can now be investigated. The crystallographic preliminaries were given in Sect. 3.3.2. Since most open channels in a crystal are bordered by the most closely spaced atomic rows, the crystal directions of most interest are those listed in Tables 3.3–3.8. An element of a collision sequence along an isolated [$uvw$] row is shown in Fig. 3.6. A *focusing parameter* may be defined as

$$A(\vartheta_j) \equiv \sin \vartheta_{j+1}/\sin \vartheta_j = \operatorname{ctn} \vartheta_j \cos \Theta_j/2 - \sin \Theta_j/2, \tag{3.4.21}$$

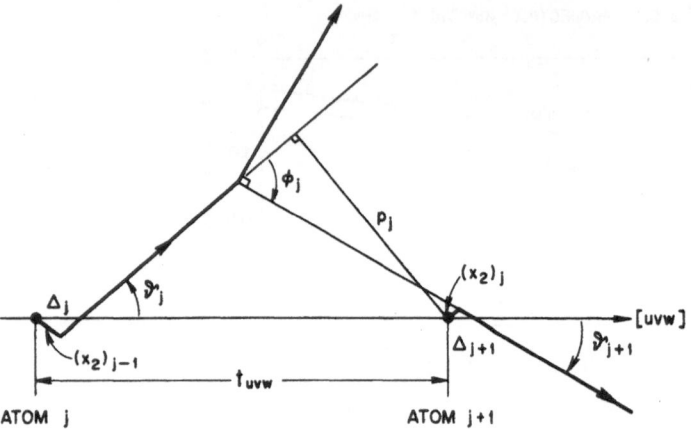

**Fig. 3.6.** One unit in a linear collision sequence along an isolated atomic row

a result that may be obtained from Fig. 3.6, using (3.4.9). Equation (3.4.21) differs from the usual definition of $\Lambda(\vartheta_j)$ [3.6], which employs the angles themselves rather than their sines. For small angles, of course, the difference is negligible and (3.4.21) is more convenient in calculations. The impact parameter in the collision is

$$p_j = (t_{uvw} - \Delta_j) \sin \vartheta_j, \tag{3.4.22}$$

where

$$\Delta_j = (x_2)_{j-1} \sin(\vartheta_j + \vartheta_{j-1})/\sin \vartheta_j$$

is the (small) displacement along the row, resulting from the previous collision in the sequence. If $\Lambda(\vartheta_j) < 1$, each particle moves at a smaller angle with the row than its predecessor and the collision is said to be *focusing*. In the contrary case, the successive angles increase and the sequence is *defocusing*. A *focusing energy* may be defined as the energy at which $\Lambda(0) = 1$. From (3.4.21), this condition may be written as

$$\lim_{\vartheta_j \to 0} \cos \vartheta_j \cos \Theta_j / 2 = 2$$

or, using (3.4.1, 5), as

$$\int_{R_f}^{\infty} (dr/r^2)[1 - 2V(r)/E_f]^{-1/2} = \lim_{p \to 0} (\pi - \Theta)/2p = 2/t_{uvw}, \tag{3.4.23}$$

where the displacement has been neglected and where the limit of integration and the focusing energy are related through

$$E_f = 2V(R_f). \tag{3.4.24}$$

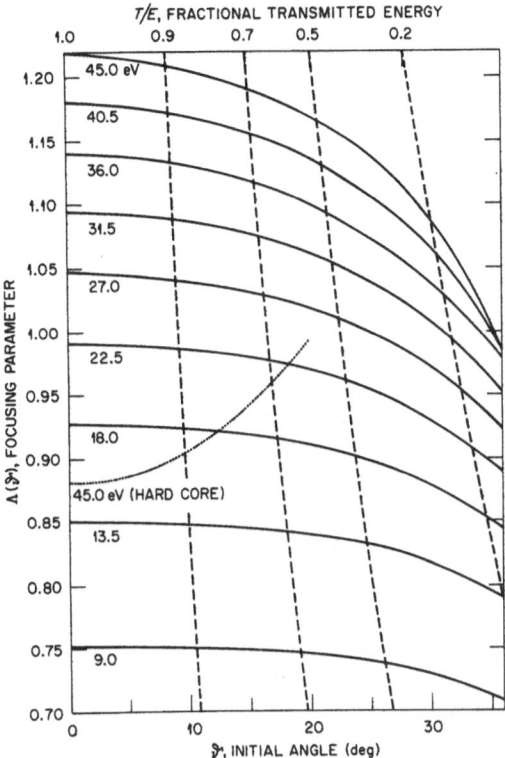

**Fig. 3.7.** The dependence of the focusing parameter for isolated Cu⟨011⟩ rows on the energy and angle of the projectile. The solid lines are exact calculations using a Born-Mayer potential with the parameters (3.4.16). The dotted line shows one calculation in the hard-core approximation. The dashed lines are loci of constant fractional transmitted energy

Exact evaluations of (3.4.21, 23) can be carried out using tabulated scattering integrals [3.112].

Figure 3.7 shows the angular dependence of $\Lambda(\vartheta)$ for a Cu ⟨110⟩ row, using a Born-Mayer potential with the parameters (3.4.16). At each energy, $\Lambda(\vartheta)$ is a maximum when the first atom moves directly along the axis and decreases somewhat as the initial angle increases. As the initial angle increases, however, the energy transmitted to the next atom is reduced. The figure shows loci that indicate the fractional energy transmission at various angles. Under circumstances where the energy transmission is large, that is, over an angular range of 20° or so in the example pictured, the focusing parameter is nearly constant. Thus, it is reasonable to regard the focusing energy (about 23 eV in the example) as applying at all angles. The neglect of the displacement $\Delta_j$ in evaluating (3.4.21) produces a small overestimate of $\Lambda(\vartheta)$, especially close to the axis. Consequently, the focusing energy given by (3.4.23) is a lower limit, but the error is neglible compared to that introduced by assuming the row to be isolated. Figure 3.8 shows the focusing energy for isolated atomic rows as a function of the atomic separation, for both the Born-Mayer and the Molière potentials. This figure may be used to deduce values of $E_f$ for a variety of crystals and axes.

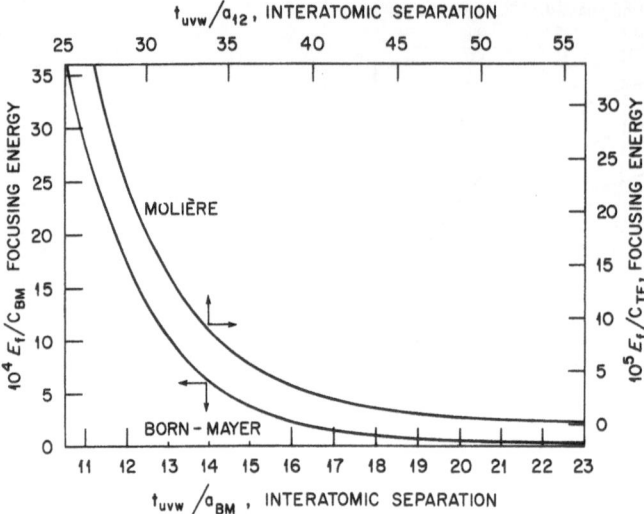

**Fig. 3.8.** The focusing energies in isolated atomic rows according to two different interatomic potentials

Focusing collision sequences have often been treated by approximate analytical methods [3.116], especially by the hard-core approximation. In this approximation, the focusing parameter is

$$\Lambda^{HC}(\vartheta) = \alpha \cos \vartheta - (1 - \alpha^2 \sin^2 \vartheta)^{1/2} \approx (\alpha - 1)[1 + \tfrac{1}{2}\alpha\vartheta^2 + \mathcal{O}(\vartheta^4)], \qquad (3.4.25)$$

where

$$\alpha = t_{uvw}/R_0 \qquad (3.4.26)$$

and the hard-core radius is defined by (3.4.20). The focusing energy is

$$E_f^{HC} = V(t_{uvw}/2). \qquad (3.4.27)$$

An evaluation of (3.4.25) is shown in Fig. 3.7 for one energy. The angular dependence has a curvature opposite to that of the exact result. Furthermore, the magnitude of $\Lambda^{HC}(\vartheta)$ is much less than the exact result. Thus, the hard-core approximation seriously overestimates the focusing energy, while at the same time giving an incorrect picture of the angular behavior of $\Lambda(\vartheta)$. More accurate estimates of the focusing energy can be obtained by matching potential methods. The angular dependence of $\Lambda(\vartheta)$ in the truncated power approximation resembles the exact result for small angles; it has not been studied in detail, however.

The focusing energies obtained by approximate methods are compared with the exact results in Table 3.11. None of them is very accurate. In view of the ease of making exact calculations from Fig. 3.8, the approximations are probably useful

**Table 3.11.** Comparison of exact and approximate focusing energies in Cu

| Method of Evaluation | Focusing Energy [eV] | | |
|---|---|---|---|
| | $\langle 011 \rangle$ | $\langle 001 \rangle$ | $\langle 111 \rangle$ |
| Born-Mayer potential with the parameters (3.4.16) | | | |
| Hard-core Approximation | 67.6 | 4.6 | 0.005 |
| Truncated Coulomb | 28.0 | 1.8 | 0.002 |
| Truncated $1/r^2$ | 27.5 | – | – |
| Exact | 23.1 | 1.5 | <0.002 |
| Molière potential with parameter (3.4.18) | | | |
| Hard-core Approximation | 36.7 | 3.0 | 0.008 |
| Truncated Coulomb | 15.2 | 1.2 | 0.003 |
| Exact | 12.6 | 1.0 | <0.002 |

only for rough estimations in new situations or beyond the limits of the figure or the tables from which it was developed. A second point emerging from the comparisons in Table 3.11 is the great importance of the atomic spacing along a row. Only for the $\langle 011 \rangle$ rows is the focusing energy for isolated rows high enough that focusing collision sequences would be expected to be a significant feature of displacement cascade development. Finally, the considerable sensitivity of the focusing energy to the interatomic potential should be noted. This point has not been much investigated, but it is clear that some caution concerning focusing calculations is called for. Table 3.12 displays the focusing energies calculated for isolated close-packed rows of atoms in several metallic elements. If focusing collision sequences are to make an important contribution to sputtering, the last member of the sequence must receive an energy greater than the surface binding energy (see Sect. 3.2.3). Thus, $E_f/U_0$ must be at least 1 to 2 (and probably much more) to permit focusing sequences of appreciable length. Comparison of the values of the focusing energy in Table 3.12 with the values of the cohesive energy in Table 3.1 shows $E_f/U_0 < 2$ for Mg, Al, and Nb. Thus, from the simple theory of the isolated row, focusing is not expected to contribute strongly to sputtering ejection in these metals.

### 3.4.3 Assisted Focusing

The approximation that collision sequences occur along atomic rows which are isolated from their neighbors is a severe one. As may be seen in Figs. 3.2 and 3.3, each row is surrounded by similar rows at distances comparable to the spacings within the rows. As an atom moves from its lattice site towards a collision with the next member of its row, it must penetrate one or more rings of more-or-less symmetrically disposed neighboring atoms. If the atom initially moves obliquely

**Table 3.12.** Focusing energies for isolated close-packed rows in crystals of several metallic elements

| Element | $\langle uvw \rangle$ | Lattice Constant, $a$ (Å) [3.117] | Focusing Energy [eV] | |
|---------|------|------|------|------|
| | | | Born-Mayer (3.4.17) | Molière (3.4.19) |
| Al | 011 | 4.0495 | 2.4 | 1.3 |
| Ni | 011 | 3.524 | 18.1 | 15.0 |
| Cu | 011 | 3.6147 | 16.6 | 13.6 |
| Ag | 011 | 4.0857 | 15.7 | 15.9 |
| Pt | 011 | 3.9239 | 43.9 | 58.0 |
| Au | 011 | 4.0783 | 34.5 | 45.5 |
| Pb | 011 | 4.9502 | 8.8 | 11.9 |
| Fe | 111 | 2.8664 | 16.6 | 13.3 |
| Nb | 111 | 3.3006 | 13.8 | 13.0 |
| W | 111 | 3.1650 | 43.7 | 56.8 |
| Mg | 11$\bar{2}$0 | 3.2094 | 1.0 | 0.5 |
| Zn | 11$\bar{2}$0 | 2.6649 | 13.5 | 11.2 |

**Table 3.13.** The effects of neighboring rows on focusing energies

| Crystal | Potential | Method | Focusing Energy [eV] | | |
|---------|-----------|--------|------|------|------|
| | | | $\langle 011 \rangle$ | $\langle 001 \rangle$ | $\langle 111 \rangle$ |
| Cu | Born-Mayer (3.4.16) | Isolated Row | 23.1 | 1.5 | < 0.002 |
| | | Dynamics [3.19] | 30 | 39 | |
| | Molière (3.4.18) | Isolated Row | 12.6 | 1.0 | < 0.002 |
| | | Dynamics [3.121] | 17 | 20 | 225 |
| Fe | Mixed (3.4.28) | Isolated Row | | 8.7 | 20.7 |
| | | Dynamics [3.119, 120] | | 18.25 | 27.7 |

to its row, interactions with atoms in such rings will deflect it back towards the axis, much as a light ray is refracted by a converging lens. Because of the collision with the ring atoms, the next central collision occurs with a smaller impact parameter than would have been the case otherwise. The encounter with the ring improves the focusing and a higher focusing energy results. Such *assisted focusing* [3.19, 118] is especially important for rows that are not closely packed, such as fcc $\langle 001 \rangle$ and $\langle 111 \rangle$, but it makes a significant contribution for close-packed rows such as fcc $\langle 011 \rangle$ also. Analytical treatments of assisted focusing are possible [3.116, 118], but these have the same drawbacks that they have for the isolated row problem. Numerical treatments are available for a few cases, however, using the dynamical method first applied to atomic collision problems in crystals by *Vineyard* and coworkers [3.19, 119]. The method allows focusing problems to be studied with the inclusion of the effects of neighboring rows of atoms and the effect of the displacements as well.

Table 3.13 reports some estimates available for $E_f$ based on dynamical calculations and compares them with the corresponding values for isolated rows. The dynamical calculations for Fe [3.119, 120] used the somewhat complicated potential function

$$V(r) = \begin{cases} 8573(1.0034/r)\exp(-r/0.2189) & 0 < r < 1.0034 \\ \\ 8573\exp(-r/0.2189) & 1.0034 \leqq r < 1.9348 \\ \\ 601.8\exp(-r/0.5161) - 103.9\exp(-r/1.0322) & 1.9348 \leqq r < 3.5830 \end{cases}$$

(3.4.28)

where $r$ is measured in angstroms and $V(r)$ in eV. The isolated-row estimates in the table are based on the Born-Mayer segment of this potential only and may be somewhat inaccurate. The Molière dynamical results for Cu [3.121] are based on calculations involving only two chain atoms and the required number of ring atoms. In all cases, the focusing energy is increased by the assistance of the ring(s) of neighbors. The effect is especially dramatic for the non-close-packet rows. Thus, dynamical calculations predict the possibility of significant focusing effects in these additional directions as well.

As a collision sequence propagates down a row of atoms, energy is lost from it by two mechanisms. One of these, present even for isolated rows, results from the non-head-on nature of the collisions. Its magnitude can be seen from the contour lines in Fig. 3.7. The other mechanism is the energy deposited in the encounter with the focusing rings. Because this loss is present even for perfectly focused sequences, it limits the range of collision sequences. Such losses can only be calculated in a rigorously correct manner by the dynamical method, especially in the case of close-packed rows. Their estimation by approximate methods will help to show, however, the essentially many-body nature of focusing.

In estimating the energy lost from a perfectly focused collision sequence, a first approximation is the separation of the central collisions, in which no energy is lost, from the encounters with the rings. The projectile moves in a straight line along the [$uvw$] axis, through a symmetric ring of $m$ neighbors. Andersen and Sigmund [3.122] have made a detailed study of the dynamics of such encounters. The central projectile atom interacts with the $m$ ring atoms through repulsive potentials. By the symmetry of the problem, the impact parameters $p_j$ ($j = 1, 2, ..., m$) are equal. Hence, the target scattering angles $\phi_j$ are also equal and the net deflection of the projectile must be zero. Andersen and Sigmund show that this problem may be reduced to an equivalent one of the scattering of a particle in an elliptical field of force. Their treatment to this point is exact, but solution of the elliptical equations of motion requires numerical methods. They found, however, that the problem could be approximated by using appropriately chosen spherically symmetrical potentials, matched to the exact elliptical one. For Born-Mayer potentials, these spherical matching potentials are also simple exponential functions. Thus, the elliptical problem can be solved by conventional methods, in particular by the use of tabulated scattering integrals [3.112]. When the impact parameter in the ring encounter is large, the integral tables are used with the effective impact parameter

$$\hat{p} = p_j/(1+m) \tag{3.4.29}$$

and the effective relative energy

$$\hat{E}_r = [E/(1+m)] \exp(m\hat{p}/a_{\mathrm{BM}}). \tag{3.4.30}$$

The energy transferred to the ring atoms is

$$\Delta E_{uvw}/E = 4\hat{y}(1-\hat{y}), \tag{3.4.31}$$

where

$$\hat{y} = [m/(1+m)] \sin^2 \Theta/2 \tag{3.4.32}$$

is evaluated using the tabulated values of $\sin^2 \Theta/2$.

Another estimate of $\Delta E_{uvw}$ can be based on the binary collision approximation following a procedure used in the collision cascade simulation program MARLOWE [3.110, 123]. In this program, it is necessary to deal with events in which a projectile collides almost simultaneously with several target atoms, a situation dealt with by superposing the independent binary collisions of the projectile with each target atom as discussed in detail later. For the case of perfectly focused collision sequences, the energy lost to the ring is

$$\Delta E_{uvw}/E = y/(1-y+y^2), \tag{3.4.33}$$

where

$$y = m \sin^2 \Theta/2 \tag{3.4.34}$$

and $\Theta$ is the usual barycentric scattering angle for one of the component binary collisions.

Figure 3.9 shows the energy loss calculated by each of these procedures for Cu $\langle 011 \rangle$ rows, using a Born-Mayer potential. Two dynamical results are shown [3.121], one of which employed only the projectile and the four ring atoms. The excellent agreement between this calculation and the approximation by *Andersen* and *Sigmund* is obvious. The other dynamical calculation included the next atom along the row in addition to the projectile and the four ring atoms. Above about 8 eV, the energy loss in this calculation is larger and less energy-dependent than in any of the "ring-only" calculations, in agreement with the more elaborate dynamical calculations of *Gibson* et al. [3.19]. It is clear from the dynamical results that the separation of the ring encounter from the following central collision is not

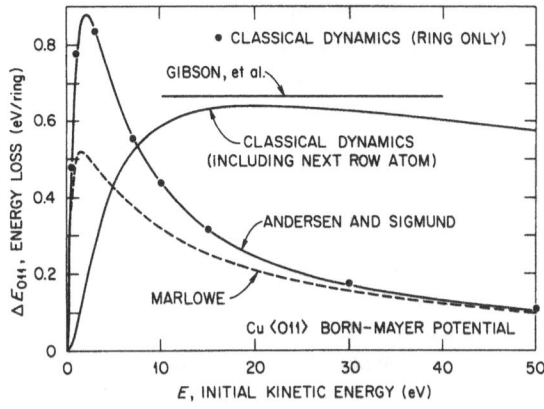

**Fig. 3.9.** Energy losses calculated by several methods for perfectly focused linear collision sequences along Cu⟨011⟩ rows

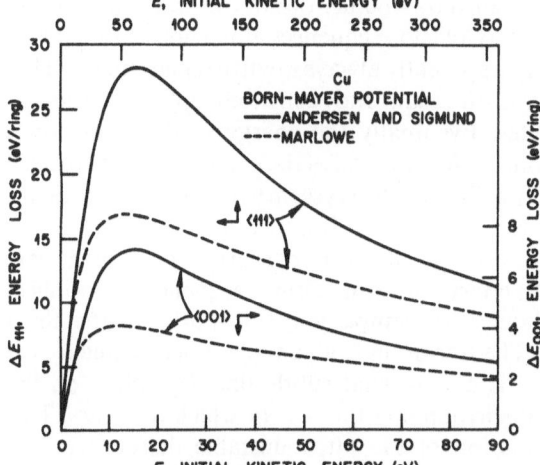

**Fig. 3.10.** Energy losses calculated for linear collision sequences along Cu⟨001⟩ and ⟨111⟩ rows

well-founded. This result can be understood in binary collision terms by noting that the time integral for the central collision locates the classical turning point for this encounter. From tables [3.112], it is found that this turning point *precedes* the plane of the ring at projectile energies below 17 eV and that, even at 90 eV, it is only one-fourth of the way from the ring to the next atomic site on the row. Thus, there is an essential many-body character to collision sequences on close-packed rows. The projectile transfers more energy to the ring atoms in this circumstance since it moves much more slowly through the plane of the ring than it would do in the absence of the next row atom. This effect should be less important along more widely spaced atomic rows. In Cu ⟨001⟩ rows, for instance, the classical turning point in the central collision is always well beyond the plane of the intervening ring and the separation of the ring encounter from the central collision should be a better approximation. Figure 3.10 compares the energy losses estimated by (3.4.31, 33) for the ⟨001⟩ and ⟨111⟩ axes in Cu. The only dynamical results available for ⟨001⟩ [3.19] show an energy loss between 7 and 8 eV, about the peak of the Andersen-Sigmund curve in Fig. 3.10. For the ⟨111⟩ axis, the calculations give an energy to penetrate a single ring (there are two rings between the row atoms in this case) between 25 and 40 eV; the energy needed to penetrate both rings is between 60 and 100 eV. These results are consistent with the Andersen-Sigmund curve in Fig. 3.10. Thus, some essential dynamical character is still present for ⟨001⟩ sequences, but probably not for ⟨111⟩. More thorough investigation of these points is needed. The MARLOWE approximation worsens as the rings become smaller, but, for a given ring, becomes better as the projectile energy increases. Its advantages over Andersen's and Sigmund's procedure are its easy applicability to potentials other than the Born-Mayer and the fact that it can be applied to cases where the projectile does not pass exactly through the center of the ring.

Linear collision sequences have been classified so far as either focusing or defocusing. They may also be classified by whether or not the projectile *replaces* the next row atom on its lattice site. Collision sequences along non-close-packed rows such as fcc ⟨001⟩ and ⟨111⟩ apparently always involve replacements. That is, the first site in such a sequence is left vacant and each atom replaces the next as long as the sequence propagates. Eventually the energy in the sequence is dissipated and an interstitial atom is left at the end of the chain. In closely packed rows such as fcc ⟨011⟩, defocusing collisions always seem to involve replacements, but focusing collisions need not do so. At the end of an encounter, a projectile may be directed back towards its original lattice site by the concerted action of the next row atom and the members of the intervening ring. Thus, a sequence is possible in which energy is transmitted without any accompanying transport of matter. Such a sequence is termed a *focuson*. The transition from a replacement sequence (or *dynamic crowdion*) to a focuson is a somewhat subtle one depending on the behavior of many atoms in and surrounding the row in which it occurs. This transition, and the further evolution of the often unstable defect structure produced by a linear collision sequence, can only be studied in detail by dynamical methods. For sputtering ejection problems, the distinction between focusons and replacement sequences is probably not important. The turning back of each projectile which is characteristic of focuson propagation results from the presence of the next atom along the row. When the row intersects a surface, this atom will be missing, the turning back cannot occur, and the last atom in the row can be ejected if its energy exceeds the appropriate surface binding energy. On the other hand, in experiments involving collision sequences in ordered systems [3.100], only replacement events are of significance, since only these result in the changes in local order which are measured. Such sequences may be defocusing over much of their length, since the energy losses along the sequence may bring it into the focusing regime before it ceases to propagate.

Two mechanisms for the loss of energy from linear collision sequences have been identified, namely the loss within a row because of less-than-perfect focusing and the loss to neighboring rows, even in perfectly focused sequences. In addition, there are important losses because the atoms in a crystal at any temperature do not occupy their ideal lattice sites except on the average. Such thermal displacements are very effective in decreasing the lengths of collision sequences [3.120, 121, 124–128]. *Tenenbaum's* calculation [3.128] for Cu using a Born-Mayer potential with the parameters (3.4.16) shows a maximum range for $\langle 110 \rangle$ replacement sequences of about 15 $t_{110}$ at 0 K and about 8 at 293 K. Similar shortening also occurs for $\langle 100 \rangle$ sequences. It is clear therefore that long linear collision sequences cannot be expected, in spite of some claims to the contrary (for instance [3.129]).

### 3.4.4 The Interaction of Linear Collision Sequences with Surfaces

If a linear collision sequence reaches a surface with sufficient energy, an atom can be ejected. However, even if the sequence is perfectly focused, the direction taken by the ejected particle may deviate from the axis of the atomic row along which the sequence propagates. Among the factors which contribute to such deviations are the following.

i) If the axis of the row passes obliquely through the surface, the ejected atom must usually pass a group of neighboring atoms lying on only one side of its path [3.18, 130]. An alternative description is that the last focusing ring is incomplete, one or more of its members being missing. The arrangement of neighbors is such that the *polar* angle of ejection is reduced, that is, the atom is deflected towards the surface normal. Alteration of the *azimuthal* angle, that is, rotation of the direction of ejection around the surface normal, may also occur.

ii) In the planar surface binding model (see Sect. 3.2.3), only the normal component of the ejected particle velocity is available to surmount the surface binding barrier. This produces a deflection away from the surface normal which will at least partially compensate the first effect [3.18].

iii) Relaxation from their substrate lattice positions of atoms in the crystal selvage can alter the ejection direction [3.18]. Small relaxations outward deflect the ejection direction towards the surface normal; inward relaxations have the opposite effect. If the relaxations are large, the collision sequence could be disrupted altogether. More drastic reordering of the surface would produce more drastic effects on ejection patterns. Altered thermal vibrations near a target surface could contribute to alterations in ejection directions by changing the local symmetries of atoms about the ejection site.

iv) The effects of radiation damage and surface roughening must be mentioned. Their effects on collision sequences would be more-or-less arbitrary and not subject to precise analysis.

The first two effects mentioned above are illustrated here by two examples. In both cases, atoms of Cu are assumed to interact through a Born-Mayer potential. The surface binding energy $U_s$ is taken equal to the cohesive energy (Table 3.1). Figure 3.11 illustrates a perfectly focused $\langle 001 \rangle$ sequence passing through a $\{111\}$ surface at an angle from the surface normal of $\cos^{-1} 3^{-1/2} = 54.74°$. Two atoms of

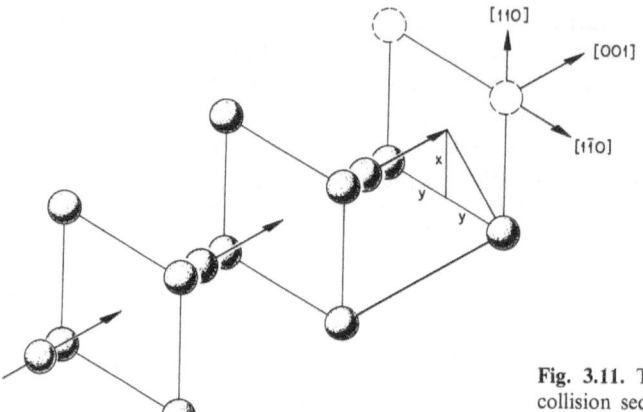

**Fig. 3.11.** The interaction of a linear collision sequence along an fcc ⟨001⟩ row with a {111} crystal surface

the last focusing ring are missing. The general situation is similar for a ⟨011⟩ sequence passing through a {001} surface at an angle of $\cos^{-1} 2^{-1/2} = 45°$. In both cases, a simultaneous encounter occurs with two ring atoms lying on one side of the axis. If this encounter is treated in the modified binary collision approximation [3.123], the resultant scattering angle and energy loss are

$$\cos \vartheta = [(x^2 + y^2) \cos^2 \Theta / (x^2 \cos^2 \Theta + y^2)]^{1/2} \tag{3.4.35}$$

$$\Delta E / E_0 = [2(x^2 + y^2) \sin^2 \Theta / 2] / [x^2 + y^2 + 2(y^2 - x^2) \sin^2 \Theta / 2$$

$$+ 4x^2 \sin^4 \sin^4 \Theta / 2], \tag{3.4.36}$$

where $\Theta$ is the barycentric scattering angle for one of the component collisions and where $x$ and $y$ are defined in Fig. 3.11. The deflection and energy loss predicted by these formulas are substantially greater than those resulting from the mere vectorial addition of the laboratory deflections of the projectile used in an earlier calculation [3.130]. The effects of surface scattering were evaluated using (3.4.35, 36); the results were corrected for binding energy effects using (3.2.1, 2). The final ejection angles are shown in Fig. 3.12 as functions of the initial kinetic energy of the projectile as it leaves its lattice site. The two axes behave rather differently. In the ⟨011⟩ case, where the neighbor atoms are rather far from the axis, the surface scattering is overwhelmed by the binding energy effect at all ejection energies. That this need not always be true is shown by the behavior of the ⟨001⟩ axis at the {111} surface. Here surface scattering plays the major role except at low energies. The binding energy begins to be important at ∼15 eV and is the major factor at the lowest energies. These calculations appear to be consistent with experimental ejection patterns. Assuming the Wehner spots to result from linear collision sequences, the curves in Fig. 3.12 can be used to predict that the ⟨011⟩ spots from {001} Cu crystals will be centered at polar angles somewhat greater than the ideal 45°, the exact amount depending on averaging over the

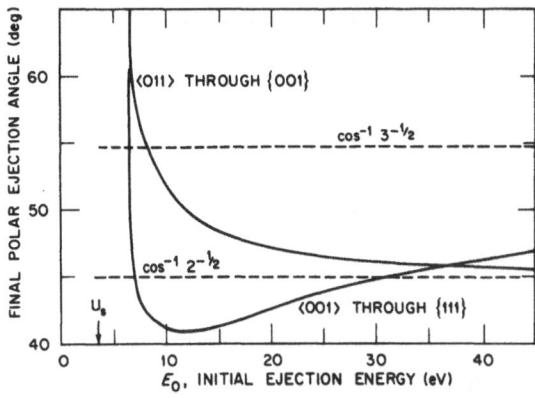

**Fig. 3.12.** The combined effects of surface scattering and binding energy on the directions of ejection of linear collision sequences passing through various surfaces of Cu

ejection energy spectrum. *Weijsenfeld*'s experiments [3.18] show spot centers at 50°. The other curve shows that the ⟨001⟩ spot from {111} crystals should be observed much closer to the surface normal than the ideal 54.74°, probably near 42°. This result is also consistent with experiment [3.18, 68, 93].

Surface scattering can not only shift a Wehner spot closer to the surface normal, but can also sharpen it. Atoms originally ejected farther from the normal than the ideal direction (below the axis in Fig. 3.11) will be more strongly scattered than those on the axis. The contrary effect occurs for atoms starting closer to the normal than the ideal direction. Thus, the distribution of initial directions will be narrowed. On the other hand, the binding energy refraction will tend to broaden the distribution, since particles starting farther from the normal are more strongly refracted.

### 3.4.5 The Role of Linear Collision Sequences in Sputtering Ejection

When *Silsbee* first proposed [3.6] the idea of focusing collision sequences along closely spaced rows of atoms in crystals, it seemed obvious that this was the origin of the recently discovered Wehner spots [3.4, 5]. This interpretation was questioned by *Lehmann* and *Sigmund* [3.17], who proposed an alternate mechanism for forming the spots. Among the grounds for doubting the focuson theory are the two following points.

i) Wehner spots are observed in low energy sputtering experiments. The original observations [3.4] were made with 150 eV $Hg^+$ ions on a crystal of W and with $Hg^+$ ions of energies as low as 50 eV on Ag crystals. Penetration of the target by the incident ions in such circumstances is very slight. In fact, if tabulated range calculations [3.131] are extrapolated to the low energies of Wehner's first experiments, penetration depths of less than 1 Å are found. In such cases, there is no plausible way that collision sequences along close-packed rows can contribute to ejection into the Wehner spots.

ii)Wehner spots are observed for directions not associated with long, evenly-spaced rows of atoms. Such spots have been observed from the $\langle 111 \rangle$ directions of crystals with the diamond and zinc-blende structures, such as Ge and InSb [3.5, 45, 132]. In these rows, the atomic spacings alternate between a long and a short value (see Table 3.6) and focusing should be significant only for the latter. The observation and interpretation of ejection patterns in these materials is complicated by radiation damage, however (see Sect. 3.2.2). Probably a more important example is the observation of Wehner spots corresponding to the $\langle 20\bar{2}3 \rangle$ rows of the hcp metals [3.95, 133, 134]. Again, (see Table 3.8) a short and a long interatomic spacing alternate. No plausible model ever seems to have been suggested for collision sequences of more than two members in these rows.

To avoid such difficulties with the focuson model, *Lehmann* and *Sigmund* argued that the spot structure could result from the regularity of the surface structure, coupled with the energy spectrum of the sputtered particles, which varies approximately as $E^{-2}$ in random cascade theory (see Chap. 2). Their model can be illustrated by a simple example. Consider a solid surface perpendicular to a nearest neighbor direction, such as an {011} surface of an fcc metal. The atoms in the second layer are assumed to move isotropically, with an $E^{-2}$ energy spectrum. Their distribution may be written as

$$p_0(\mu_0, E_0)d\mu_0 dE_0 = (U_s/E_0^2)d\mu_0 dE_0, \ U_s \leqq E_0 \leqq \infty, 0 \leqq \mu_0 \leqq 1, \qquad (3.4.37)$$

where $\mu_0 = \cos\vartheta_0$ is the initial projectile direction cosine, $E_0$ is its initial kinetic energy, and $U_s$ is the surface binding energy. Equation (3.4.37) is normalized to unity over the specified ranges. This distribution impinges on an atom in the crystal surface, as shown in the inset portion of Fig. 3.13. The direction of motion of the target atom is

$$\mu_T = \cos\vartheta_T = [1 - \Lambda^2(1 - \mu_0^2)]^{1/2}, \qquad (3.4.38)$$

where $\Lambda$ is the focusing parameter defined in (3.4.21). As Fig. 3.7 shows, it is an excellent approximation to take $\Lambda$ independent of $\vartheta_0$. The energy of the target atom is

$$E_T = E_0[\mu_0\mu_T - \Lambda(1 - \mu_0^2)]^2. \qquad (3.4.39)$$

After scattering, the direction and energy of the projectile are

$$\mu_p = \Lambda(1 - \mu_0^2)^{1/2} \qquad (3.4.40)$$

$$E_p = E_0(1 - \mu_0^2)(\mu_T + \Lambda\mu_0)^2. \qquad (3.4.41)$$

For simplicity in the following development, $\Lambda$ is taken independent of $E_0$, although $\Lambda$ varies roughly as $E_0^{1/3}$ in Fig. 3.7. The distribution of the ejected target

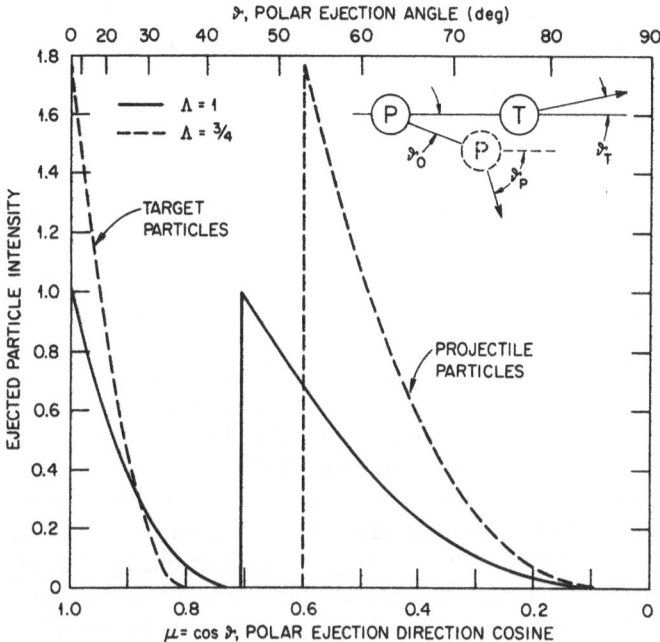

**Fig. 3.13.** The formation of *Wehner* spots by the *Lehmann-Sigmund* mechanism

atoms can be found by transforming (3.4.37) from the coordinates $(\mu_0, E_0)$ to $(\mu_T, E_T)$. The ejected particle distribution is

$$p_T(\mu_T, E_T) = (U_s/E_T^2)(\mu_T/\Lambda^3)[\Lambda^2 - (1 - \mu_T^2)]^{-1/2}$$

$$\cdot \{\mu_T[\Lambda^2 - (1 - \mu_T^2)]^{1/2} - (1 - \mu_T^2)\}^2,$$

$$U_s \leqq E_T \leqq \infty, (1 + \Lambda^2)^{-1/2} \leqq \mu_T \leqq 1. \tag{3.4.42}$$

The corresponding distribution of the scattered projectiles is

$$p_p(\mu_p, E_p) = (U_s/E_p^2)(\mu_p/\Lambda)^3(\Lambda^2 - \mu_p^2)^{-1/2}[(1 - \mu_p^2)^{1/2} + (\Lambda^2 - \mu_p^2)^{1/2}]^2,$$

$$U_s \leqq E_p \leqq \infty, 0 \leqq \mu_p \leqq (1 + \Lambda^2)^{-1/2}. \tag{3.4.43}$$

The ejected particle angular distributions can be found by integrating (3.4.42, 43) over the indicated energy ranges; the results are given by the same equations without their leading factors. These angular distributions are plotted in Fig. 3.13 for two values of $\Lambda$. Even where there is no focusing ($\Lambda = 1$), the ejected target atoms are closely grouped around the nearest neighbor direction and form a rather well-defined spot. The spot is considerably sharper when focusing occurs ($\Lambda = 3/4$). The scattered projectiles form a ring of intensity at rather large angles, completely surrounding the target "spot". Their distribution is that of a typical blocking pattern of the kind familiar at higher particle energies [3.65–67, 135, 136]. This sample calculation shows that an ejection pattern spot can indeed be produced by *Lehmann*'s and *Sigmund*'s mechanism, even in the absence of focusing. The mechanism is subject to several criticisms, however, which must be examined.

*Lehmann* and *Sigmund* [3.17] considered the distribution of target atoms, but ignored the recoiling projectiles. As has been shown, these particles add a ring of ejected material at rather large angles from the nearest neighbor row. This would constitute a serious objection to the theory except that it is unlikely that such particles could escape from a crystal without making additional collisions with neighboring atoms. These additional collisions would have major effects on the projectile ring which would either eliminate it or spread it out into a more diffuse background emission. A similar conclusion was reached by *Lenskjaer* et al. [3.137], on the basis of a rather different physical model.

A more serious objection to the theory is its neglect of the refraction of the ejected particle trajectories by the surface binding energy. Inclusion of the planar binding model (see Sect. 3.2.3) would affect the angular distributions significantly. The refraction associated with escaping the surface binding energy compensates in considerable part for the focusing effects, both that in the final collision and that stemming from the nature of the energy spectrum. Such effects are the largest for ejection spots that are distant from the surface normal.

It seems clear from this discussion that neither the focuson theory of the Wehner spots nor the model of *Lehmann* and *Sigmund* can be taken quite literally, although both contain elements which must be present in a satisfactory theory of the ejection pattern. But, perhaps because neither of the models deals with the full three-dimensional nature of the collision cascades leading to ejection, neither can be regarded as really successful in accounting for experimental observations. Furthermore, neither model deals with the origin of atoms which are sputtered, not into the Wehner spots, but into the "background" on which the spots are superimposed. Experiments on 1 to 10 keV $Hg^+$ and $Cs^+$ ion irradiations of Cu and Mo crystals [3.138, 139] have shown that $\sim 80\%$ of the total yield is in this background. A complete theory of the ejection process must account for this result. The controversy [3.140, 141] about the extent to which "random" ejection is to be attributed to phase changes brought about by the implanted ions or to radiation damage may not even be germane (see Sect. 3.2.2). It should be possible to explain the random ejection as a normal aspect of the sputtering of crystals.

## 3.5 The Computer Simulation of Monocrystal Sputtering

As the preceding discussion has shown, neither the channeling theory of the sputtering yield nor the models of the ejection process from monocrystal surfaces can give more than a qualitative or, at best, a semiquantitative picture of the experimentally observed phenomena. The limitations of these models seem to stem mainly from their neglect of the full three-dimensional structures of the target crystals and their concentration on overly-simplified one-dimensional abstractions. Such limitations can be overcome by computer simulation methods. Since sputtering is only one aspect of the more general problem of defect production by the energetic particle irradiation of solids, it is convenient to discuss the computer simulation of sputtering in the more general context of the simulation of radiation damage.

Two fundamental assumptions are common to all of the computer simulation models to be discussed. First, the constituent particles of solids are assumed to be atoms or atomic ions interacting with one another through conservative pairwise forces. A volume-dependent interaction may be included as well. This assumption restricts the models to studies of the atomic properties of solids, their overtly electronic properties being explicitly ignored. Second, model calculations are based on the methods of classical mechanics. The use of Newtonian mechanics is an aspect of the general neglect of electrons in these models. Electron effects can be important in several ways. Most importantly, in the vicinity of crystal defects such as vacant lattice sites or interstitial atoms or near crystal surfaces, electron rearrangements may significantly alter the interatomic forces among the atoms of the solid [3.142]. In addition, encounters involving energetic atoms are significantly inelastic and one or more of the atoms involved may be left in an electronically excited state, with consequent effects on its subsequent interactions. In some of the models to be discussed, approximate corrections for these electronic effects are included, but these are made in the spirit of classical mechanics and not by adding real electrons.

Computer simulation models may be classified according to the number of mutually interacting atoms which they consider and the boundary conditions that guarantee the stability of the model and represent its interactions with the surrounding matrix. Studies of the structure and other properties of defects in solids require that many atoms be considered simultaneously, often more than 1000 being included in the numerical crystallite. In such calculations, the boundary conditions applied to the model must represent both the static and dynamic responses of the matrix. The application of atomistic computer simulation methods to studies of defect properties in metals is reviewed in detail elsewhere [3.34, 143–146]. Such models can also be used to study radiation damage and sputtering. However, the calculations are often very time-consuming or require very large amounts of computer memory. Such practical computing problems limit dynamical calculations on fully stable models to studies of individual events at fairly low primary particle energies. Consequently, more approximate models must be used if the objective of the calculations is to study high energy events or to consider ensembles of primary particles large enough to permit statistically significant comparisons with experiment. The following subsections will discuss some of the approximations that can be made, their limitations, and their effects on calculated quantities. Other reviews which discuss at least some of the same topics, often from a viewpoint somewhat different to the present one, will be found in [3.147–150].

### 3.5.1 Displacement Cascades in Stable Dynamical Models

The inherent stability of the model solid is a principal feature which characterizes the calculations to be discussed here. A disturbance introduced into such a model is eventually dissipated and its atoms find an appropriate equilibrium. Because of this stability, such models may be used to study many aspects of solids, including the formation and migration of point defects, displacement threshold energy surfaces, relaxations in the vicinity of dislocations, and the development of radiation damage displacement cascades. The dynamics of low energy radiation damage has been studied [3.19, 119, 120, 127, 128, 150–168] in a variety of materials, including the fcc metal Cu, the bcc metals Fe and W, the ordered alloy $Fe_3Al$, and ionic crystals such as $PbI_2$ and KCl.

Although the various computational models differ in detail, they all closely follow the pattern established by [3.19]. The computational crystallite consists of from 500 [3.19] to 5000 [3.167] atoms which interact according to two-body central forces derived from a potential $V(r)$. Typical potentials have been the purely repulsive Born-Mayer potential of (3.4.12) [3.19, 163] and the Morse potential [3.62, 63, 119, 120, 151, 158, 159]:

$$V_{MO}(r) = C_{MO}y(y-2),\qquad\qquad(3.5.1)$$

where

$$y = \exp[-(r-r_0)/a_{MO}].\qquad\qquad(3.5.2)$$

The parameters $r_0$, $C_{MO}$, and $a_{MO}$ can be obtained from the work of *Girifalco* and *Weizer* [3.32] or elsewhere. Adjustments are usually made to the potential function so that it vanishes smoothly at an appropriate point, for example, within the second neighbor separation in the crystal [3.63]. With such restrictions on the range of the interaction potential, relatively small numbers of atoms contribute to the force at any one point, greatly reducing the numerical labor. The selection of suitable empirical potentials is discussed in some detail by [3.34, 143].

When only repulsive forces are used between the atoms, the stability of the numerical crystallite must be assured by putting special static external forces on the atoms in the crystallite boundaries. Such forces represent a contribution to the binding energy of the model which is proportional to the volume of the crystallite and thus mimic the binding due to the conduction electrons in a metal. Even when the two-body forces are partially attractive ones, external boundary forces may be needed to obtain the correct lattice spacing or to match other physical properties properly, such as the cohesive energy or the elastic constants. In addition, boundary conditions must be applied to the model to represent its embedding in a macroscopic medium. These conditions must include the reaction of the matrix to atomic displacements within the crystallite. In the work of *Gibson* et al. [3.19], the additional boundary conditions were a Hookean (spring) force to represent the elastic response of the matrix, and a viscous force to permit excess kinetic energy within the crystallite eventually to drain away. Another possibility is to make the matrix completely rigid [3.62], that is, to fix the boundary atoms permanently to their lattice sites. Such a boundary can only be suitable if it can be shown not to influence the results of the calculations. Two problems can occur: since the elastic response of the medium is ignored, the static behavior of defects will be altered unless they are much farther from the boundary than the range of the interatomic forces. In addition, kinetic energy introduced into the crystallite cannot flow away into the matrix and this may influence both the dynamic development of a collision cascade and the distribution of defects which results. A third possibility for embedding the crystallite in a matrix is to employ periodic boundary conditions [3.63, 162–167]. In this case, the model crystallite is one element of a superlattice, in each cell of which identical events are occurring. Energy flowing out through one face of the model re-enters through the opposite face with possible consequences for the dynamics. It is possible that long-range interactions will occur between defects in one cell and their images in other cells. Intermediate types of boundary conditions are also possible, such as cyclic boundaries with dissipation to reduce energy flow. Ideally, the processes being studied within the crystallite will not approach the boundaries closely enough for the precise choice of constraints to be significant. The extent to which such independence of the boundaries is achieved in particular calculations is not always clear. Since efficient use of the computing machinery demands that the numerical model be kept as small as possible, boundary effects must always be of concern.

The simulation of a radiation damage event in a dynamical model proceeds as follows. A selected atom is set moving in a particular direction with the desired kinetic energy. The classical equations of motion are integrated for all the particles in the crystallite as long as a significant fraction of the original kinetic energy remains. Methods of integration are beyond the scope of this article, but are discussed fully elsewhere [3.169, 170]. Time steps for the integration are typically 1 to 3 fs and a whole calculation may require 100 to 1000 such steps. The calculations are very time-consuming even on the fastest contemporary machines. For example, the program COMENT [3.164–167], when executed on a Control Data Corporation 7600 machine, requires approximately one second for each computational time step in a crystallite of 1000 atoms. (The remarkable development of computers in the past twenty years is attested by the fact that this time is 1/100 of that required on the IBM 704 machine used in the first work of *Gibson* et al. [3.19].) Because of their computing requirements, dynamical calculations are mainly restricted to studies of individual events at fairly low initial kinetic energies. They are indispensable for identifying the significant anatomical features of low energy collision cascades. Such features may often then be modelled in a simple manner for inclusion in more approximate model calculations.

Figure 3.14 shows a typical dynamical calculation for a low-energy collision cascade in bcc Fe, using the potential (3.4.28) [3.119]. The strongly marked $\langle 111 \rangle$ linear collision sequences are evident. A somewhat more detailed account of the energy loss mechanism in such sequences can be given here than was possible in Sect. 3.4.3. As the first atom in the sequence moves forward, it squeezes between

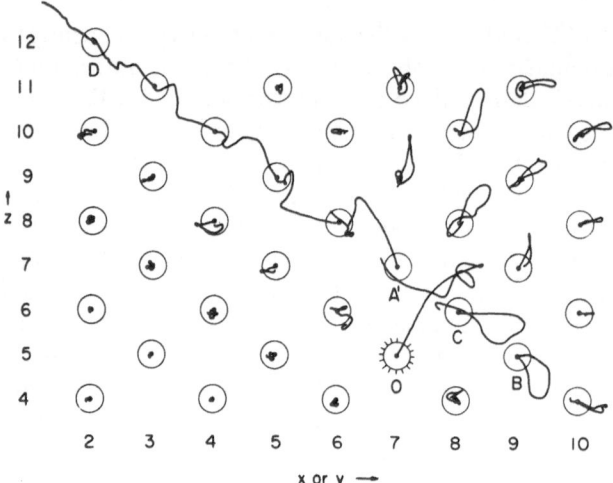

**Fig. 3.14.** Trajectories of atoms in an event initiated by a 65 eV primary Fe atom directed 25 degrees from [001] in a (1$\bar{1}$0) plane of an Fe crystal. The figure is reproduced with the permission of *Erginsoy* et al. [3.119]

the atoms of the adjacent focusing ring which move aside, but then encounter their own farther neighbors and turn back towards the sequence axis, returning to their original sites after the main impulse has passed by. In a replacement sequence, their return supplies an extra nudge to the original projectile, helping it on its way; in a focusing sequence, the extra impulse assists the motion of the next sequence atom instead. In either event, part of the energy transferred to the ring atoms is not lost from the sequence, but returns to be transmitted to the next focusing ring. As a result, the energy lost in the first step of the sequence is greater than in subsequent steps, since the "returnable" energy is stored only once [3.19, 119, 120]. It is clear from Fig. 3.14 that a collision sequence cannot be decoupled from its surroundings and that, furthermore, these surroundings include many more atoms than merely those in the adjacent rows. Such behavior again emphasizes the essential many-body character of collision sequences.

When the initial kinetic energy of the primary recoil atom has been largely dissipated, after 0.1 to 1 ps of cascade development, the numerical crystallite contains vacant lattice sites, interstitial atoms, and a considerable residual atomic agitation. In fcc metals, the isolated interstitial adopts the $\langle 100 \rangle$ split or dumbbell configuration [3.19], first suggested by *Huntington* and *Seitz* [3.171]. The interstitial and an adjacent lattice atom form (in effect) a diatomic molecule centered on the lattice site, with the molecular axis parallel to a $\langle 100 \rangle$ direction. In bcc metals, the interstitial has a similar structure, but with its axis along $\langle 110 \rangle$. These configurations, first proposed on theoretical grounds, have now, after some considerable controversy, been firmly established experimentally as the correct ones [3.172, 173]. Experiment has also established a $\langle 0001 \rangle$ dumbbell as the stable interstitial configuration in the hcp metals. In all metals, the vacancy has the expected simple configuration of a single unoccupied lattice site. The neighboring atoms relax somewhat around both vacancies and interstitials. The computer simulations show further that around either kind of point defect (vacancy or interstitial) there is a region in which the other kind of point defect (interstitial or vacancy) is unstable [3.19]. If an interstitial atom is produced within the instability volume around a vacancy, the two defects will recombine (annihilate) within about 1 ps [3.157]. In the case of Cu interacting through a Born-Mayer potential with the parameters of (3.4.16), the instability volume was found to contain 74 lattice sites.

The concept of a threshold energy for producing atomic displacements has long been used in theories of radiation-induced defect production in solids [3.174, 175]. While it was early realized [3.176] that the threshold energy, $E_d$, depended on the initial direction of motion of the recoiling atoms, it was not until the introduction of computer simulation that accurate calculations of such dependence could be made and its physical basis properly established. Both linear collision sequences and the instability volume play a part in determining $E_d$. When an atom recoils from a lattice site with an energy near $E_d$, it may become an interstitial itself or it may initiate a replacement sequence at the end of which another atom becomes the interstitial member of a Frenkel pair. In either case, if the interstitial lies outside the instability volume around the vacant primary lattice site, the Frenkel pair will be stable; if not, its members will recombine. Thus, allowing for the effects of replacement events, the surface of the instability volume gives the range of recoils with energy $E_d$. It is clear from this description that $E_d$ describes the conditions for Frenkel pair *survival* and not the conditions for recoil atom motion. Thus, $E_d$ has nothing to do with sputtering problems, which are governed instead by surface binding energies. Detailed calculations of the directional dependence of $E_d$ have been made for fcc Cu [3.19, 160] and for bcc Fe [3.119] in the absence of thermal vibrations. Calculations including thermal effects have been made for fcc Fe [3.161] and for bcc W [3.162]. There are several recent reviews of experimental determinations of displacement threshold energy surfaces [3.177–179].

### 3.5.2 Sputtering Studies with Stable Dynamical Models

A few attempts have been made to use stable dynamical models to study aspects of sputtering [3.62–64, 180–186]. An important advantage of such models is that there is no need to make assumptions concerning the binding of atoms at and near the target surface, since such binding is automatically incorporated in the interatomic forces and the boundary constraints. In all cases to date, a Morse potential (3.5.1, 2) was used, with various choices of parameters. The studies of the ejection of atoms from monocrystal surfaces [3.62] were discussed in Sect. 3.2.3. A similar, though more limited, calculation is that of *Bespalova* and *Gurvich* [3.186]. More recently, *Cherns* et al. [3.63] have studied the transmission sputtering of {111} Au foils by 0.5 to 1.1 MeV electrons, both experimentally and theoretically. Depending on the energies of the electrons, the primary Au recoils had maximum initial kinetic energies from 8.30 to 25.45 eV. The recoil energy was maximal when the atoms moved initially along [111] and decreased as $\cos^2 \vartheta$ for more oblique directions. The numerical crystallite contained nine (111) layers, each consisting of 36 atoms. The two parallel (111) surfaces of the crystallite were free, while periodic boundary conditions were imposed on its sides. Beyond the nearest neighbor separation in the crystal, a cubic spline was used to reduce the potential smoothly to zero within the second neighbor distance. The potential parameters were chosen to fit the cohesive energy and the bulk modulus of Au. The program was used to follow selected Au primaries, mainly directed near $\langle 110 \rangle$, starting in the last four layers of target atoms. Several features of the results are noteworthy.

i) The threshold energy for sputtering was $\sim 6.7$ eV. This value is higher than the surface binding energy, 5.0 eV, suggested in [3.49]. Experiment [3.187] also suggests a value near 5 eV. The higher value from the dynamical calculation doubtless reflects inadequacies in the interaction potential, perhaps particularly near the target surface.

ii) Detailed trajectories show the behavior expected in the planar surface binding model, namely, that low energy ejected atoms are refracted away from the surface normal. The opposite effect occurs for fast particles which are mainly influenced by surface scattering (see Sect. 3.4.4).

iii) At the higher primary energies, much of the ejection is caused by momentum transport along $\langle 110 \rangle$ rows. While this result implies that *transmission* sputtering is closely connected to linear collision sequences, it does not imply that such is also the case for *reflection* sputtering.

iv) The model calculations are in reasonable quantitative agreement with the experiments, especially with respect to angular distributions.

More extensive sputtering calculations with a stable dynamical model have been reported by *Harrison* et al. [3.64, 180–185]. This work was an extension of earlier calculations using a metastable model [3.20, 188–191] which will be discussed below. In the stable model calculations, the interactions of the Cu target atoms are represented by a Born-Mayer potential with the parameters (3.4.16) for interatomic separations less than 1.5 Å, by a Morse potential for separations greater than $\sim 2.0$ Å, and by a cubic spline which interpolates smoothly between them. The Morse potential parameters were either those of [3.32] or one of the sets of truncated Morse potentials proposed by *Anderman* [3.192]. The model contained about 250 atoms arranged in four to six layers, depending on orientation. In addition to the sputtering calculations, a few static calculations of surface relaxations were made. The results were generally similar to [3.31], with the additional feature that the surface planes were not atomically smooth, but were rumpled, the atoms lying $\pm 0.036$ Å from the average surface plane. Whether this effect is real or an artifact of the model is not clear. The rumpling should produce an observable LEED effect, since the size of the surface mesh would be altered. Such effects are not observed from such fcc metals as Al and Cu. On the other hand, since the individual layers of the crystallite are not much larger than the range of the potential used, boundary constraint effects are likely. Surface binding energy values are reported [3.180] which are about half the values found by *Jackson* [3.49] and others (see Sect. 3.2.3) using very similar potential functions. The source of this discrepancy is not clear.

Dynamical calculations [3.180] were made of the sputtering of Cu {100} surfaces by normally incident 5 keV Ar ions. The Ar–Cu interactions were described by a Born-Mayer potential with the parameters [3.189]

$$C_{BM} = 71.3 \text{ keV} \quad \text{and} \quad a_{BM} = 0.218 \text{ Å}. \tag{3.5.3}$$

The screening length is approximately the same as that proposed by *Andersen* and *Sigmund* [3.108, 109], but the energy parameter is more than an order of magnitude larger than theirs [see (3.4.17)]. The numerical crystallite was completely free (no matrix embedding), but edge and corner atoms were not included in the sputtering results. Calculations were performed for 36 incident particles, uniformly distributed over the unit mesh. The sputtering yield obtained was 4.7 atoms/ion, so that about 170 sputtered particles were generated. Since Poisson statistics should apply approximately to this calculation, the uncertainty of the calculated yield is about ±0.4. The result is therefore in satisfactory agreement with the experimental values 4.2 ± 0.2 [3.68] and 4.1 ± 0.2 [3.193]. The results also agreed well with earlier calculations [3.20, 188–190] using a metastable model. Consequently, the significantly faster metastable model calculations were used for other ion energies and orientations. These results will be considered further below. In a later report [3.64], yields were calculated for 600 eV Ar normally incident on the three principal Cu surfaces. The *relative* yields are in reasonable accord with experiment [3.68, 94, 98, 193], but the *absolute* yields calculated were about twice the observed values. It appears probable that such high yields are a consequence of the very "hard" interaction potential, which restricts the displacement cascade to the surface region more effectively than would a potential with more reasonable parameters.

Another application of the Harrison stable model is to the simulation of the sputtering of molecular species [3.181–185]. This calculation was essentially the same as the foregoing, with the addition of a determination of whether or not pairs of sputtered atoms were bound together. This determination was originally [3.181] based on the same interatomic potential used in the crystal, but in later work [3.182–185] an appropriate diatomic molecule function was used instead. An important finding was that the atoms comprising an ejected molecule need not be (in fact, usually were not) sputtered in a closely correlated manner. Instead, a relatively fast atom (or collision sequence) from one part of a cascade could overtake a relatively slow atom ejected from the surface earlier on. This overtaking process was materially assisted by the lattice structure of the solid, including effects of atomic rows and of focusing rings.

### 3.5.3 Studies of Sputtering in a Metastable Dynamical Model

For many applications, dynamical calculations with energetically stable numerical models are so time-consuming that approximation procedures must be sought to make the computations feasible. One approach, introduced by *Harrison* and his colleagues [3.20, 180, 181, 188–191] for studies of sputtering, is to abandon the requirement that the numerical crystallite be stable. If the interaction potential is suitably restricted in range and if the phenomena being studied are rapid enough, little error should be introduced by this approximation. The lifetime of a collision cascade in a structureless medium is roughly 0.1 to 1.0 ps, depending on the primary energy and the particle masses [3.194]. At the end of this time, the lattice is left with a substantial residual agitation which dissipates by processes related to ordinary heat conduction and diffusion. Since the periods of the thermal motions of the lattice atoms which participate in these processes are $\sim 0.1$ ps or more, cascade evolution may often be regarded as decoupled from the thermal and elastic responses of the solid. Special procedures may then be supplied which emulate the behavior of the neglected long-time responses. Such special models are probably best founded on fully stable dynamical models, although this is not always practicable.

Harrison's procedure is to limit the size of the numerical crystallite and to "erode" the interatomic potential so that it vanishes at $r_0$, the nearest neighbor separation in the crystal. The eroded and infinite potentials differ very little and the forces derived from them are identical, except that a discontinuity in the force at $r_0$ introduces a numerical problem requiring somewhat careful attention [3.20, 190]. Because of the eroded potential, the numerical crystallite is metastable: in the initial condition with all atoms on their lattice sites, the total energy is zero. Once the system is disturbed, however, no return to it is possible, since there are no restoring forces. Surface binding was included in the model by applying an energy-dependent escape probability to atoms leaving the crystallite through its surface (atoms leaving through its back were dropped from the calculation; the few leaving through the sides were examined to see if they could cause sputtering and were then treated appropriately). Three models were explored. In the notation of Fig. 3.1, the escape probabilities may be written as

$$p_1(\mu^2 E_0) = \begin{cases} 1 & \mu^2 E_0 > U_s \\ 0 & \mu^2 E_0 < U_s \end{cases} \tag{3.5.4}$$

$$p_2(\mu^2 E_0) = \begin{cases} 1 - \exp(1 - \mu^2 E_0/U_s) & \mu^2 E_0 > U_s \\ 0 & \mu^2 E_0 < U_s \end{cases} \tag{3.5.5}$$

$$p_3(\mu^2 E_0) = 1 - \exp(-\mu^2 E_0/U_s). \tag{3.5.6}$$

In each case, the direction and energy of the sputtered particle were modified according to (3.2.1, 2). The step function, (3.5.4), is the planar binding model described in Sect. 3.2.3. No physical justification has been provided for the other two functions. Equation (3.5.5) lowers the probability of emission of particles with energies near $U_s$, especially if ejected at large angles from the normal. It thus tends to eliminate those atoms most strongly affected by the binding energy refraction.

Most of Harrison's work was done with the third form, (3.5.6). According to this equation, particles with $\mu^2 E_0 < U_s$ may still be emitted from the target. Such particles should be totally reflected from the surface binding barrier. How these two conflicting statements can be reconciled is unclear. Equation (3.5.6) also tends to favor particles ejected with little binding refraction.

All of Harrison's metastable model calculations used an eroded Born-Mayer potential with the parameters (3.4.16) to represent the interaction of the Cu atoms in the numerical crystallite. The interactions of the incident $Ar^+$ ions with the lattice atoms were also described by Born-Mayer potentials. The original work [3.20] has been criticized [3.195] mainly for the parameters used in these latter potentials. The very hard potentials used were certainly responsible for many of the quantitative aspects of the results obtained. More recently, partly in response to this criticism, the work has been repeated using somewhat less objectionable potential parameters [3.180, 189] although the set (3.5.3), which has been mostly used, still appears stronger than warranted. The argument that these strong potentials are semiempirical ones inferred [3.196] from ion-induced secondary electron emission yields appears to be without merit. Such potentials only reflect inadequacies in the secondary electron theory. Before the quantitative aspects of Harrison's calculations warrant detailed analysis, several points require clarification or correction:

i) the ion-lattice atom interaction potential must be convincingly justified or replaced by one more suitable;

ii) the low surface binding energies used must be explained or corrected;

iii) the surface emission probability, (3.5.6), must be justified physically or replaced with a model that can be justified.

The calculated sputtering yields and ejected atom energy distributions are too sensitive to these points to be discussed usefully in the absence of the necessary clarifications. On the other hand, the ejected particle angular distributions appear somewhat insensitive to these points. Thus, the calculations seem to offer useful insights into possible sputtering mechanisms and it is to this aspect of Harrison's calculations that the following discussion is restricted. It should be noted, however, that the calculated ejected atom energy distributions [3.20, 180] seem to contain a somewhat large fraction of energetic particles ($E > U_0$) as compared both with experiment (see [3.55, 57] and [3.1a] Chap. 2) and with transport theory (see Chap. 2). This probably reflects the bias against low energy emissions implicit in (3.5.6), although the poor statistics achievable in these calculations makes this interpretation uncertain.

Several mechanisms have been identified [3.20] whereby atoms are ejected from a crystal surface. All of these must be distinguished from both the focuson mechanism and the Lehmann-Sigmund mechanism (see Sect. 3.4.5). Collision sequences of the Silsbee type are found very frequently in the calculations, but these are always directed inward or parallel to the surface. Thus, under the circumstances of these calculations, while long collision sequences might contribute to transmission sputtering, there was no evidence that focusons contribute to ejection into the ordinary Wehner spots. The mechanisms identified in this work fall into two groups: surface events and deep events. In the latter, the incident ion or an energetic lattice recoil penetrates several layers into the target, is there reflected and initiates a short collision sequence back towards the surface. Events of this kind were found to contribute only a small amount to the yield (up to ~25 % for 20 keV $Ar^+$ on Cu {100} at some angles of incidence) [3.180]. It seems likely that such mechanisms will become more important at higher energies and for lower ion/target mass ratios where reflection becomes more probable. Furthermore, the use of an overly-strong ion-atom interaction potential will result in an underestimate of the relative contribution of such deep events. The surface events, which dominate the ejection processes in *Harrison*'s calculations [3.20, 180, 181], are either incipient defocusing collision sequences along rows lying in the crystal surface or subsurface channeling [3.197] events. Ejection by surface collision sequences has also been seen in dynamical calculations [3.62]. It appears quite probable that events of this general character are a major factor in sputtering ejection. The fact that such processes lead to angular distributions with the appearance of Wehner spots means that such spots are not only attributable to one-dimensional mechanisms such as linear collision sequences, but also to essentially three-dimensional crystal structure effects. Unfortunately, the doubtful features in the calculations of *Harrison* et al. make it impossible to establish the quantitative role of such mechanisms, in spite of their suggestive character.

### 3.5.4 Collision Cascades in Quasistable Dynamical Models

Another technique for accelerating classical dynamics calculations involves the use of what may be termed *quasistable* models [3.61, 198–200]. In contrast to both stable and metastable model calculations, no fixed crystallite is defined inside of which a collision cascade develops. Instead, the required lattice atoms are generated along the tracks of the various recoils using criteria concerning particle energies and separations to decide which atoms are to be included in the integration procedures. *Schlaug* [3.61] used a technique of this kind to deal with low energy encounters in his sputtering calculation. He considered any recoil with energy less than 300 eV to interact with small groups of atoms in its vicinity, the number and geometry of the group depending on the recoil direction. The equations of motion of this small set of particles were integrated until the particle velocities reached their asymptotic values. Those particles with sufficient energy to cause sputtering were added to the cascade. All other particles were ignored. This model enabled *Schlaug* to deal in a realistic manner with low energy collision problems, especially with linear collision sequences. Because his model was used in connection with an otherwise binary collision model of sputtering, further discussion of it is deferred to Sect. 3.5.7.

*Torrens* [3.198] introduced a more general class of quasistable models. As in Schlaug's calculations, regions of crystal were generated as needed to follow the developing collision cascade. The classical equations of motion were integrated for any particle whose total energy exceeded a minimum value, $E_m$, typically 1 eV.

As the cascade develops, the membership of the set of moving particles changes, increasing as atoms are set into motion, decreasing as they are brought to rest. Because the motions of particles with energies less than $E_m$ are ignored, the basic instability of the model is circumvented. *Torrens* used this model to study collision sequences near the displacement threshold in Cu. More recently [3.199, 200], it has been made the basis of a collision cascade simulation program and its application to other problems, including sputtering, should be straightforward. Similar models have been used to study linear collision sequences [3.121] and the scattering of energetic ions from crystals [3.149, 201, 202]. Conceptually, there are close similarities between quasistable classical dynamical models and binary collision models.

The moving particles in a quasistable classical dynamics calculation interact with each other and with any stationary lattice atoms within a cut-off distance $p_c$. Beyond this distance, typically between the first and second neighbor separations in crystals, the potential energy and interatomic force are set to zero. It was found empirically [3.199] for calculations in the fcc metal Cu using the Molière potential (3.4.13), that the results were sensitive to changes in $p_c$ if it was less than the nearest neighbor separation, $a/2^{1/2}$, but were rather insensitive to changes in $p_c$ when it was larger than this. The value $p_c = 0.885\,a$ was adopted. This is sufficient to ensure that a particle at the center of an octahedral hole in the fcc lattice will interact with all the atoms in the surrounding unit cube (the most distant atoms are at the cube corners, $3^{1/2}\,a/2$ away). The minimum energy criterion was $E_m = 1.5$ eV. This was a compromise designed to reduce the computing time without introducing too much error, as judged by comparisons with the stable classical dynamics code COMENT [3.164–167]. Just as atoms are added to the moving set when their total energy exceeds $E_m$, they are dropped from this set when their energies fall below an energy $E_c$. In calculations on Cu with the machine code ADDES [3.199, 200], the cut-off criterion required the total energy of a particle to be less than 5 eV and its potential energy to be less than 3.5 eV for several time steps; when this criterion was met, the particle was dropped from the moving set. In addition, if the atom did not receive additional energy exceeding $E_m$ for several more time steps, and if it remained within $a/8^{1/2}$ of its initial lattice site, the atom was returned to that site. This last point illustrates the use of special modelling to emulate the long-time behavior of a stable classical dynamics program.

The use of ADDES to study the displacement threshold energy surface of Cu [3.199] provides an interesting illustration of what such programs can do. The Molière potential was used for this calculation with the screening length $a_{12} = 0.078$ Å. This value was selected so that the threshold, $E_d$, in the [001] direction agreed with the experimentally measured value, 20 eV [3.183, 184]. [The parameter (3.4.18) gave a value near 15 eV.] Of course, as was indicated in Sect. 3.5.1, the displacement threshold energy is dominated by the defect instability volume, and the absence of restoring forces in a quasistable model misses out just the required behavior. However, the missing feature can again be emulated in a simple way by assuming a spherical instability volume of radius $r_v$. If an interstitial atom is separated by a distance less than $r_v$ from a vacancy, the pair is recombined; otherwise it is counted as a stable Frenkel pair. The value $r_v = 1.5\ a$ was selected for the subject calculation. A sphere of this radius, centered on a lattice site in an fcc crystal, contains 54 other sites; in addition 24 sites lie just beyond its surface in $\langle 013 \rangle$ directions. Thus, a sphere of radius 1.5 $a$ is quite comparable to the instability volume of 74 sites found by stable dynamical methods [3.157]. When this model is used in ADDES to calculate the displacement threshold energy surface of Cu, satisfactory agreement is achieved for directions near $\langle 001 \rangle$ and $\langle 011 \rangle$ and the connecting $\{001\}$ plane. The deviations are greater for directions near $\langle 111 \rangle$ and $\{111\}$. The sense is that the calculated thresholds in such directions are greater than the experimental ones, up to about twice as large. Somewhat similar deviations were also found by the *Beelers* [3.160] using a stable dynamical model of fcc Fe. The origin of the discrepancy is presently unknown, but any of the following factors (among others) could be responsible in part: the form of the interatomic potential or its parameters; the neglect of the thermal motions of the lattice atoms; insufficient running time to achieve true stability in the stable model; use of a simple spherical instability volume in the quasistable model calculations. Nevertheless, the comparatively satisfactory results achieved with ADDES encourage the use of such quasistable model calculations in a variety of applications, as long as carefully designed features can be supplied to emulate the relevant long-time responses of stable models.

### 3.5.5 Collision Cascades in the Binary Collision Approximation

Although dynamical calculations can be accelerated significantly by the use of metastable or quasistable numerical models, they are nevertheless so time-consuming as to be limited to studies of collision cascades in small numbers at low kinetic energies. As has been indicated, such calculations are particularly well-adapted to the elucidation of detailed mechanisms. On the other hand, they are usually not capable of dealing effectively with the statistics of cascades, that is, with the probabilities with which the various mechanisms occur. Furthermore, most practical applications demand consideration of primary kinetic energies too high for dynamical studies. In these circumstances, the so-called *binary collision approximation* (BCA) may be used, in which the trajectories of the cascade particles are constructed as sequences of two-body encounters. The BCA may be considered either as an extreme version of the quasistable dynamical technique in which the particles are considered mainly in pairs, or as proceeding from the idea of collisions that is familiar in the kinetic theory of gases. It thus grows naturally out of the methods that have long been applied in theoretical treatments of collision cascades and energetic particle ranges in structureless (that is, gas-like) media. The earliest applications of computer simulation based on the BCA were to cascades [3.203] and particle ranges [3.204–206] in structureless media. Application of similar models to particle ranges in crystals [3.12, 13] were responsible for the discovery of the channeling effect. Cascade calculations in crystals were commenced independently by *Beeler* and his co-workers [3.207–212]. The early work in this area has been reviewed elsewhere [3.147, 148]. The present section describes the computer simulation of collision cascades using the BCA, including techniques for dealing with nearly simultaneous collisions and other special model features. This discussion is mainly concerned with the MARLOWE program [3.91, 110, 123, 213–221] which has been applied to a variety of radiation damage, sputtering, and other atomic collision phenomena in solids. No attempt will be made to review the extensive literature on BCA calculations at high energies, where applications have been mainly to channeling, surface scattering, and other aspects of ion crystallography. The references [3.66, 67, 90, 147, 148, 222–224] provide access to this material.

The BCA method has several features in common with a dynamical calculation based on a quasistable model. A particle is considered to be set in motion if it receives a kinetic energy in excess of a minimum amount $E_m$. Projectiles are followed, collision by collision, until their kinetic energies fall below an amount $E_c$. Collisions are evaluated for all encounters with impact parameters less than a value $p_c$. On the other hand, only interactions between projectiles and initially stationary target atoms are included. Since the projectile trajectories are considered to be sequences of two-body encounters, integration of the classical equations of motion reduces to evaluation of the two scattering integrals, (3.4.1, 2). As a result, calculations in the BCA are much faster than dynamical calculations and, because the number of atoms involved is severely limited, they also require much less storage capacity. These features enable BCA calculations to be extended to almost any energy and to be used to generate reasonably large ensembles of cascades. However, the degree of approximation involved may not be neglected. First, the isolation of the various two-body encounters from each other is a severe approximation, especially in crystalline media where lattice correlations lead to groups of more-or-less simultaneous collisions and at low energies where significant deflections can occur in large impact parameter collisions. Second, in the two-body encounters, it is assumed that the particles involved start and finish the collision far enough apart to be regarded as moving along their asymptotic trajectories in laboratory coordinates (see Fig. 3.4). This approximation allows the use of the infinite upper limits in the scattering integrals and influences the values of the impact parameter (angular momentum) and relative energy. Third, no explicit time variable occurs so that the concept of a time step in the dynamical codes is replaced by a scheme for ordering the various collisions.

Collision cascade calculations are often required at energies where the individual atomic encounters are significantly inelastic on account of electron excitation effects. Part of the inelasticity may be regarded as local, that is, as depending on the impact parameter in each collision. The remainder is nonlocal, intended to represent distant interactions between energetic ions and target conduction electrons. The theories of *Firsov* [3.225] and of *Lindhard* and *Scharff* [3.226] are used to supply the parameters of the models. The applicability of these theories to collision cascade problems has been reviewed elsewhere [3.227] (see also Chap. 2). Both theories are restricted to fairly low energies, but the limits are not normally approached very closely in MARLOWE calculations.

It is characteristic of the development of collision cascades in crystalline targets that the projectiles often encounter the target atoms in more-or-less symmetrical groups. An example is the motion of a particle down a crystal channel. Such situations are detected in MARLOWE and each group of nearly simultaneous collisions is then combined into a whole. In fcc crystals, up to four "simultaneous" target atoms may be present when the code is used with typical parameters. The scattering integrals are evaluated for each collision partner as if it alone were involved and the momentum of each target atom is found. The momentum of the scattered projectile is then obtained from the conservation law. Although the total linear momentum has been conserved, the total energy will be conserved only if there is a single target atom. In other cases, the kinetic energies of the scattered particles and the inelastic energy loss are all scaled to ensure energy conservation. Thus, the final result of an event involving several simultaneous target atoms is a set of energies which are exactly conserved and of momenta which are approximately conserved.

This approximation procedure is conceptually somewhat akin to the impulse approximation and, like the impulse approximation, is most accurate when the barycentric scattering angles in the individual collisions are small. This in turn implies that the approximation is best at large impact parameters and high energies. Numerical illustrations of this approximation were discussed in Sect. 3.4.3 and displayed in Figs. 3.11, 3.12 for the encounter of a projectile with a symmetric ring of neighbors. Comparisons of the MARLOWE approximation with the essentially exact treatment of this problem [3.122] show the former to increase in accuracy as the projectile energy is increased. The accuracy is also greater for $\langle 011 \rangle$ focusing rings than for the smaller $\langle 001 \rangle$ and $\langle 111 \rangle$ rings. This behavior agrees with expectations, but like all procedures which separate the ring and central encounters, it seriously underestimates the energy losses, especially along close-packed atomic rows.

The modelling of linear collision sequences may be improved in MARLOWE by requiring that each new lattice atom be required to surmount an energy barrier, $E_b$, before being added to the cascade. Each atom is in effect bound to its lattice site by this amount. Ignoring any dynamical effects of such binding, a target atom receiving the energy $T$ in a collision commences to move with kinetic energy $T - E_b$. As a result, the energy losses for perfectly focused sequences will be increased over those shown in Figs. 3.11, 3.12 by an amount $E_b$ per central collision. The value $E_b = 0.5$ eV for Cu is an appropriate choice as may be seen by comparing the MARLOWE and classical dynamical results in Fig. 3.11. A further improvement is that the transition from a mass-transporting replacement sequence to a non-mass-transporting focuson is recognized in MARLOWE by noting whether or not the projectile penetrates the focusing ring.

Dynamical calculations [3.120, 128, 150, 160, 161] have shown that the thermal displacements of lattice atoms are significant factors in linear collision sequences. Thermal effects can be included in BCA calculations rather easily, since the time for the development of a single collision cascade [3.194, 195] is typically less than the periods of the thermal motions. In this circumstance, the cascade may be regarded as developing in a lattice whose atoms, although initially at rest, are slightly displaced from their average positions. In MARLOWE, each cartesian component of the thermal displacement is taken from a Gaussian distribution with mean value zero and variance $u_1^2$ given by the *Debye* model [3.228]. The individual atomic displacements are completely uncorrelated. The possible importance of correlations among the thermal displacements of nearby atoms and a method for including them in BCA calculations have been discussed by *Jackson* and *Barrett* [3.229].

To permit BCA calculations to be performed for polycrystalline and amorphous media, MARLOWE provides the possibility of disordering the target crystal structure. This is done by constructing a random rotation matrix [3.230] using standard aleatory techniques. Three combined rotations produce a target in which the density of lattice points is preserved, but in which the directional correlations are destroyed. A *polycrystalline* medium is constructed by generating a random rotation matrix once only, at the beginning of each cascade. An *amorphous* medium is simulated by generating a new random rotation matrix at every collision. Linear collision sequences are still possible in the former, but their orientations with respect to fixed features such as the crystal surface are arbitrary and change from one cascade to another.

In BCA calculations, it is necessary to include special models to emulate the long-time behavior of the stable dynamical models. Such procedures include modelling of the surface binding in sputtering applications and modelling of the stability of vacancy-interstitial (Frenkel) pairs in radiation damage applications.

### 3.5.6 Validity of the Binary Collision Approximation

It is generally recognized that the BCA is a high energy approximation and that it becomes inaccurate at low energies, but a considerable variety of opinion has been expressed as to the energy range in which it may be applied. It must be noted that "binary collision" in the sense that it is used in the BCA is not equivalent to the sense in which it is used in surface scattering (see, for instance, [3.40, 202, 231–233]). The latter use refers to an isolated encounter between two atoms in which no other particles play any role at all. This leads, for instance, to certain specific energy values for scattered particles and deviations from these values are considered deviations from "binary collision values". Such deviations occur quite naturally in BCA model calculations of surface scattering [3.234] which construct trajectories from binary collision segments, but which are rarely dominated by a single two-body encounter. These differing uses of the terminology are apparently responsible for some confusion in the literature concerning BCA calculations.

At sufficiently low energies, the time of a single collision may be so long that particles other than the two collision partners become involved before the pair separates. It was shown in Fig. 3.5 that the quantity $x_2$ is always very small, that is,

**Table 3.14.** Energies at which projectile displacements in Cu and Au become comparable to lattice spacings

| | Projectile Kinetic Energy [eV] | |
|---|---|---|
| | Cu | Au |
| Born-Mayer Potential Parameters | (3.4.16) | (3.4.17) |
| $x_1 = t_{110}/2$ at | 17.6 | 26.3 |
| $x_1 = t_{100}/2$ at | 1.1 | 1.5 |
| Molière Potential Parameter | (3.4.18) | (3.4.19) |
| $x_1 = t_{110}/2$ at | 9.0 | 33.5 |
| $x_1 = t_{100}/2$ at | 0.7 | 2.4 |

**Table 3.15.** Energies of atoms moving at the velocity of sound in metal crystals

| Metal | Longitudinal wave velocity [$10^{13}$ Å/s] [3.235] | Atom kinetic energy [eV] |
|---|---|---|
| Al | 6.26 | 5.5 |
| Cu | 4.70 | 7.3 |
| Pt | 3.96 | 15.9 |
| Au | 3.24 | 10.7 |
| Fe | 5.85 | 9.9 |
| W | 5.46 | 28.4 |

target atoms move rather little before the two-body apsis is reached, so that there is never much concern with the possibility that the colliding atoms recoil into their surroundings while still close together. On the other hand, $x_1$ becomes comparable to interatomic spacings in crystals at low energies. Estimates are given for Cu and Au in Table 3.14. It is clear from this table that at energies near 10 eV, isolation of collisions from the surrounding crystal becomes very difficult since the significant parts of the collision are occuring over large distances.

Another way to estimate the energy at which many-body effects become important [3.110] is to consider the possibilities for interaction between a projectile and relatively distant parts of the crystal. Such interactions involve particle motions resembling phonons and are transmitted at the velocity of sound. Table 3.15 shows the kinetic energies which various atoms have when moving at the velocity of sound in metal crystals. At lower energies, it may be anticipated that strong many-body effects could occur. It is interesting that the kinetic energies derived in this way are about the same as those given above.

Several comparisons of the BCA with dynamical calculations have been reported. The earliest of these, and one which illustrates some of the difficulties of such comparisons very clearly, is the work of *Gay* and *Harrison* [3.236]. They studied the interaction between 25 eV to 10 keV Cu atoms and small Cu crystallites using techniques generally similar to those in Harrison's later

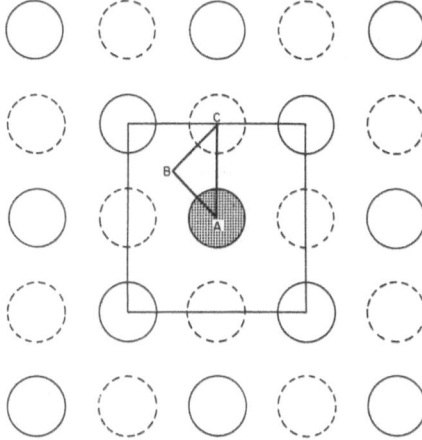

Fig. 3.15. The representative area on a {001} surface of an fcc crystal

sputtering work (see Sect. 3.5.3). The Cu projectile was normally incident upon a selected crystal face, near a particular particle designated the target. Figure 3.15 illustrates the geometry used for the fcc {100} surface. Calculations were performed for a number of impact points in the representative area, the triangle ABC. The calculation was continued until the kinetic energy of the target atom was a maximum. At this point various quantities were compared as evaluated by integrating the equations of motion for the whole crystallite of 35 to 63 atoms, plus the projectile, or, alternatively, by integrating the equations of motion of the target and projectile atoms only. Most of the calculations used the Born-Mayer potential with parameters (3.4.16). The target atom behaved rather similarly in both calculations, but the kinetic energy remaining to the projectile differed strongly in two parts of the impact triangle. The BCA kinetic energies were about twice the dynamical values for impacts near the point B and were very much greater near C, where the projectile in the dynamical case was nearly at rest. But these results are a consequence of the comparison made and cannot be considered as reflecting on the BCA in any significant way. When the projectile is directed at point C, it is brought to rest in the dynamical calculation by a central collision with an atom in the second layer of the crystallite, but this interaction was not considered in the BCA calculation, although it would of course be included in any meaningful BCA model. Similarly, the projectile incident at B makes a simultaneous encounter with two target atoms, only one of which is included in the BCA of [3.236], although both would be included by MARLOWE and also by the range programs [3.13] from which it descended. Thus, most of the shortcomings attributed to the BCA by *Gay* and *Harrison* are in fact shortcomings of their comparison procedure. Their conclusion that "all theoretical research below approximately 500 eV must consider a more complex model than the binary collision approximation" is without foundation in their work and must be rejected.

A comparison of the BCA and dynamical calculations has been made in the context of surface scattering by *Yurasova* [3.149]. She integrated the equations of motion for 50- to 500-eV Ar$^+$ ions incident on a Cu {001} surface, either at normal incidence or at 70° from the normal in a {110} plane. Interactions between target Cu atoms were described by a Born-Mayer potential with the parameters of (3.4.16). Interactions of the ion with the target atoms were described by a Born-Mayer potential for separations greater than 0.5 Å; for smaller separations a screened Coulomb potential of the *Thomas-Fermi* type [3.107] with the Firsov screening length, (3.3.12) [3.89], was used. Three models were studied: the integrations included all interactions (Ar–Cu and Cu–Cu); or, they included only the Ar–Cu interactions; or, they included interactions between the Ar ion and only the nearest Cu atom of the moment. The whole trajectory of the ion was evaluated. The first two models gave closely similar results. That is, the motion of the Ar ion was influenced very little by interactions of the Cu target atoms with each other. In view of what was said above, such influences would not be expected to be very significant at energies as high as 50 eV. It was also found that the reflected ion energies and directions in the binary encounter calculation agreed moderately well with the many-body results (within a few percent in energy and about half a degree in scattering angle). Larger effects were found for ions which penetrated the target, presumably because the simple binary encounter model used neglected nearly simultaneous collisions.

Several comparisons have been made between BCA and stable dynamical model calculations. Unfortunately, one of these [3.151] considered only the most obvious binary encounter possibility and neglected the weak distant collisions that would be included in MARLOWE. The other comparisons [3.199, 200], however, are part of a developing effort to make detailed comparisons of various collision cascade simulation models. In this work, the stable classical dynamics code COMENT [3.164–167], the quasistable classical dynamics code ADDES [3.199], and the BCA code MARLOWE (see Sect. 3.5.5) were compared using the Molière potential with $a_{12} = 0.078$ Å to describe interactions in a Cu crystal. The comparisons to date are limited and no attempt has been made to optimize the choices of the many model parameters. When MARLOWE was run with $E_m = E_c = 5$ eV and $E_b = 0$, very long collision sequences occurred along $\langle 110 \rangle$, a result of the low energy losses in this model. To eliminate these, further calculations were made with $E_m = E_c = 25$ eV. Comparisons of ADDES and MARLOWE for 0.25 to 1 keV primary recoils then gave similar numbers of Frenkel pairs and similar distributions of pair separations. Further comparisons of these three programs are in progress.

### 3.5.7 Monocrystal Sputtering in the Binary Collision Approximation

Little work has been published on the computer simulation of monocrystal sputtering using the BCA. *Schlaug* [3.61] made a study of the sputtering of Cu {001} surfaces by normally incident Cu, Ar, and Cs atoms in the energy range up

**Table 3.16.** Calculated sputtering yields for normally incident atoms on Cu {001} [3.61]. Potential parameters refer to incident atom – Cu interaction

(a) 1 keV Ar ($Y_{exp}$ = 3.4 [3.70], 3.0 [3.68], 2.4 [3.94])

| $A_{BM}$ [keV] | 13.5 | | 22.5 | | 30.0 | |
|---|---|---|---|---|---|---|
| $a_{BM}$ [Å] | 0.2118 | | 0.1966 | | 0.1382 | |
| $U_s$ [eV] | 3.52 | 7.00 | 3.52 | 7.00 | 3.52 | 7.00 |
| $Y_{calc}$ | 6.84 | 3.68 | 7.24 | 3.88 | 3.52 | 1.96 |

(b) 1 keV Cs ($Y_{exp}$ = 2.1 [3.138])

| $A_{BM}$ [keV] | 50.0 | | 22.5 | | 100.0 | |
|---|---|---|---|---|---|---|
| $a_{BM}$ [Å] | 0.1751 | | 0.1966 | | 0.1111 | |
| $U_s$ [eV] | 3.52 | 7.00 | 3.52 | 7.00 | 3.52 | 7.00 |
| $Y_{calc}$ | 5.76 | 2.92 | 5.24 | 2.44 | 2.72 | 1.60 |

to 10 keV. His calculation used a simple BCA (that is, with no allowance for simultaneous collisions) for projectiles with energies greater than 300 eV. Below this energy, the quasistable method described in Sect. 3.5.4 was used. No inelastic losses were included. Particles passing through the target surface were considered to be sputtered if their energies exceeded the surface binding energy $U_s$, irrespective of their directions of motion. Values of $U_s$ ranging from about 2/3 $U_0$ to about 2$U_0$ were used in exploratory calculations. The interactions between Cu atoms were described by a Born-Mayer potential with the parameters of (3.4.16). For the incident Ar and Cs atoms, Born-Mayer potentials were also used. Initially, the screening length was scaled from that for Cu–Cu interactions using (3.3.10) and the energy parameter was found by making the Born-Mayer potential tangent to a Thomas-Fermi potential as evaluated by *Firsov* [3.89]. When the results proved unsatisfactory, several other parameter sets were also tried. Some of *Schlaug*'s results are collected in Table 3.16. For both Ar and Cs irradiations, agreement between the calculations and experiment was achieved only through use of either a very large surface binding energy or of a rather hard interaction potential. *Schlaug* made the latter choice for a series of calculations at other energies.

Similar results were obtained for Ar on Cu {001} using an early version of MARLOWE [3.237]. In this calculation, the Molière potential was used with the parameter (3.4.18) for Cu–Cu collisions and a value $a_{12}$ = 0.08916 Å for Ar–Cu collisions. The latter value is the harmonic mean of the Cu–Cu value and the Ar–Ar value given by *Firsov* [3.89], (3.3.12). Inelastic energy losses were included using a local model and the isotropic surface binding model was used. In agreement with *Schlaug*'s work, it was found necessary to use a value of $U_s$ near 2$U_0$ to obtain agreement with the experimental data from 1 to 5 keV. However, for structureless media, the theoretical yield for the isotropic binding model is about twice that for the planar binding model, other things being equal. Thus, the need to use a value of $U_s$ near 2$U_0$ can be attributed in both calculations to use of

**Table 3.17.** Model parameters used for MARLOWE investigations of the sputtering of Au {001} surfaces by 700 eV Xe

| | |
|---|---|
| Lattice constant $a$ | 4.0783 Å [3.117] |
| Debye temperature $\theta_D$ | 180 K [3.239] |
| Minimum energy for motion $E_m$ | 4.5 eV |
| Particle cut-off energy $E_c$ | 4.0 eV |
| Lattice binding energy $E_b$ | 0.5 eV |
| Maximum impact parameter $p_c/a$ | 0.62 |
| Surface binding energy $U_s$ | 4.0 eV |
| Molière potential parameters $a_{12}$ | |
| Xe – Au | 0.0731 Å |
| Au – Au | 0.0750 Å |

the isotropic model. Had the planar model been used, both calculations would have produced resonable agreement with experiment using a value of $U_s$ near $U_0$ as expected from the discussion in Sect. 3.2.3. Since the planar binding model is the preferred one, there seems to be no need to adopt the hard potentials that *Schlaug* used.

Sputtering problems have been investigated sporadically with MARLOWE, partly in the course of developing improvements in the speed of the program and in the details of the underlying model. A brief description of some of the results obtained with the most recent version of this program will be given here, using the example of 700 eV Xe atoms normally incident on {001} Au monocrystal targets [3.238]. The model parameters used are listed in Table 3.17. The only parameter requiring comment is the surface binding energy, $U_s$. Following [3.49], the surface binding energy at the Au {001} surface would be expected to be about 30 % greater than the cohesive energy or about 4.94 eV. The value of $U_s$ in the model calculation would then be this quantity reduced by the amount of the local binding energy $E_b$. The value used in the calculation may therefore be somewhat too small and the resulting sputtering yields somewhat too high. In view of the general uncertainty concerning surface binding, however, as discussed in Sect. 3.2.3, more precision is probably not possible. The incident Xe atoms were distributed uniformly over the entire unit surface mesh (see Fig. 3.15).

Figure 3.16 shows the dependence on target thickness of the sputtering yields calculated for a static lattice using the planar binding model. The reflection (ordinary) yield reaches a limiting value for targets thicker than about $2a$ (5 atomic layers, 8.16 Å). The transmission yield, on the other hand, reaches a maximum at about the target thickness where the reflection yield becomes constant and then decreases. The upper portion of the figure shows the distribution of Xe penetration depths. About one atom in five hundred penetrates beyond $4a$ and about one in twenty thousand beyond $6a$. Some 22 % of the Xe atoms are reflected; these rarely penetrate beyond the third or fourth layer of atoms. Thus, the reflection sputtering yield and the reflection of the primary particles are determined in the same near-surface region of the crystal.

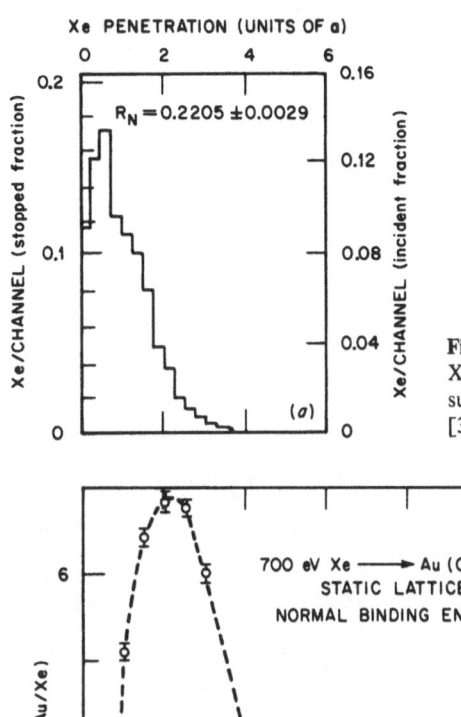

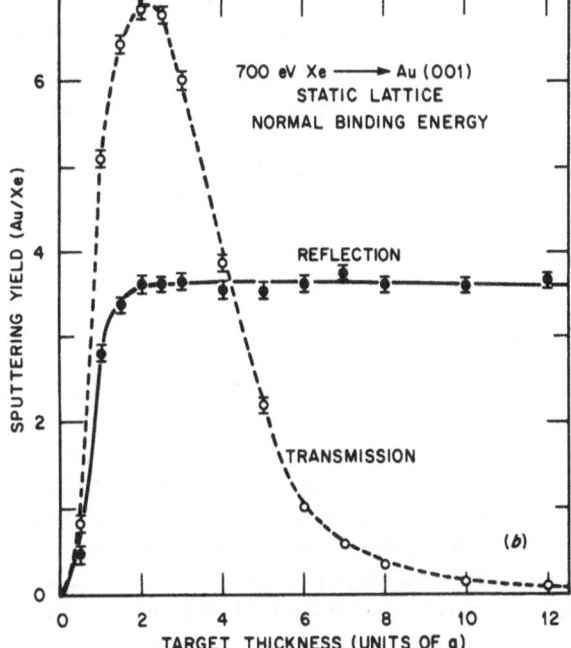

**Fig. 3.16.** Sputtering yields calculated for 700 eV Xe atoms normally incident on an Au {001} surface, using the BCA program MARLOWE [3.238]

Table 3.18 summarizes the calculated depths of origin of the ejected Au atoms and the frequency of ejection by linear collision sequences. Less than 1 % of the ejected atoms originate beyond the second layer of the target and less than about one atom in $10^5$ originates beyond the depth at which the reflection yield reaches its limiting value. This result is in agreement with the conclusions of the transport theory of the sputtering of structureless media [3.50] (see also Chap. 2). The table also shows that, although linear collision sequences do contribute to the ejection,

**Table 3.18.** Some properties of atoms ejected from Au {001} surfaces by normally incident 700 eV Xe atoms [3.238]

(a) Depths of Origin of Sputtered Atoms

| Atom Layer | 1 | 2 | 3 | 4 | 5 | 6 |
|---|---|---|---|---|---|---|
| Yield (Au/Xe) | 3.291 | 0.285 | 0.018 | 0.005 | 0.006 | 0.001 |
| Fraction of emissions | 0.9127 | 0.07912 | 0.0048 | 0.0014 | 0.0018 | 0.0001 |

(b) Ejection of Atoms by Linear Collision Sequences

| Sequence length | 0 | 1 | 2 | 3 | 4 | 5 | >5 |
|---|---|---|---|---|---|---|---|
| Yield (Au/Xe) | 2.459 | 0.837 | 0.219 | 0.054 | 0.026 | 0.006 | 0.005 |
| Fraction of emissions | 0.6821 | 0.2321 | 0.0608 | 0.0150 | 0.0071 | 0.0016 | 0.0013 |

**Table 3.19.** Sputtering yields for 700 eV Xe normally incident on Au targets [3.238]

| Target | Inelastic Loss | Temperature | Calculated Sputtering Yield (Au/Xe) | |
|---|---|---|---|---|
| | | | Isotropic Binding | Planar Binding |
| Polycrystalline | yes | static | $5.161 \pm 0.059$ | $3.221 \pm 0.044$ |
| (001) | yes | static | $4.809 \pm 0.015$ | $3.606 \pm 0.013$ |
| (001) | yes | 120 K | $4.845 \pm 0.027$ | $3.679 \pm 0.023$ |
| (001) | no | 120 K | $5.093 \pm 0.076$ | $3.661 \pm 0.065$ |

Experimental yields for $Xe^+$ ions on polycrystalline Au [3.240]:

| Ion Energy (eV) | 400 | 500 | 600 | 700 |
|---|---|---|---|---|
| Observed Yield (Au/Xe) | $2.5_2$ | $3.2_2$ | $3.4_1$ | $3.7_0{}^a$ |

[a] Extrapolated.

they do so with only a small likelihood. Two-thirds of the atoms are ejected by events not involving any replacement and nearly a quarter are ejected in a single isolated replacement. Nevertheless, linear collision sequences do occur and some of these involve as many as ten collisions and originate about as far within the target as the incident atoms penetrate. The low probability of these events is probably responsible for their absence in other calculations (especially [3.20]). Although no attempt has been made to isolate particular ejection mechanisms in MARLOWE sputtering calculations, they must resemble *Harrison*'s near-surface mechanisms [3.20, 180]. Again in agreement with *Harrison*'s work, many linear collision sequences occurred, mainly along $\langle 110 \rangle$, but with lesser numbers along $\langle 001 \rangle$ and a few along $\langle 111 \rangle$. Such sequences were mostly directed inward or parallel to the target surface. Transmission sputtering, except in the thinnest targets, was dominated by $\langle 110 \rangle$ and $\langle 001 \rangle$ sequences. For targets thicker than about $12a$ these were the only modes of ejection.

Calculated sputtering yields for several different conditions are collected in Table 3.19, which also lists the only available experimental data [3.240]. A considerable difference (30 to 60 %) will be noted between the calculations based on the different surface binding models. Since the discussion in Sect. 3.2.3

indicated that the planar binding model was to be preferred, it is gratifying that this model also gives the better agreement with the experimental data (see also [3.220]). While adjustment of some of the model parameters could improve this agreement, the present accuracy seems sufficient until more calculations and more experiments are available. The insensitivity of the results to temperature supports the conclusion that linear collision sequences are not a major contributor to sputtering under the conditions of these calculations. It should be noted, however, that at 120 K the rms thermal displacement of Au atoms from their lattice sites is little more than the zero-point value. The effect of higher temperatures has not been investigated.

The sputtering calculations with the BCA program MARLOWE that have been described here show considerable promise. They are nevertheless somewhat preliminary and need to be extended to systems where more experimental observations are available before too much is made of them. Furthermore, in spite of the comparative speed of BCA calculations, they still require rather large amounts of computing time. On an IBM System/360, Model 195 machine, a calculation for 700 eV Xe atoms on Au 001 # requires about ten minutes per thousand particles and this time increases roughly proportionally to the incident energy.

## 3.6 Conclusions and Prospects

This chapter has reviewed several theoretical aspects of the sputtering of single crystal targets, with results that are summarized briefly below.

i) The crystallographic orientation dependence of the sputtering yield results mainly from the channeling of the incident ions. However, a satisfactory quantitative theory of such effects cannot be based on the theory [3.82] of axial channeling because of the low energies of main interest in sputtering experiments. The required isolation of individual rows of atoms fails in these circumstances, leading to important planar effects and to a general inability to isolate the behavior of the incident particles from the collision cascades which they produce. Unfortunately, the available experiments are not sufficient to distinguish axial from planar effects in many circumstances.

ii) The structure observed in the angular distributions of the sputtered material cannot be accounted for by linear collision sequences, whether or not these exhibit focusing. Although such sequences are an important feature of collision cascade development, their contribution to reflection sputtering is a minor one in most circumstances. No satisfactory theory of ejection patterns has emerged which is based on one-dimensional abstractions. Experimentally, there are rather few investigations available which report such quantitative data as accurate *Wehner* spot locations and ejection pattern intensity contours. Such data are required for the testing of theoretical models.

iii) Computer simulation methods show considerable promise of being able to account for most features of experimental sputtering observations, but only a beginning has been made so far. Development of these methods is significantly hampered by the lack of accurate models of crystal surfaces, both with respect to the locations of the atoms and with respect to surface binding energetics. Such methods require the largest and fastest computing machines to produce statistically reliable quantitative results, but somewhat qualitative information can be developed more easily.

Several topics have been omitted from this chapter, among which the following may be mentioned.

i) Sputtering threshold energies have not been discussed, except for a brief mention in connection with electron sputtering (see Sect. 3.5.2). No satisfactory theory of such thresholds is available. The work of *Hotston* [3.241] indicates a simple connection between the surface binding energy and the sputtering threshold energy which is adequate for very light and for very heavy incident particles, but for the most important situation, where the primary and target masses are comparable, the formula is not of much value. Experimental observations [3.242] indicate some rather interesting crystallographic effects on threshold energies.

ii) The dependence of sputtered atom energy spectra on the direction of ejection has not been investigated theoretically, although a considerable body of experimental data is available (see [3.1a] Chap. 2). Such investigations must be made without strong assumptions concerning ejection mechanisms. Computer simulation models can be used for this work in principle, but such calculations may be prohibitively expensive, even in the BCA.

iii) The effects of temperature on monocrystal sputtering have been investigated only a little. Such effects can be studied by computer simulation, at least within the framework of the Debye model, without undue difficulty. Rather few experimental results are available for testing such calculations, however.

iv) The computer simulation of the sputtering of polycrystalline and amorphous media can be carried out by methods similar to those discussed here. This topic lies outside the scope of this chapter. See, however, [3.220] and other references cited there.

*Acknowledgement.* In preparing this review, I have received indispensable assistance from many of my colleagues. I am particularly grateful to H. L. Davis and D. M. Zehner for their patient instruction in numerous aspects of modern surface science. It is a pleasure to acknowledge many valuable discussions with J. H. Barrett and O. S. Oen, especially concerning ion channeling and related matters. I am indebted to them, as well as to J. B. Roberto and F. W. Young, Jr., for reading the manuscript and making numerous useful suggestions about it. I also owe a debt of gratitude to Ms. Velma Hendrix for her skill and patience in typing many versions of this manuscript.

# References

3.1     J. Stark, G. Wendt: Ann. Phys. (Leipzig) **38**, 921–940 (1912)
3.1a    R. Behrisch (ed.): *Sputtering by Particle Bombardment*, III, Topics in Applied Physics (Springer, Berlin, Heidelberg, New York 1981) (to be published)

3.2    J.Stark, G.Wendt: Ann. Phys. (Leipzig) **38**, 941–957 (1912)

3.3    J.Stark: Phys. Z. **13**, 973–977 (1912)

3.4    G.K.Wehner: J. Appl. Phys. **26**, 1056–1057 (1955)

3.5    G.K.Wehner: Phys. Rev. **102**, 690–704 (1956)

3.6    R.H.Silsbee: J. Appl. Phys. **28**, 1246–1250 (1957)

3.7    P.K.Rol, J.M.Fluit, F.P.Viehbock, M.de Jong: In *Proc. 4th Intern. Conf. Ionization Phenomena in Gases*, ed. by N.R.Nilsson (North-Holland, Amsterdam 1960) pp. 257–279

3.8    O.Almén, G.Bruce: Nucl. Instrum. Methods **11**, 257–278 (1961)

3.9    V.A.Molchanov, V.G.Tel'kovski, V.M.Chicherov: Doklady Akad. Nauk S.S.S.R. **137**, 158–159 (1961) [Engl. transl.: Sov. Phys.-Doklady **6**, 222–223 (1961)]

3.10   P.K.Rol, J.M.Fluit, J.Kistemaker: Physica **26**, 1009–1011 (1960)

3.11   M.T.Robinson: Appl. Phys. Lett. **1**, 49–51 (1962)

3.12   M.T.Robinson, O.S.Oen: Appl. Phys. Lett. **2**, 30–32 (1963)

3.13   M.T.Robinson, O.S.Oen: Phys. Rev. **132**, 2385–2398 (1963)

3.14   G.R.Piercy, F.Brown, J.A.Davies, M.McCargo: Phys. Rev. Lett. **10**, 399–400 (1963)

3.15   H.Lutz, R.Sizmann: Phys. Lett. **5**, 113–114 (1963)

3.16   R.S.Nelson, M.W.Thompson: Philos. Mag. **8**, 1677–1689 (1963)

3.17   C.Lehmann, P.Sigmund: Phys. Stat. Sol. **16**, 507–511 (1966)

3.18   C.H.Weijsenfeld: "Yield, Energy, and Angular Distribution of Sputtered Atoms"; Thesis, Utrecht, (1966); Philips Research Reports Supplements 1967, No. 2

3.19   J.B.Gibson, A.N.Goland, M.Milgram, G.H.Vineyard: Phys. Rev. **120**, 1229–1253 (1960)

3.20   D.E.Harrison, Jr., N.S.Levy, J.P.Johnson, III, H.M.Effron: J. Appl. Phys. **39**, 3742–3761 (1968)

3.21   E.A.Wood: J. Appl. Phys. **35**, 1306–1312 (1964)

3.22   J.A.Strozier, Jr., D.W.Jepsen, F.Jona: In *Surface Physics of Materials*, Vol. 1, ed. by J.M.Blakeley (Academic, New York 1975) pp. 1–77

3.23   G.A.Somorjai: In *Treatise on Solid State Chemistry*, Volume 6A, Surfaces I, ed. by N.B.Hannay (Plenum, New York 1976) pp. 1–55

3.24   E.G.McRae, H.D.Hagstrum: In *Treatise on Solid State Chemistry*, Vol. 6A, Surfaces I, ed. by N.B.Hannay (Plenum, New York 1976)

3.25   T.Rhodin, D.Adams: In *Electronic Structure and Reactivity of Metal Surfaces*, ed. by E.G.Derouane, A.A.Lucas (Plenum, New York 1976) pp. 1–34

3.26   G.E.Rhead: In *Electronic Structure and Reactivity of Metal Surfaces*, ed. by E.G.Derouane, A.A.Lucas (Plenum, New York 1976) pp. 229–266

3.27   S.Andersson: In *Electronic Structure and Reactivity of Metal Surfaces*, ed. by E.G.Derouane, A.A.Lucas (Plenum, New York 1976) pp. 289–339

3.28   F.Jona: J. Phys. C**11**, 4271–4306 (1978)

3.29   D.M.Zehner, B.R.Appleton, T.S.Noggle, J.W.Miller, J.H.Barrett, L.H.Jenkins, O.E.Schow, III: J. Vac. Sci. Technol. **12**, 454–457 (1975)

3.30   P.Wynnblatt, N.A.Gjostein: Surf. Sci. **12**, 109–127 (1968)

3.31   D.P.Jackson: Can. J. Phys. **49**, 2093–2097 (1971)

3.32   L.A.Girifalco, V.G.Weizer: Phys. Rev. **114**, 687–690 (1959)

3.33   M.W.Finnis, V.Heine: J. Phys. F**4**, L37–L41 (1974)

3.34   R.A.Johnson, P.J.White: Phys. Rev. B**13**, 5293–5302 (1976)

3.35   M.G.Lagally: In *Surface Physics of Materials*, Vol. 2, ed. by J.M.Blakeley (Academic, New York 1975) pp. 419–473

3.36   J.F.Nicholas: *An Atlas of Models of Crystal Surfaces* (Gordon and Breach, New York, 1965)

3.37   H.E.Farnsworth, R.E.Schlier, T.H.George, R.M.Burger: J. Appl. Phys. **26**, 252–253 (1955)

3.38   H.E.Farnsworth, K.Hayek: Surf. Sci. **8**, 35–56 (1967)

3.39   J.J.Bellina, H.E.Farnsworth: J. Vac. Sci. Technol. **9**, 616–619 (1972)

3.40   W.Heiland, E.Taglauer: Radiat. Eff. **19**, 1–6 (1973)

3.41   A.Chutjian: Phys. Lett. **24**A, 615–616 (1967)

3.42   P.J.Estrup, E.G.McRae: Surf. Sci. **25**, 1–52 (1971)

3.43   D.M.Zehner: Oak Ridge National Laboratory (unpublished results)

3.44   J.W.Mayer, L.Eriksson, J.A.Davies: *Ion Implantation in Semiconductors* (Academic, New York 1970) pp. 76–97

3.45   G.S.Anderson, G.K.Wehner: Surf. Sci. **2**, 367–375 (1964)

3.46   G.S.Anderson: J. Appl. Phys. **37**, 2838–2840 (1966)

3.47   R.Hultgren, P.D.Desai, D.T.Hawkins, M.Gleiser, K.K.Kelley, D.D.Wagman: *Selected Values of the Thermodynamic Properties of the Elements* (American Society for Metals, Metals Park, Ohio, USA, 1973)

3.48   G.M.Rosenblatt: In *Treatise on Solid State Chemistry*, Volume 6A, Surfaces I, ed. by N.B.Hannay (Plenum, New York 1976) pp. 165–240

3.49   D.P.Jackson: Radiat. Eff. **18**, 185–189 (1973)

3.50   P.Sigmund: Phys. Rev. **184**, 383–416 (1969)

3.51   J.A.Appelbaum: In *Surface Physics of Materials*, Vol. 1, ed. by J.M.Blakely (Academic, New York and London, 1975) pp. 79–119

3.52   G.Allan: In *Electronic Structure and Reactivity of Metal Surfaces*, ed. by E.G.Derouane, A.A.Lucas (Plenum, New York 1976) pp. 45–79

3.53   N.D.Lang: In *Electronic Structure and Reactivity of Metal Surfaces*, ed. by E.G.Derouane and A.A.Lucas (Plenum, New York 1976) pp. 80–111

3.54   H.J.Leamy, G.H.Gilmer, K.A.Jackson: In *Surface Physics of Materials*, ed. by J.M.Blakely (Academic, New York 1975) Vol. 1, pp. 121–188

3.55   M.W.Thompson: Philos. Mag. **18**, 377–414 (1968)

3.56   J.Politiek, J.Kistemaker: Radiat. Eff. **2**, 129–131 (1969)

3.57   H.Oechsner: Z. Phys. **233**, 433–451 (1970)

3.58   E.B.Henschke: Phys. Rev. **106**, 737–753 (1957)

3.59   E.Langberg: Phys. Rev. **111**, 91–97 (1958)

3.60   D.E.Harrison, G.D.Magnuson: Phys. Rev. **122**, 1421–1430 (1961)

3.61   R.N.Schlaug: "Sputtering Calculations from a Realistic Model", Thesis, University of California, Berkeley 1965 (University Microfilms, Inc., Ann Arbor, Michigan 1966)

3.62   D.P.Jackson: Can. J. Phys. **53**, 1513–1523 (1975)

3.63   D.Cherns, M.W.Finnis, M.D.Mathews: Philos. Mag. **35**, 693–714 (1977)

3.64   D.E.Harrison, Jr., P.W.Kelly, B.J.Garrison, N.Winograd: Surf. Sci. **76**, 311–322 (1978)

3.65   S.Datz, C.Erginsoy, G.Leibfried, H.O.Lutz: Ann. Rev. Nucl. Sci. **17**, 129–188 (1967)

3.66   D.V.Morgan (ed.): *Channeling: Theory, Observations, and Applications* (John Wiley & Sons, London 1973)

3.67   D.S.Gemmel: Rev. Mod. Phys. **46**, 129–227 (1974)

3.68   A.L.Southern, W.R.Willis, M.T.Robinson: J. Appl. Phys. **34**, 153–163 (1963)

3.69   J.M.Fluit, P.K.Rol, J.Kistemaker: J. Appl. Phys. **34**, 690–691 (1963)

3.70   G.D.Magnuson, C.E.Carlston: J. Appl. Phys. **34**, 3267–3273 (1963)

3.71   D.D.Odintsov: Fiz. Tverd. Tela **5**, 1114–1116 (1963) [Engl. transl.: Sov. Phys.-Solid State **5**, 813–815 (1963)]

3.72   Yu.V.Martynenko: Fiz. Tverd. Tela **6**, 2003–2009 (1964) [Engl. transl.: Sov. Phys.-Solid State **6**, 1581–1585 (1965)]

3.73   H.P.Smith, Jr.: J. Appl. Phys. **35**, 2522–2524 (1964)

3.74   D.Onderdelinden: Appl. Phys. Lett. **8**, 189–190 (1966)

3.75   D.Onderdelinden, F.W.Saris, P.K.Rol: In *Proc. 7th Intern. Conf. Phenomena in Ionized Gases*, Vol. 1 (Gradevinska Knjiga, Beograd 1966) pp. 157–161

3.76   D.Onderdelinden: *Proc. Conf. Applications of Ion Beams to Semiconductor Technology* Grenoble, ed. by P.Glotin (Editions Orphrys, Gap, France 1967) pp. 389–395

3.77   D.Onderdelinden: Can. J. Phys. **46**, 739–745 (1968)

3.78   D.Onderdelinden: "Sputtering of F.C.C. Metals"; Thesis, Leiden (1968)

3.79   J.J.P.Elich, H.E.Roosendaal: Radiat. Eff. **10**, 175–184 (1971)

3.80   J.J.P.Elich, H.E.Roosendaal, D.Onderdelinden: Radiat. Eff. **14**, 93–100 (1972)

3.81   J.Lindhard: Phys. Lett. **12**, 126–128 (1964)

3.82   J.Lindhard: K. Dan. Vidensk. Selsk. Mat. Fys. Medd. **34**, No. 14 (1965)

3.83  J.D.H.Donnay, G.Donnay: In *International Tables for X-Ray Crystallography*, ed. by J.S.Kaspar, K.Lonsdale (International Union of Crystallography, Kynoch Press, Birmingham, U.K. 1959) Vol. 2, pp. 99–158

3.84  A.L.Patterson, J.S.Kaspar: In *International Tables for X-ray Crystallography*, ed. by J.S.Kaspar, K.Lonsdale (International Union of Crystallography, Kynoch Press, Birmingham, U.K. 1959) Vol. 2, pp. 342–354

3.85  J.F.Nicholas: Acta Crystallogr. **21**, 880–881 (1966)

3.86  C.Lehmann, G.Leibfried: Z. Phys. **172**, 465–487 (1963)

3.87  C.Lehmann, G.Leibfried: J. Appl. Phys. **34**, 2821–2836 (1963)

3.88  J.Lindhard, V.Nielsen, M.Scharff: K. Dan. Vidensk. Selsk. Mat. Fys. Medd. **36**, No. 10 (1968)

3.89  O.B.Firsov: Zh. Eksp. Teor. Fiz. **33**, 696–699 (1957) [Engl. transl.: Sov. Phys.-JETP **6**, 534–537 (1958)]

3.90  J.H.Barrett: Phys. Rev. B**3**, 1527–1547 (1971)

3.91  M.Hou, M.T.Robinson: Appl. Phys. **17**, 371–375 (1978)

3.92  M.Balarin, V.A.Molchanov, V.G.Tel'kovskii: Dokl. Akad. Nauk S.S.S.R. **147**, 331–333 (1962) [Engl. transl.: Sov. Phys.-Doklady **7**, 1005–1007 (1963)]

3.93  M.T.Robinson, A.L.Southern: J. Appl. Phys. **38**, 2696–2973 (1967)

3.94  T.W.Snouse, L.C.Haughney: J. Appl. Phys. **37**, 700–704 (1966)

3.95  M.T.Robinson, A.L.Southern: J. Appl. Phys. **39**, 3563–3475 (1968)

3.96  J.H.Barrett: Phys. Rev. Lett. **31**, 1542–1545 (1973)

3.97  G.J.Ogilvie, J.V.Sanders, A.A.Thomson: J. Phys. Chem. Solids **24**, 247–259 (1963)

3.98  E.J.Zdanuk, S.P.Wolsky: J. Appl. Phys. **36**, 1683–1687 (1965)

3.99  T.H.Blewitt, M.A.Kirk, T.L.Scott: In *Fundamental Aspects of Radiation Damage in Metals*, Vol. 1, ed. by M.T.Robinson, F.W.Young, Jr. (U.S. Energy Research and Development Administration, Oak Ridge, Tenn. 1975) pp. 156–170

3.100  M.A.Kirk, T.H.Blewitt, T.L.Scott: Phys. Rev. B**15**, 2914–2922 (1977)

3.101  D.N.Seidman: In *Radiation Damage in Metals*, ed. by N.L.Peterson, S.D.Harkness (Amer. Soc. for Metals, Metals Park, Ohio 1976) pp. 28–57

3.102  K.H.Ecker: Radiat. Eff. **23**, 171–180 (1974)

3.103  G.Ayrault, R.S.Averback, D.N.Seidman: Scr. Metall. **12**, 119–123 (1978)

3.104  H.Goldstein: *Classical Mechanics* (Addison-Wesley, Reading, Mass. 1959) pp. 58–89

3.105  M.Born, J.E.Mayer: Z. Phys. **75**, 1–18 (1932)

3.106  G.Molière: Z. Naturforsch. **2a**, 133–145 (1947)

3.107  P.Gombás: In *Handbuch der Physik*, ed. by S.Flügge (Springer-Verlag, Berlin, 1956), Vol. 36, pp. 109–231

3.108  H.H.Andersen, P.Sigmund: Nucl. Instrum. Methods **38**, 238–240 (1965)

3.109  H.H.Andersen, P.Sigmund: "On the Determination of Interatomic Potentials in Metals by Electron Irradiation Experiments", Danish Atomic Energy Commission Report Riso-103 (1965)

3.110  M.T.Robinson, I.M.Torrens: Phys. Rev. B**9**, 5008–5024 (1974)

3.111  D.J.O'Conner, R.J.MacDonald: Radiat. Eff. **34**, 247–250 (1977)

3.112  M.T.Robinson: "Tables of Classical Scattering Integrals", U.S. Atomic Energy Commission Report ORNL-4556 (1970)

3.113  G.Leibfried: *Bestrahlungseffekte in Festkörpern* (G.B.Teubner, Stuttgart 1965) pp. 46–52

3.114  G.Leibfried, O.S.Oen: J. Appl. Phys. **33**, 2257–2262 (1962)

3.115  C.Lehmann, M.T.Robinson: Phys. Rev. **134**, A37–A44 (1964)

3.116  G.Leibfried: J. Appl. Phys. **30**, 1388–1396 (1959) and reference 3.113, pp. 214–244. There is an error in Eq. (7.4) of the latter which is corrected in Eq. (3.4.25) of the present text

3.117  W.Hume-Rothery, K.Lonsdale: In *International Tables for X-Ray Crystallography*, ed. by C.H.MacGillavry, G.D.Rieck (Intern. Union of Crystallography, Kynoch Press, Birmingham, U.K. 1968) Vol. 3, pp. 277–285

3.118  R.S.Nelson, M.W.Thompson: Proc. Roy. Soc. (London) A**259**, 458–470 (1961)

3.119  C.Erginsoy, G.H.Vineyard, A.Englert: Phys. Rev. **133**, A595–A606 (1964)

3.120  V.M.Agranovich, V.V.Kirsanov: Fiz. Tverd. Tela **12**, 2671–2682 (1970) [Engl. transl.: Sov. Phys.-Solid State **12**, 2147–2155 (1971)]

3.121 D.K.Holmes, M.T.Robinson: Solid State Division Ann. Prog. Report Dec. 31, 1975, U.S. Energy Research and Development Administration Report ORNL-5135 (1976) pp. 16–18

3.122 H.H.Andersen, P.Sigmund: K. Dan. Vidensk. Selsk. Mat. Fys. Medd. **34**, No. 15 (1966)

3.123 M.Hou, M.T.Robinson: Nucl. Instrum. Methods **132**, 641–645 (1976)

3.124 R.S.Nelson, M.W.Thompson, H.Montgomery: Philos. Mag. **7**, 1385–1405 (1962)

3.125 J.B.Sanders, J.M.Fluit: Physica **30**, 129–143 (1964)

3.126 T.S.Pugacheva: Fiz. Tverd. Tela **9**, 102–105 (1967) [Engl. transl.: Sov. Phys.-Solid State **9**, 75–77 (1967)]

3.127 A.Tenenbaum: Phys. Lett. **63**A, 155–157 (1977)

3.128 A.Tenenbaum: Philos. Mag. A**37**, 731–748 (1978)

3.129 R. von Jan, R.S.Nelson: Philos. Mag. **17**, 1017–1032 (1968)

3.130 M.T.Robinson: J. Appl. Phys. **40**, 4982–4983 (1969)

3.131 K.B.Winterbon: *Ion Implantation Range and Energy Deposition Distributions*, Vol. 2, Low Incident Ion Energies (IFI/Plenum, New York 1975)

3.132 R.J.MacDonald: Philos. Mag. **21**, 519–553 (1970)

3.133 B.Perovic: *Proc. 5th Intern. Conf. Ionization Phenomena in Gases*, Vol. 2, ed. by H.Maecker (North-Holland, Amsterdam 1962) pp. 1172–1178

3.134 W.O.Hofer: Radiat. Eff. **19**, 263–270 (1973)

3.135 O.S.Oen: Phys. Lett. **19**, 358–359 (1965)

3.136 O.S.Oen: *Proc. 7th Intern. Conf. Atomic Collisions in Solids* (Moscow, 1977) to be published

3.137 T.Lenskjaer, F.Nyholm, S.D.Pedersen, N.B.Petersen: Phys. Lett. **47**A, 63–65 (1974)

3.138 N.T.Olson, H.P.Smith, Jr.: Phys. Rev. **157**, 241–245 (1967)

3.139 R.G.Musket, H.P.Smith, Jr.: J. Appl. Phys. **39**, 3579–3586 (1968)

3.140 R.S.Nelson: J. Appl. Phys. **40**, 3859 (1969)

3.141 H.P.Smith, Jr., R.G.Musket: J. Appl. Phys. **40**, 3859–3860 (1969)

3.142 M.J.Stott: J. Nucl. Mater. **69/70**, 157–175 (1978)

3.143 R.A.Johnson: J. Phys. F**3**, 295–321 (1973)

3.144 R.A.Johnson: In *Computer Simulation for Materials Applications*, ed. by R.J.Arsenault, J.R.Beeler, Jr., J.A.Simmons, Nuclear Metallurgy **20**, 1–38 (1976)

3.145 P.H.Dederichs, C.Lehmann, H.R.Schober, A.Scholz, R.Zeller: J. Nucl. Mater. **69/70**, 176–199 (1978)

3.146 F.C.Gehlen, J.R.Beeler, Jr., R.I.Jaffee (eds.): *Interatomic Potentials and Simulation of Lattice Defects*, (Plenum, New York 1972)

3.147 D.P.Jackson, D.V.Morgan: Contemp. Phys. **14**, 25–48 (1974)

3.148 D.P.Jackson: In *Atomic Collisions in Solids*, Vol. 1, ed. by S.Datz, B.R.Appleton, C.D.Moak (Plenum, New York 1975) pp. 185–197

3.149 V.E.Yurasova: *Physics of Ionized Gases 1974*, ed. by V.Vujnovic (Inst. of Phys., Univ. of Zagreb, Yugoslavia, 1974) pp. 427–476

3.150 V.M.Agranovich, V.V.Kirsanov: Usp. Fiz. Nauk. **118**, 3–51 (1976) [Engl. transl.: Sov. Phys.-Usp. **19**, 1–25 (1976)]

3.151 C.Erginsoy, G.H.Vineyard, A.Shimizu: Phys. Rev. **139**, A118–A125 (1965)

3.152 R.O.Jackson, H.P.Leighly, Jr., D.R.Edwards: Philos. Mag. **25**, 1169–1193 (1972)

3.153 L.T.Chadderton, D.V.Morgan, I.M.Torrens: Phys. Lett. **20**, 329–331 (1966)

3.154 L.T.Chadderton, I.M.Torrens: Proc. R. Soc. (London) A**294**, 93–111 (1966)

3.155 I.M.Torrens, L.T.Chadderton: Phys. Rev. **159**, 671–682 (1967)

3.156 L.T.Chadderton, I.M.Torrens: *Fission Damage in Crystals* (Methuen, London 1969) pp. 144–189

3.157 A.Scholz, C.Lehmann: Phys. Rev. B**6**, 813–826 (1972)

3.158 P.H.Dederichs, C.Lehmann, A.Scholz: Phys. Rev. Lett. **31**, 1130–1132 (1973)

3.159 P.H.Dederichs, C.Lehmann, A.Scholz: Z. Phys. B**20**, 155–163 (1975)

3.160 J.R.Beeler, Jr., M.F.Beeler: *Atomic Collisions in Solids*, Vol. 1, ed. by S.Datz, B.R.Appleton, C.D.Moak (Plenum, New York 1975) pp. 105–114

3.161 J.R.Beeler, Jr., M.F.Beeler: *Fundamental Aspects of Radiation Damage in Metals*, Vol. 1, ed. by M.T.Robinson, F.W.Young, Jr. (U.S.E.R.D.A. Report CONF-751006, 1975) pp. 21–27

3.162 R.N.Stuart, M.W.Guinan, R.J.Borg: Radiat. Eff. **30**, 129–133 (1976)

3.163  M.W.Guinan, R.N.Stuart, R.J.Borg: Phys. Rev. B**15**, 699–710 (1977)
3.164  J.O.Schiffgens, K.E.Garrison: J. Appl. Phys. **43**, 3240–3254 (1972)
3.165  J.O.Schiffgens, D.H.Ashton: J. Appl. Phys. **45**, 1023–1039 (1974)
3.166  J.O.Schiffgens, R.D.Borquin: J. Nucl. Mater. **69/70**, 790–796 (1978)
3.167  J.O.Schiffgens, to be published
3.168  A.Tenenbaum, N.V.Doan: J. Nucl. Mater. **69/70**, 771–775 (1978)
3.169  C.W.Gear: *Numerical Initial Value Problems in Ordinary Differential Equations* (Prentice-Hall, Englewood Cliffs, N. J. 1971)
3.170  L.F.Shampine, M.K.Gordon: *Computer Solution of Ordinary Differential Equations: The Initial Value Problem* (W.H.Freeman, San Francisco 1975)
3.171  H.B.Huntington, F.Seitz: Phys. Rev. **61**, 315–325 (1942)
3.172  P.Ehrhart: J. Nucl. Mater. **69/70**, 200–214 (1978)
3.173  W.Schilling: J. Nucl. Mater. **69/70**, 465–489 (1978)
3.174  F.Seitz: Discuss. Faraday Soc. **5**, 271–282 (1949)
3.175  G.H.Kinchin, R.S.Pease: Rep. Prog. Phys. **18**, 1–51 (1955)
3.176  H.B.Huntington: Phys. Rev. **93**, 1414–1415 (1954)
3.177  P.Jung: In *Atomic Collisions in Solids*, Vol. 1, ed. by S.Datz, B.R.Appleton, C.D.Moak (Plenum, New York 1975) pp. 87–104
3.178  P.Lucasson: *Fundamental Aspects of Radiation Damage in Metals*, Vol. 1, ed. by M.T.Robinson, F.W.Young, Jr. (U. S. Energy Research and Development Administration Report CONF-751006, 1975) pp. 42–65
3.179  P.Vajda: Rev. Mod. Phys. **49**, 481–521 (1977)
3.180  D.E.Harrison, Jr., W.L.Moore, Jr., H.T.Holcombe: Radiat. Eff. **17**, 167–183 (1973)
3.181  D.E.Harrison, C.B.Delaplain: J. Appl. Phys. **47**, 2252–2259 (1976)
3.182  N.Winograd, B.J.Garrison, D.E.Harrison, Jr.: Phys. Rev. Lett. **41**, 1120–1124 (1978)
3.183  B.M.Garrison, N.Winograd, D.E.Harrison, Jr.: J. Chem. Phys. **69**, 1440–1444 (1978)
3.184  N.Winograd, D.E.Harrison, Jr., B.J.Garrison: Surf. Sci. **78**, 467–477 (1978)
3.185  B.J.Garrison, N.Winograd, D.E.Harrison, Jr.: Phys. Rev. B**18**, 6000–6010 (1978)
3.186  N.S.Bespalova, L.G.Gurvich: Dokl. Akad. Nauk SSSR **202**, 804–806 (1972) [Engl. transl.: Soviet Physics-Doklady **17**, 123–125 (1972)]
3.187  D.Cherns, F.J.Minter, R.S.Nelson: Nucl. Instrum. Methods **132**, 369–376 (1976)
3.188  D.E.Harrison, Jr., J.P.Johnson, II, N.S.Levy: Appl. Phys. Lett. **8**, 33–36 (1966)
3.189  D.E.Harrison, Jr.: J. Appl. Phys. **40**, 3870–3872 (1969)
3.190  D.E.Harrison, Jr., W.L.Gay, H.M.Effron: J. Math. Phys. **10**, 1179–1184 (1969)
3.191  P.V.Mundkur: "Computer Simulation of Sputtering", U. S. Atomic Energy Commission Report LBL-3172 (1974)
3.192  A.Anderman: "Computer Investigation of Radiation Damage in Crystals", Air Force Cambridge Research Laboratory Report AFCRL-66-688 (1966) [also known as Atomics International Report AI-66-252 (1966)]
3.193  G.D.Magnuson, C.E.Carlston: J. Appl. Phys. **34**, 3267–3273 (1963)
3.194  O.S.Oen, M.T.Robinson: J. Appl. Phys. **46**, 5069–5071 (1975)
3.195  M.T.Robinson: J. Appl. Phys. **40**, 2670 (1969)
3.196  D.E.Harrison, Jr., C.E.Carlston, G.D.Magnuson: Phys. Rev. **139**, A737–A745 (1965)
3.197  R.Sizmann, C.Varelas: Nucl. Instrum. Methods **132**, 633–638 (1976)
3.198  I.M.Torrens: J. Phys. F**3**, 1771–1780 (1973)
3.199  D.M.Schwartz, J.O.Schiffgens, D.G.Doran, G.R.Odette, R.G.Ariyasu: In *Computer Simulation for Materials Applications*, ed. by R.J.Arsenault, J.R.Beeler, Jr., J.A.Simmons, Nucl. Metall. **20**, 75–88 (1976)
3.200  H.L.Heinisch, J.O.Schiffgens, D.M.Schwartz: J. Nucl. Mater. **85/86**, 607–610 (1979)
3.201  A. van Veen, J.Haak: Phys. Lett. **40**A, 378–380 (1972)
3.202  B.Poelsma, L.K.Verhey, A.L.Boers: Surf. Sci. **55**, 445–466 (1976)
3.203  M.Yoshida: J. Phys. Soc. Jpn. **16**, 44–50 (1961)
3.204  M.T.Robinson, D.K.Holmes, O.S.Oen: *Le Bombardement Ionique* (Centre Nat. Rech. Sci., Paris 1962) pp. 105–117

3.205  O.S.Oen, D.K.Holmes, M.T.Robinson: J. Appl. Phys. **34**, 302–312 (1963)
3.206  O.S.Oen, M.T.Robinson: J. Appl. Phys. **34**, 2515–2521 (1964)
3.207  J.R.Beeler, Jr., D.G.Besco: *Radiation Damage in Solids*, Vol. 1, (Intern. Atomic Energy Agency, Vienna 1962) pp. 43–63
3.208  J.R.Beeler, Jr., D.G.Besco: J. Phys. Soc. Jpn. **18**, Suppl. III, 159–164 (1963)
3.209  J.R.Beeler, Jr., D.G.Besco: J. Appl. Phys. **34**, 2873–2878 (1963)
3.210  J.R.Beeler, Jr., D.G.Besco: Phys. Rev. **134**, A530–A532 (1964)
3.211  J.R.Beeler, Jr.: J. Appl. Phys. **37**, 3000–3009 (1966)
3.212  J.R.Beeler, Jr.: Phys. Rev. **150**, 470–487 (1966)
3.213  I.M.Torrens, M.T.Robinson: *Interatomic Potentials and Simulation of Lattice Defecta*, ed. by P.C.Gehlen, J.R.Beeler, Jr., R.I.Jaffee (Plenum, New York 1972) pp. 423–438
3.214  I.M.Torrens, M.T.Robinson: *Radiation-Induced Voids in Metals*, ed. by J.W.Corbett, L.C.Ianniello (U. S. Atomic Energy Commission, Oak Ridge, Tenn. 1972) pp. 397–429
3.215  M.T.Robinson, K.Rössler, I.M.Torrens: J. Chem. Phys. **60**, 680–688 (1974)
3.216  O.S.Oen, M.T.Robinson: *Application of Ion Beams to Materials* 1975, ed. by G.Carter, J.S.Colligon, W.A.Grant (Institute of Physics, London 1976) pp. 329–333
3.217  O.S.Oen, M.T.Robinson: Nucl. Instrum. Methods **132**, 647–653 (1976)
3.218  O.S.Oen, M.T.Robinson: J. Nucl. Mater. **63**, 210–214 (1976)
3.219  M.Hou, M.T.Robinson: Appl. Phys. **17**, 295–301 (1978)
3.220  M.Hou, M.T.Robinson: Appl. Phys. **18**, 381–389 (1979)
3.221  R.Behrisch, G.Maderlechner, B.M.U.Scherzer, M.T.Robinson: Appl. Phys. **18**, 391–398 (1979)
3.222  D.V.Morgan, D. van Vliet: Can. J. Phys. **46**, 503–516 (1968)
3.223  D.V.Morgan, D.P.Jackson: Nucl. Instrum. Methods **132**, 153–161 (1976)
3.224  H.J.Pabst: Radiat. Eff. **31**, 197–202 (1977)
3.225  O.B.Firsov: Zh. Eksp. Teor. Fiz. **36**, 1517–1523 (1959) [Engl. transl.: Sov. Phys.-JETP **36**, 1076–1080 (1959)]
3.226  J.Lindhard, M.Scharff: Phys. Rev. **124**, 128–130 (1961)
3.227  M.T.Robinson: *Radiation Damage in Metals*, ed. by N.L.Peterson, S.D.Harkness (Amer. Soc. Metals, Metals Park, Ohio 1976) pp. 1–57
3.228  M.Blackman: "The Specific Heat of Solids", in *Crystal Physics I*, ed. by S.Flügge, Encyclopedia of Physics, Vol. 7/1 (Springer, Berlin, Göttingen, Heidelberg 1955) p. 377
3.229  D.P.Jackson, J.H.Barrett: Comput. Phys. Commun. **13**, 157–166 (1977)
3.230  H.Goldstein: *Classical Mechanics* (Addison-Wesley, Reading, Mass. 1959) pp. 107–109
3.231  E.Taglauer, W.Heiland: Surf. Sci. **33**, 27–34 (1972)
3.232  E.S.Mashkova, V.A.Molchanov: Radiat. Eff. **16**, 143–187 (1972); 23, 215–270 (1974)
3.233  W.Heiland, E.Taglauer: Surf. Sci. **68**, 96–107 (1977)
3.234  W.Heiland, E.Taglauer, M.T.Robinson: Nucl. Instrum. Methods **132**, 655–660 (1976)
3.235  C.Kittel: *Introduction to Solid State Physics* (Wiley, New York 1953) p. 57
3.236  W.L.Gay, D.E.Harrison, Jr.: Phys. Rev. **135**, A1780–A1790 (1964)
3.237  M.T.Robinson, I.M.Torrens: "A Computer Simulation of Sputtering in the Binary Collision Approximation", unpublished paper presented at Intern. Conf. on Ion-Surface Interaction, Sputtering, and Related Phenomena, Garching bei München, West Germany, 24–27 Sept. 1972
3.238  M.T.Robinson: unpublished research
3.239  K.Lonsdale: *International Tables for X-Ray Crystallography*, ed. by C.H.MacGillavry, G.D.Rieck (Intern. Union of Crystallography, Kynoch Press, Birmingham, U.K. 1968) Vol. 3, p. 237
3.240  G.K.Wehner, R.V.Stuart, D.Rosenberg: unpublished data cited by R.Behrisch, Ergeb. Exakten Naturwiss. **35**, 319 (1964)
3.241  E.Hotston: Nucl. Fusion **15**, 544–547 (1976)
3.242  L.L.Tongson, C.B.Cooper: Radiat. Eff. **24**, 187–193 (1975)

# 4. Sputtering Yield Measurements

Hans Henrik Andersen and Helge L. Bay

With 50 Figures

The conditions for performing reproducible sputtering measurements are a well-defined ion beam with high enough current and uniform current density, a low enough vacuum and a well characterized target. The different methods to determine total and differential sputtering yields, i.e., the measurement of the loss of target material and the flux of sputtered atoms, are outlined. All available yield data measured for different ions on different materials at normal incidence are depicted on a set of graphs and some are compared with the results of *Sigmund*'s theory. The dependence of the yields on ion-mass, energy and angle of incidence and on target structure and temperature are discussed. For light ion sputtering an empirical analytical formula is given. Energy reflection coefficients, also named sputtering efficiencies are discussed.

## 4.1 Experimental Conditions and Methods

Energetic atomic or molecular particles impinging on a solid (or liquid) cause ejection of target atoms or clusters of target atoms from their surfaces. By far the most often measured parameter characterizing such processes is the so-called sputtering yield, which is defined as the ratio of the average number of ejected to the number of incoming particles. Usually, we shall specify this definition strictly to mean the ratio of the average number of ejected to incoming *atoms* as this is the parameter relevant for estimating the erosion of the target. Partial ion yields or cluster yields may also be measured, as discussed in Chap. 2 and [Ref. 4.1a, Chap. 2]. With incoming molecules, the yield per molecule is usually measured. These yields may often not be obtained by the addition of atomic yields, as discussed below in Sect. 4.2. Sputtering may occur both from the bombarded side of the target and, provided it is sufficiently thin, from the side where the beam exits from the target. The corresponding yields are called backsputtering and transmission-sputtering yields. The former is often simply called the sputtering yield.

Sputtering-yield measurements have now been performed, at least qualitatively, for more than 125 years [4.1, 2], and a vast amount of data have been published. A number of earlier experiments have had a critical impact on the development of sputtering theory, but these early data may now be discarded as obsolete, particularly because of the development of experimental techniques.

Over the years, sputtering-yield measurements have been marred by bad reproducibility, as is clearly seen in the extended set of data presented in the review by *Behrisch* [4.3] fifteen years ago.


### 4.1.1 The Characterization of Experimental Conditions

The lack of reproducibility apparent among sputtering-yield measurements stems mainly from insufficient characterization of experimental conditions. This statement holds for a missing precise characterization of the impinging-particle beam as well as of the irradiated target. Apart from its properties before the irradiation starts, the latter is heavily influenced by both vacuum conditions and by the implantation of the impinging-beam particles.

Until approximately twenty years ago, nearly all sputtering yields were measured in *plasma discharges*. Large current densities could be obtained up to a few keV ion energy, and surfaces could usually be kept clean, which made discharges superior to ion beams in the low-energy region except when vacuum conditions were very good. Apart from that, the irradiation conditions were largely undefined. Impurities and molecular compounds in the plasma could contribute heavily to measured yields and, in particular at energies close to threshold, different charge states of the ions could smear essential features of the energy dependence. Dose measurements were difficult as it was not possible to suppres secondary electron currents. Rough corrections could be made by use of known secondary electron yields, but such yields will often change drastically during heavy-ion bombardment. Hence, experimental methods involving plasma discharges will generally not be discussed here. For applications, however, "yields" measured as erosion per current unit per unit area are of dominating importance as most applications use plasma discharges both for thin-film deposition ([Ref. 4.1a, Chap. 6] and *Christensen* [4.4]) and ion milling [Ref. 4.1a, Chap. 5].

A necessary condition for obtaining reproducible data is that the irradiation is performed with a *well-defined ion beam* both with regard to ion species and energy. Such beams are mostly obtained from accelerators of the ion-implantion type, which have been discussed at length in recent monographs (see, e.g., [4.5, 6]), but also ion beams up to much higher energies, as supplied by tandem accelerators, have been used [4.7, 8]. Ions of the desired species must be produced as either positives or negatives, accelerated to the necessary energy, mass-separated to obtain a clean beam, and focussed onto the target with a well-defined direction of incidence. The necessary technology is presented by *Freeman* [4.9]. If the beam is not mass-separated, it may contain impurities as well as charge states other than that desired. The latter possibility may also imply several different energies. Low-energy, non-mass separated beams of hydrogen and helium may, in particular, contain heavier neutrals that decisively influence measured sputtering yields, as shown in a check experiment by *Weiss* et al. [4.10]. The work of *Bhattacharya* et al. [4.11] illustrates some of the other difficulties that may be encountered while making experiments with non-mass separated beams.

For many types of measurements, it is essential that the entire target area investigated is homogeneously irradiated. This is usually achieved by defocussing the beam (resulting in a heavy loss of intensity) or by sweeping the incident beam [4.12]. The beams to be swept must be well focussed on the target to ensure that a large fraction of the current is utilized. Preferentially, the focussed beam must have its halo scraped off by a set of apertures in front of the sweeping system. Sweeping techniques are described in the literature on ion-implantation technology, and *Freeman* [4.9] is also here the most thorough source. Space-charge effects in the focussed beam may cause trouble at low energies, where homogeneous irradiation conditions are more easily obtained in a plasma discharge. Space-charge interactions may also give rise to intensity problems at low energies. In such cases, it is possible to decelerate a more energetic beam immediately in front of the target [4.13].

At least as important as a precise characterization of the incoming beam are the parameters describing the *target conditions*. The influence of target structure (crystallinity, texture) will be discussed separately below as will target purity, surface topography, and beam-induced changes in both of these parameters. However, target properties may also change during an experiment due to contamination by *background gas* in the vacuum. As a (too) crude rule-of-thumb, it has often been stated that a sufficient condition for a dynamically clean surface is that the flux of irradiating particles be larger than the flux of background gas to the target area. *Behrisch* [4.3] modified this condition substantially by emphasizing that the condition also depended on sputtering yields and he depicted experimental conditions for a large number of experiments in a flux versus background-pressure diagram. It turned out at that time, that very few experiments met the single condition that particle flux times sputtering yield be larger than background flux. The most characteristic development in sputtering-yield measurements over the last fifteen years, however, has been the steady improvement in vacuum quality. A number of measurements are now made in actual UHV (e.g., [4.14–16]), and very few measurements are performed in a vacuum inferior to a few times $10^{-7}$ Torr (whereas *Behrisch* [4.3] listed only two experiments with conditions superior to this).

The condition quoted above, viz.

$$Y \times I > \Gamma, \qquad (4.1.1)$$

where $Y$ is the sputtering yield, $I$ the projectile flux, and $\Gamma$ the background flux, is too simple for at least two reasons. Firstly, the formula does not take into account that sticking probabilities of background gas atoms are usually substantially smaller than one on most surfaces. This fact makes condition (4.1.1) stronger than necessary and was taken into account in the analysis by *Yonts* and *Harrison* [4.17]. On the other hand, sputtering yields of adsorbed atoms on surfaces may be different from the yield of the substrate. They may be higher because they are more loosely bound or because of a better mass-fit to the projectiles at low-energy irradiations. Also, they may be lower due to stronger binding because they usually constitute only a single layer, or because of incomplete energy transfer from substrate to adsorbate. However, most important, an adsorbed layer may influence the yield of the substrate atoms. Finally, there is no sharp transition from a covered surface to a clean one. For any given experimental conditions, a steady state will eventually be reached. Let the sticking probability of background-gas species, $i$, be $\gamma_{i,s}$ on a clean substrate s. Let the sticking probability further, be 0 on a completely covered surface and vary linearly with surface concentration $c_i$. If the area per absorption site is $a_i$, we have

$$\frac{1}{a_i}\frac{dc_i}{dt} = \Gamma\gamma_{i,s}(1-c_i) - IY_{i,s}c_i, \tag{4.1.2}$$

where $Y_{i,s}$ is the sputtering yield of species $i$ absorbed on substrate s. The use of (4.1.2) is based on the presence of one dominant background species. The steady-state concentration $c_{i\infty}$ is seen to be

$$c_{i\infty} = \frac{\Gamma\gamma_{i,s}}{\Gamma\gamma_{i,s} + IY_{i,s}}. \tag{4.1.3}$$

Judged from the general accuracy of sputtering measurements, it will probably be reasonable to demand that

$$c_{i\infty} < 0.1 \tag{4.1.4}$$

or

$$Y_{i,s}I \gtrsim 10 \times \Gamma\gamma_{i,s}. \tag{4.1.5}$$

As $\gamma_{i,s}$ and, in particular, $Y_{i,s}$, are known in a few cases only (see, e.g., [4.18–20]), it is difficult to judge whether (4.1.5) is more or less stringent than (4.1.1).

Recoil implantation may further complicate the situation. The irradiating particles hitting the surface may knock adsorbed atoms and, in particular, light ones [4.21–23] into the substrate. A theoretical framework for calculation of their steady-state depth distribution exists [4.24]. In steady state, as many adsorbed atoms will be implanted as will be uncovered due to sputtering. Hence, if $\gamma_{i,s}$ and $Y_{i,s}$ are not changed due to the presence of recoil-implanted atoms, $c_{i\infty}$ will not be influenced by recoil implantation, but the time required to reach $c_{i\infty}$ to within a certain fraction will be increased.

Finally, it should be mentioned that it does not help to have vacuum conditions that allow a clean surface if the surface is not cleaned before the actual yield measurements, and that in this connection, cleaning may also mean removal of stable oxides.

### 4.1.2 Determination of the Irradiation Dose

As the sputtering yield is defined as the ratio between the number of sputtered and irradiating atoms, it is equally important for a precise measurement to determine the irradiation dose and the amount of material removed. In principle, determination of the dose should be very simple as the irradiation is usually performed with charged particles, and a dose measurement amounts to a current integration only. In practice, secondaries emitted from the target and its surroundings complicate beam-current integrations considerably. For bombardments performed in a plasma, they make a precise dose determination virtually impossible. Secondaries may, however, also give rise to trouble in beam experiments. Obviously, secondary ions and electrons emitted from the target must be prevented from influencing the measurements. However, also sputtered neutrals may give rise to discernible effects as they strike grids and apertures within the target chamber and generate their own secondary electrons and ions. Detailed designs aimed at circumventing these difficulties will be given below.

The current integration necessary for a dose determination may be performed either directly or indirectly. The most direct way is to make the target part of a Faraday cup. If that is not possible, a Faraday cup may be inserted periodically into the beam in front of the target, or a secondary beam may be measured continuously while periodically being normalized to the main beam. Finally, a large number of secondary phenomena may be calibrated against the main beam and used for continuous measurement.

If the intention is not to collect the sputtered material for later investigation, the target may be made part of a Faraday cage and sufficiently accurate current integration be performed, as shown by *Blank* [4.25], *Blank* and *Wittmaack* [4.26], *Blank* et al. [4.27], *EerNisse* [4.28], and *Weissmann* and *Behrisch* [4.29]. (For experimental details, see [4.25].) Most of the above authors occasionally measure target-mass changes by means of frequency changes of a quartz-crystal oscillator, which serves as target-material substrate. It should be mentioned that the first experimentalists to combine a quartz-crystal target with direct current integration were *MacDonald* and *Haneman* [4.30]. Recently, *EerNisse* [4.31] and *Peterson* and *EerNisse* [4.32] have demonstrated that by building a very complicated setup, a careful current integration may also be combined with the collection of sputtered material.

Most often, current integration is performed indirectly. *Almén* and *Bruce* [4.33] placed a Faraday cage next to their target. As their measurements were performed in an isotope separator, the dose could be determined on a neighbouring-mass line with known intensity ratio to the mass line being utilized. *Andersen* and *Bay* [4.12] placed a retractable Faraday cup in front of their target. When the cup was removed, their quartz-oscillator target was exposed to the beam. The beam heating caused a frequency change of the oscillator, which immediately revealed any significant changes in current density during irradiation. For their transmission-sputtering experiments, *Bay*

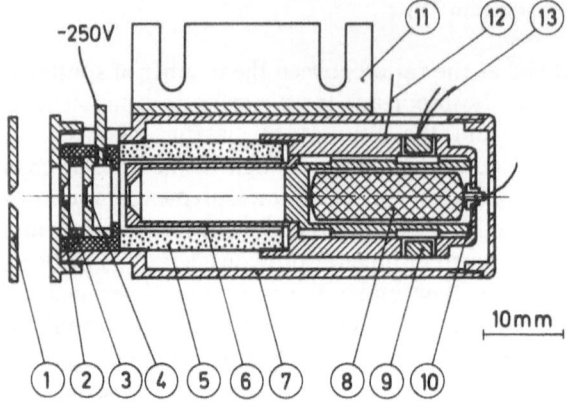

**Fig. 4.1.** Details of Faraday cup with option for calorimetric checks on projectile charge state as used by *Bay* et al. [4.7, 35]. The numbers in the figure refer to the following construction details: (*1*) beam-limiting grounded aperture, (*2*) ceramic insulator rings, (*3*) grounded aperture connected to shield surrounding the cup [this aperture is necessary to prevent particles slit-scattered at (*1*) from hitting (*4*) secondary-electron-suppression aperture], (*5*) ceramic insulator ring supporting main cup, (*6*) Faraday cup [all hatched parts touching each other are in electrical and thermal contact], (*7*) grounded shield [preventing low-energy secondaries in the target chamber from striking the cup], (*8*) heater for calibration of cup as calorimeter, (*9*) copper ring electrically insulated from cup by thin mica strips, (*10*) feed-through for heater power, (*11*) mechanical-mounting support, (*12*) current feed-through, (*13*) thermocouple. The figure does not illustrate construction details aimed at ensuring adequate vacuum performance

et al. [4.34, 35] also used a removable Faraday cup. Here, the cup was used to calibrate a channeltron detector, which was placed at a scattering angle of 135° to the incoming beam and which detected backscattered beam particles and target-atom x rays. Indirect current determination through integration of scattering or secondary-emission yields (calibrated with a Faraday cage) may also be performed by utilizing either a rotating or vibrating wing that intersects the beam for a known fraction of the time [4.36, 37]. If such a wing is covered with a thin layer of a heavy material to improve backscattering spectra, careful attention should be given to the possibility of sputtering of this layer during irradiation.

Faraday cages for direct current measurements or calibration of secondary-integration devices may take widely different shapes. Figure 4.1 shows the Faraday cup used by *Bay* et al. [4.7, 35] for their tandem-accelerator measurements. The cup demonstrates the main principle that the beam should first meet a grounded, beam-limiting aperture, then a negatively biased electron repeller that must not be hit by either the direct beam or by backscattered or secondary particles from the bottom of the cup and from the beam-limiting aperture, a principle which requires two more apertures which must also not be hit by the direct beam. Furthermore, the cup shown here is also a calorimeter. Hence, it may check the charge state of the incoming beam. This was also done calorimetrically by *Andersen* and *Bay* [4.12]. Performing the measurements at an isotope separator, they could set an upper limit of 5% to the content of neutrals in the beam, while *Bay* et al. [4.13] suppressed their positive current ($\sim$4-keV H$^+$ and He$^+$) and measured secondary-electron emission from transmitted neutral hydrogen and helium.

It may be concluded that by taking sufficient precautions against the influence of secondaries, current integration may be performed to better than 5% during sputtering measurements. With no precautions, large systematic errors may appear, even if corrections are attempted. The content of unwanted charge states may also be measured to better than 5% and, if necessary, corrected.

Finally, if the sputtered material is measured through a thickness (or mass/area) change, the fluence rather than the dose is needed. The first precaution will be to ensure a homogeneous flux over the irradiated area, as mentioned in Sect. 4.1.2. Second, if the dose is determined through current integration, an accurate measurement of the irradiated area must be performed. This will not usually be difficult, but if the irradiated area is small, a measuring microscope to determine aperture size or a scanning electron microscope to determine crater size [4.34] may be required. Crater sizes can also be measured with a tallysurf stylus, which might be used simultaneously to check the beam homogeneity [4.38, 39]. *Gittings* et al. [4.40] demonstrate both kinds of determination of crater sizes and precautions to ensure a homogeneous fluence.

Sputtering-yield measurements on insulator materials present special problems. The insulating surface will charge up, causing the beam distribution to be very inhomogeneous and rendering current measurements (if attempted during sputtering) very difficult. *Anderson* et al. [4.41] first described the now widely employed ac method for sputtering insulators in a plasma discharge (see also [4.4]), while *Edwin* [4.42] and *Navinšek* [4.43] demonstrated that beam experiments could be performed on insulators provided the target was flooded simultaneously with low-energy electrons supplied by a tungsten filament.

### 4.1.3 Yield Measurements on the Sputtered Target

#### a) Mass-Change Measurements

The classical method for determining the amount of removed material is to weigh the target before and after irradiation and deduct the change in mass of the target. Irrespective of the particular method used for mass-change determinations, two complications arise. The first is concerned with the amount of implanted material retained within the target. We may write

$$Y = \frac{\Delta m}{M_2 n_1} N_0 + \gamma_1 M_1 / M_2, \tag{4.1.6}$$

where $\Delta m$ is the mass change, $M_1$ is the projectile mass, $M_2$ is the target-atom mass, $N_0$ is Avogadro's number, $n_1$ is the number of incoming projectile atoms, and $\gamma_1$ is their trapping coefficient. For large doses, trapping may be neglected, and we have

$$Y = \frac{\Delta m}{M_2 n_1} N_0, \tag{4.1.7}$$

which is a common approximation, particularly in early works. For small doses, on the other hand, we may often assume that all projectiles not directly reflected from the target are retained, and we have

$$Y = \frac{\Delta m}{M_2 n_1} N_0 + (1 - R_0) M_1 / M_2, \tag{4.1.8}$$

where $R_0$ is the reflection coefficient [4.44, 45]. For intermediate cases where neither method works, the detailed function $\gamma_1(n_1)$ is needed. Detailed trapping (or retention) functions are known only for a few systems [4.25, 27, 33, 46–49].

Helium and hydrogen irradiations present special problems. The trapping in the near-surface region has been studied by nuclear-reaction analysis [4.50–52]. If both the solubility and diffusivity of the light-gas atoms are low, $\gamma_1(n_1)$ may be deduced from the above measurements [4.53, 54], but the behaviour of hydrogen is complicated. In nonreactive metals, rapid diffusion may occur, e.g., in nickel [4.55], and reactive materials (e.g., C, Nb, Ti, Zr, Er) show trapping coefficients that vary strongly with temperature [4.56]. The situation was reviewed by *McCracken* [4.57] and *Scherzer* ([Ref. 4.57a, Chap. 6] and [4.58]). It may be stated that particularly when hydrogen-sputtering yields are measured by the weight-loss method, very careful individual assessments of the trapping behaviour for each individual system must be made.

Most weight-loss measurements have been performed with a microbalance placed outside the vacuum chamber. The balance must have a sensitivity of the order of 1 µg [4.13]. The classical case of a large, systematic series of weight-loss measurements is that of *Almén* and *Bruce* [4.33] (see also [4.59, 60]). As more recent examples, we may mention *Behrisch* et al. [4.61], *Holmén* [4.15], *Pearmain* and *Unvala* [4.38], and *Bohdansky* et al. [4.62]. These measurements all have the drawback that the samples must be removed from the vacuum to be weighed. *Akaishi* et al. [4.63], on the other hand, built an extremely sensitive torsion microbalance which operated in vacuum, as did *Aizentson* and *Kapukhin* [4.64]. In both experiments, it was possible to follow dose effects of the yield. Unfortunately, the setups were provided with a simple ion gun without mass separation. Also *Narusawa* et al. [4.65] and *Ohtsuka* et al. [4.66], and *Hart* and *Cooper* [4.67] applied a vacuum microbalance, but in their cases for weighing of sputtered, collected material.

The fastest and most sensitive way to measure mass changes within the vacuum and to follow the sputtering yield as a function of dose is a quartz-oscillator crystal. Before the sputtering measurements, the target material has been deposited on the crystal surface. Mass changes during sputtering are reflected in changes of the resonant frequency of the crystal. The sensitivity of the commercial equipment is a few nanograms [4.12] and hence two orders of magnitude higher than torsion microbalances. The first examples of the use of quartz-oscillators in sputtering work are those of *McKeown* [4.68] and *Hayward* and *Wolter* [4.69], who used non-mass separated beams. Systematic use of the technique with many ion-target combinations and separated beams was introduced by *Andersen* and *Bay* [4.12, 70–73] while *Andersen* [4.74] demonstrated how the technique could be used to follow dose effects. The above authors used commercially available crystals with nonsmooth surfaces. Influence of surface roughness on measured yields will be discussed later in the present chapter, but it appears that the few absolute yields given by *Andersen* and *Bay* are systematically high (cf. [4.25, 31]), but the surface topography does not influence relative yields, which were their main target of study. Also, it appears that commercial crystals may be safely polished without parasitic resonances being introduced [4.25]. The irradiation of a deposited thin film on the crystal will stress the film. This stress will give rise to frequency changes [4.28, 75–78] which may be studied by using two different cuts of the quartz crystal. A detailed discussion of the use of such techniques is given by *EerNisse* [4.78]. If the oscillator is used for sputtering-yield measurements only, it appears simpler to use a rather thick metal film between the sputter-target material and the crystal. Both silver [4.12] and aluminum [4.25] apparently work well. If the layer is made too thick, the damping of the oscillations may be too strong, but for most recent commercial oscillators, frequency changes up to 1 MHz ($\sim 30$ μC) are tolerable.

### b) Thickness-Change Measurements

Thickness changes may be measured in three different ways, i.e., as a change in area density of a number of atoms of the target, as a change in geometrical thickness of the target, and as the change given by the total removal of a premeasured thickness of material. Methods based on detection of the total removal of a thin film will be discussed in Sect. 4.1.3d.

To investigate changes in area density, scattering of fast, light ions has been used extensively, mostly as Rutherford backscattering. The recent, extensive advances in the analysis of thin films by this method may be utilized [4.79–82]. In the standard form, the method requires the target to be a thin film [4.7, 29, 34, 35, 37, 83–86] which may or may not be supported. As first demonstrated by *Behrisch* and *Weissmann* [4.84], measurements on such thin films may give detailed information on the sputtering mechanisms. Reliable bulk backsputtering yields may, however, be measured only if the target is thicker than approximately half the sputtering-projectile range [4.34, 87]. This fact is clearly illustrated in Fig. 4.2, showing sputtering yields for 500-keV argon bombardment of Au/Be sandwiches as a function of gold-layer thickness [4.88]. The absolute yields are high due to the rough surface of the beryllium substrate, but that will not influence the thickness variation appreciably.

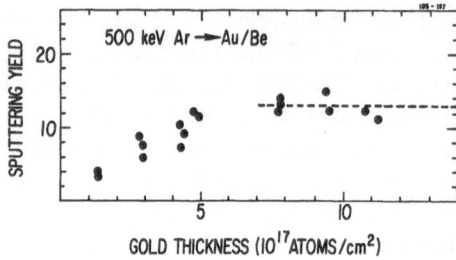

**Fig. 4.2.** Sputtering yield of Au vacuum evaporated onto a Be substrate as a function of Au-film thickness. The sputtering was performed by *Andersen* [4.88] with 500-keV Ar and the thickness analysis with 2-MeV He. The dashed line represents the theoretical thick-target yield [4.89]

Other methods of measuring changes in area density could be envisaged, for example, ion-induced x ray analysis [4.90], nuclear-reaction analysis [4.91] or changes in previously induced radioactivity as used to study neutron sputtering of gold [4.92]. Among the few examples known, it may be mentioned that *Anderson* [4.93] used x ray analysis to study the build-up of a copper-rich layer on the surface of a sputtered Ag/Cu compound target, and *Fritzsche* and *Rothemund* [4.94] used an electron microprobe to determine changes in target thickness. Similarly, *Kirschner* and *Etzkorn* [4.95] used electron-induced x rays to monitor film thicknesses when depth-profiling a Pd–Al–Si sandwich. Finally, *Tardy* et al. [4.96] studied the sputtering of LiF films by means of the resonant nuclear reactions $^7\mathrm{Li}(p, \alpha)\,^4\mathrm{He}$ and $^{19}\mathrm{F}(p, \alpha)\,^{16}\mathrm{O}$. They could thus separately measure the yields of the two components. Also the energy loss of a charged-particle beam, or multiple scattering of an ion beam or an electron beam could be used as thickness gauges for thin films. Indirectly, heavy ions were used in such a manner by *Mertens* [4.97] for transmission-sputtering studies, while *Medved* and *Poppa* [4.98] used a transmission-electron microscope which allowed the checking of changes over very small lateral distances (2 μm). In such measurements, carbon contamination of the investigated area may be important.

Thickness changes may be investigated by measuring the movement of a low-concentration marker implanted into the sample or by measuring the changes in the profile of the sputtering-beam particles implanted into the target [4.26, 27, 95, 99, 100]. Such measurements may be very sensitive, but possible diffusion of the implanted projectiles or the marker must be carefully checked [4.27, 74].

Measurements of changes in geometrical thickness (often utilizing Tolansky multiple-beam interference) have been used for a longer time but also suffer from systematic experimental errors. In particular, thickness changes may not directly imply sputtering but could be related to radiation damage below the surface. Usually, the damaged layer will expand [4.101, 102] but in cases of amorphous targets, it may also be compacted [4.103, 104]. Corrections for such effects have usually not been reported in the sputtering literature. To make them unimportant, layers large compared to projectile ranges, should be sputtered away. The method was first used by *Ogilvie* and *Ridge* [4.105] for measurements on polycrystalline silver and by *Hines* and *Waller* [4.106] and

*Edwin* [4.42] for silica. Silica and other glasses were also the materials studied by *Bach* [4.107–109] and *Bach* et al. [4.110], while *Tsunoyama* et al. [4.111, 112] sputtered iron. The latter authors also measured crater depths with a scanning electron microscope. Density changes are probably not important for iron but apart from that, reservations concerning such effects also remain for SEM work. Optical step-height measurements were further performed by *Fritzsche* and *Rothemund* [4.94].

Crater depths may also be measured with a mechanical microstylus, the so-called Tallysurf. Examples relevant for sputtering work may be found in *Schulz* et al. [4.39] and *Kimura* et al. [4.113].

For particular systems, changes in the geometrical thickness may be measured through changes in interference colours. For sputtering studies, such effects were first utilized by *Kelly* [4.114] to measure the sputtering yield of 2–10 keV $O_2$ and Kr bombarding $TiO_2$. The technique may be used for other oxide films also [4.115], and, surprisingly enough, also for some metals, as demonstrated by *Guthrie* and *Blewer* [4.116] for scandium films. Readers interested in this technique are urged to get hold of a color print of Fig. 2 of the above article to judge the striking sensitivity of the effect.

For thin metal wires or foils, thickness changes during irradiation may be estimated from changes in electrical resistivity provided point-defect generation and, probably more important, electrical size effects [4.117] may be neglected. The method has been used by *Fetz* [4.118], *Scott* [4.119], *Teodosić* [4.120], and *Navinšek* and *Carter* [4.121].

Finally, *Kelly* and associates [4.122, 123] have used specialized methods to study oxide sputtering. Marker movement, for example, was studied via a chemical stripping technique.

### c) Microscopic Measurements on the Target

The methods mentioned up till now are macroscopic in the sense that they measure yields as a ratio between, e.g., weight change and projectile current. Hence, measured yields are average values for a large number of cascades. Similarly, theoretical estimates are also based on averaging over cascades. However, yields are expected to fluctuate strongly from cascade to cascade, and these fluctuations may carry information on cascade dynamics and, in particular, be important for nonlinear cascade effects [4.70, 71, 124, 125].

Measurements of fluctuations are difficult. Scanning electron microscopes do not have sufficient resolution to measure craters, due to individual cascades, except for the very largest ones. Transmission electron microscopy has the lateral resolution but usually not sufficiently precise depth information. Only if thin layers of heavy materials are sputtered (or evaporated) from a light substrate by ion impact may the result of individual impacts be seen [4.83]. One instrument, however, presents unique possibilities, namely the field-ion microscope. Publications on field-ion microscopy and sputtering exist only from the group of *Walls* et al. [4.126–128] and these papers do not give much direct sputtering information. However, with the field-ion microscope, it should be possible to measure the number of atoms sputtered from each collision cascade directly and hence measure complete statistics for the sputtering process as well as a possible splitting of cascades into subcascades. The measuring time required for such a program may, however, be forbiddingly large.

### d) Yield Determination Through Analysis of Surface Compositions

Preferential sputtering effects on alloy targets will not be discussed in the present section of the book, but to give meaningful data for alloy sputtering yields, the possible influence of preferential sputtering must be assessed. Hence, it appears important to mention briefly a number of surface-analytical tools and their application to sputtering studies. Extensive discussion of various experimental methods will be found in recent conference proceedings, e.g., *Andersen* et al. [4.79], *Mayer* and *Ziegler* [4.80], *Meyer* et al. [4.81], *Dobrozemsky* et al. [4.129], and *Benninghoven* et al. [4.130].

Ion scattering has been used extensively both as low-energy-ion surface scattering (ISS) [4.131, 132] and as high-energy Rutherford backscattering (RBS) [4.85, 86, 95, 133, 134] to study changes in surface composition. However, the main technique appears to be Auger-electron spectroscopy (AES), either in combination with RBS [4.85, 86, 134, 135] or as a technique in itself [4.135–153]. The surface enrichment in copper of an Ag/Cu compound target was simply seen as a color change by *Anderson* [4.93], and similar qualitative information has been obtained by numerous other authors.

Provided the target consists of a thin film of known thickness, either unsupported or supported by a substrate of different material, and provided the irradiation is laterally homogeneous, any surface-analysis technique may be used to reveal when the target is sputtered away. Hence, these methods may be used to measure absolute sputtering yields. The warning issued in Sect. 4.1.3b concerning the influence of a substrate of mass widely different from the thin-film target material, is very appropriate here, in particular for light-ion sputtering. Hence the hydrogen-sputtering yields of *KenKnight* and *Wehner* [4.154], who watched the breakthrough of self-supporting films, are of limited value, while the measurements of *Kelly* [4.114] (2–10 keV $O_2$ and Kr on $Al_2O_3$), which also utilized the breakthrough of self-supporting films, will not be appreciably influenced by systematic errors. *Smith* et al. [4.147–149] used AES to indicate the sputtering-through. In their first [4.147] publication, they studied argon sputtering for silver and niobium films on glass substrates, and the methods appear to be fast, reliable, and without systematic errors. In the following [4.148] work, they extended the targets to carbon and tungsten and the projectiles to hydrogen. Unfortunately, the specific combinations of multisandwich targets and substrate material were not stated, but in the latter [4.149] publication, hydrogen sputtering of carbon on platinum was measured.

This combination is bound to give too high yields due to backscattering from the heavy substrate, particularly at higher energies, an effect which is clearly seen in their data (see Fig. 4.5).

Sputtering-through of thin films has also been used as a yield-measuring technique by *Benninghoven* [4.155], who utilized secondary-ion mass spectrometry (SIMS). (Strictly speaking, this and the following technique is concerned with the sputtered material and hence belong in the next section. It is treated here together with other measurements utilizing breakthrough.) He did not discuss the influence of the substrate, but this was carefully taken into account by *Hofer* and *Liebl* [4.156] and *Hofer* and *Martin* [4.157]. *Oechsner* et al. [4.158] used neutral mass spectrometry among other cases for $Ta_2O_5/Ta$ films as did *Hofer* and *Martin*. Here, the influence of the substrate should be negligible. Finally, *Braun* et al. [4.159] used optical emission as an indication of sputtering-through, but they again neglected the influence of different target and substrate mass. The above detection methods will be discussed in more detail in the following section.

A number of systematic errors, apart from those already mentioned, may influence yield measurements based on the sputter-through of thin films. Not only inhomogeneous beam distributions may lead to an uneven erosion rate. The sputter process in itself may create microscopic inhomogeneities; cone formation [4.150, 151, 4.57a, Chap. 2] in particular, may be troublesome [4.156, 160]. Preferential sputtering, in connection with the unavoidable atomic mixing at the interface [4.161] may give rise to an apparent shift of the interface. Similar effects may appear if the sputtered particles are used as indicators. Both in SIMS and with optical emission, matrix effects in connection with atomic mixing may cause an apparent shift of the position of the interface.

*Winters* and *Sigmund* [4.20] studied the sputtering of very thin layers, namely of chemisorbed nitrogen on tungsten. To find the amount of sputtered nitrogen, they compared the amount of nitrogen flash-evaporated from identical tungsten wires in the sputtered and unsputtered state. The method appears to be reliable and reproducible.

### 4.1.4 Measurements on the Sputtered Material

#### a) Dynamical Techniques

The main fraction of the particles sputtered from solid surfaces are neutrals of low energy [4.162] and thus defer direct detection. Only for the alkalis and earth alkalis may a hot tungsten surface (Langmuir-Taylor) be used to ionize the neutrals with a high efficiency. Such a technique has been used by *Stein* and *Hurlbut* [4.163] to measure angular distributions of sputtered potassium atoms, by *Fluit* et al. [4.164] to study isotopic fractionation in lithium sputtering, and by *Können* et al. [4.165] for measurements of energy spectra in connection with a time-of-flight mass spectrometer.

Attempts to ionize the neutral, sputtered material by electron beams were by and large unsuccessful as they only achieved ionization rates of $10^{-4}$ [4.166]. Much higher ionization rates may be achieved by passage through a low-pressure plasma discharge [4.167–174]. These early versions were mainly used to study energy distributions of sputtered neutrals, but recently, methods involving postionization in a plasma have been developed into a regular analytical technique [4.175–180]. Such instruments may be used for studies of preferential alloy sputtering and for measurements of relative sputtering yields. Recently, the use for absolute-yield determinations [4.158] as well as for determination of sputtered-ion fractions [4.181] has been demonstrated.

Sputtered species may also be postexcited. Tunable lasers allow high-precision optical fluorescence spectroscopy to be performed on a number of elements. The Doppler shift of the absorption line allows determinations of velocity distributions [4.182–184]. Measurements of different transitions allow a determination of ground-state versus excited-state population [4.185–187]. Both setups are extremely sensitive and well-suited for relative-yield measurements. For elements where transitions with sufficiently well-known oscillator strengths exist, it should also be possible to use the technique for absolute-yield measurements (even for very low yields) along with its other possibilities, which includes its being used under actual fusion-reactor conditions in a Tokamak [4.188].

If it is attempted to study only a limited fraction of the sputtered particles, the charged fraction may be directly investigated (SIMS) or species in excited states may be detected via their optical radiation. The different techniques for detecting sputtered ions will not be discussed here as the physics of ion production during ion bombardment has been reviewed recently by *Wittmaack* [4.162, 189]. For experimental determinations, see *Oechsner* et al. [4.181]. It will suffice here to say that it is usually very difficult to predict the absolute fraction of sputtered species appearing as ions because this fraction depends sensitively on a small amount of oxygen on the surface. Hence, if measurements are not performed under UHV conditions, absolute yields may not usually be extracted from measured dates except when the technique of sputtering-through of a known layer is used, as discussed in the previous section [4.156, 157]. Further recent examples of applications for sputtering studies may be found in *Ishitani* and *Shimizu* [4.190], *Krauss* and *Gruen* [4.191], and *Mertens* [4.192]. Finally, attention should be called to the possibility of obtaining dominating fractions of the sputtered atoms to appear as negative ions through sputtering with cesium or another alkali atom [4.193–195]. The possibilities of such techniques have up till now not been utilized for sputtering-yield studies.

The study of light emission from excited sputtered species goes back to *Sporn* [4.196]. To a large extent, the same reservations as those made for ion emission concerning matrix and surface effects on the excitation probabilities will hold. The field is treated in detail in several reviews in a recent book [4.197] and in a separately published review [4.198] (see also [Ref. 4.1a, Chap. 3]). In contrast to what has been believed previously, the emission of continuous bands has been shown to be unrelated to the species implanted through the irradiation. It is now believed to stem either from oxygen-matrix atom clusters or pure matrix-atom clusters [4.199]. The intensity of the continuous bands may possibly be used to investigate neutral-cluster emission. As discussed in Sect. 4.1.3a, optical radiation may be used as an indicator for sputtering-through of thin layers and hence for measurements of absolute yields [4.159].

If the sputtered species are gaseous, they may be detected by a mass spectrometer of known efficiency. Hence, absolute yields may be measured. The technique has been used *Erents* and *McCracken* [4.200, 201] to study sputtering (impact desorption) of chemisorbed layers but may also be useful for investigations of sputtering of oxide and nitride targets.

### b) Collection of the Sputtered Material

For many years, analysis of sputtered collected material was the dominating experimental technique in sputtering investigations and still is, as far as the angular distribution of the sputtered material from polycrystalline, as well as from single-crystal target materials, is concerned. In all collector experiments, sticking probabilities different from one may give rise to systematic errors. Generally, it is found that sputtered atoms have a higher probability of sticking to a glass or metal substrate than have evaporated atoms. *Thompson* et al. [4.202] measured the sticking probability of copper and gold on stainless steel as a function of particle energy. They could not detect differences from one for energies ranging from 2 to 100 eV and could set lower limits to the sticking probabilities only. These limits were 0.90 at 2 eV and 0.98 above 10 eV (see also Sect. 4.1.3a). Similar experiments were done for niobium and rhenium on $Al_2O_3$. The sticking probabilities were found to be 0.97 and 0.95, respectively [4.203].

Resputtering of the collected material must be considered a possibility. Nearly all sputtered particles have energies below the self-sputtering threshold, and their effect may usually be considered negligible [4.204]. If the sputtering experiment is performed in a plasma discharge and if negative ions are abundant among the sputtered species, these negative ions may be accelerated and cause substantial sputtering of the collected material (see [4.205] for the significance for technical applications). Reflected projectiles may often be energetic enough to cause considerable sputtering [4.206]. Their effects may be seen in single-crystal experiments where blocking patterns from the reflected particles may be clearly discerned on the collector [4.207–209]. The structure of the substrate as well as diffusion on the substrate surface may give rise to trouble when angular distributions are measured [4.210, 211]. Diffusion may be counteracted by strong cooling of the collectors as shown by *Hofer* [4.212].

When the sputtered material is collected, any microanalytical technique may be used for quantitative analysis. We shall here mention and briefly discuss a number of methods that have actually been used in sputtering experiments.

The most widely used method for measuring angular distributions is the collection on a glass plate and determination of the absorption of light passing through the collected layer. The absorption roughly follows an exponential law, but for thicker films, one should be aware of the problem of multiple-reflection resonances within the film. The exponential-absorption law was confirmed in a sputtering connection by *Koedam* [4.213]. The absorption measurements are based on the approximation that reflection coefficients do not change as a function of film thickness. That they actually do so in most cases has been utilized by *Chapman* and *Kelly* [4.214] for film-thickness measurements. Finally, impurities in the film and oxidation of the collected material, both of which depend on the arrival rate and hence on the film thickness, may strongly influence the optical properties. Some oxidation effects may be counteracted by optical measurements *in vacuo* [4.215]. It is concluded that although optical absorption has been used extensively, it must now be considered a qualitative and largely outdated method for measurements of collected layers.

The quartz-crystal microbalances discussed in connection with measurements on the target may also be used for absolute measurements of collected material, for yield measurements [4.68, 216] as well as for differential measurements of angular distributions [4.217].

Collected layers may be analyzed by Rutherford backscattering as was also the case for thin-film targets [4.31, 206, 218–220]. Whenever the yield is very low, RBS may be attractive. This is particularly the case for attempts to measure neutron-sputtering yields or high-energy proton yields [4.221–225]. The technique has further been found useful in connection with transmission-sputtering measurements [4.124, 226]. The high sensitivity of RBS may be used to obtain depth (or dose) resolution as was done by *Andersen* et al. [4.227] to study dose effects on the yield and transients in preferential sputtering of metal alloys.

Light ions may also be used for analysis through their excitation of characteristic x rays in the collected material. This method was used by *Rödelsperger* and *Scharmann* [4.228] for measurements of angular distributions. Characteristic x rays may further be excited by electrons in the electron microprobe [4.229–233]. Such methods are particularly useful for the determination of angular distributions.

In neutron-sputtering experiments [Ref. 4.57a, Chap. 5], the scanning electron microscope coupled with an energy-dispersive x-ray analyzer has been used extensively for studies of the so-called chunk emission, i.e., emission of micron-size lumps of material during irradiation [4.8, 221, 222, 225, 234, 235].

Sensitive surface-analysis techniques such as AES and SIMS are used both for determinations of very low yields [4.8, 65, 221, 225, 234] or for determination of the elemental or isotopic composition of the sputtered and collected material [4.236]. In a scanning mode, SIMS may be used to study the chunk emission discussed in the last paragraph. *Braun* and *Emmoth* [4.237] measured relative sputtering yields by comparing the collected amount through optical excitation by ion bombardment. Although admittedly of high sensitivity, it appears that this method demands a fair amount of testing before it may be considered generally applicable as it is well known that the optical-excitation probability shows very strong matrix effects and, in particular, depends sensitively on the oxygen content of the collected material.

A rather specialized technique for yield measurements was introduced by *Gruen* et al. [4.238] and applied for yield measurements by *Bates* et al. [4.239]. They collected the sputtered material on a cooled substrate in a matrix of noble-gas atoms. The individual collected atoms are isolated from each other (hence the name "Matrix Isolation Spectroscopy") and known oscillator strengths may be used to determine absolute amounts of collected material through optical-fluorescence measurements. The above method may also, of course, be used for measurements of the relative composition of the collected material. Such a determination may as well be performed by wet-chemistry analytical techniques [4.240], but the sensitivity is generally lower than for explicit surface-analytical techniques.

The isotopic composition of the collected material may be used for absolute-yield measurements. *Cassignol* and *Rang* [4.241] sputtered copper of natural isotopic composition with $^{65}Cu$. When equilibrium is reached, collected material contains an excess of $^{65}Cu$ over the neutral abundance from which the sputtering yield may be determined. With the present-day knowledge of isotopic effects in preferential sputtering and of transient effects during sputtering, the method deserves a revival.

The high sensitivity of radiotracer methods has been utilized both for measurements of absolute yields and angular distributions. Neutron activation is the preferred technique. Preactivated targets were used for yield measurements by *Moore* et al. [4.242], *Weiss* et al. [4.10], *Hasseltine* et al. [4.230], *Robison* [4.243], *Tischenko* [4.244], and *Agranovich* et al. [4.245], for the determination of angular distributions by *Patterson* and *Tomlin* [4.246] and for scanning of a rotor collector used for time-of-flight measurements by *Thompson* and co-workers [4.202, 247, 248]. Target materials used were Al, Co, Cu, Mo, Ag, In, Sb, Ta, and Au. *Cuderman* and *Brady* [4.249] diffused the tracer isotope $^{63}Ni$ into copper before bombardment and measured the collected active nickel to determine copper-sputtering yields. It was not discussed whether the preferential sputtering of copper in the copper/nickel system might influence the measured yields. Because of the large activation cross section of gold, it is usually preferred to post-activate the collectors when this material is investigated. Such a procedure has been used for a number of angular-distribution measurements [4.83, 250–252], but was also used for Ag and Cu [4.251, 253] and Mo and W [4.254]. Neutron and high-energy proton-sputtering measurements on gold also utilized neutron post-activation [4.8, 222]. To overcome the limited number of materials suitable for neutron preactivation, *Switkowski* et al. [4.255] used a tandem-accelerator proton beam for preactivation of Mo and V, and *Ollerhead* et al. [4.256] used alpha beams for Nb. Accelerator preactivation has the additional advantage that the activated region is easily limited to the area to be sputtered at a later time.

Uranium (and other fissionable materials) present special possibilities. If mica is used as a collector and the collector is post irradiated with thermal neutrons, the fragments from the induced fission will produce tracks in the mica, which may be counted individually. The technique has been used for $UO_2$ [4.257, 258]. Yields as low as $10^{-4}$ may be measured with confidence and angular distributions may also be determined.

Finally, macroscopic parameters such as electrical resistivity may be used to determine the thickness of deposited metallic layers. Both oxidation of the films and the electrical size effect [4.117] may give rise to systematic errors, but *Weijsenfeld* [4.259, 260] demonstrated that the difficulties could be confidently coped with. His technique was utilized by *Drentje* [4.261].

In conclusion, it may be said that collection methods are superior to measurements on the target when very low yields are measured. They are virtually the only possibility for determination of angular distributions of sputtered materials.

## 4.2 Polycrystalline and Amorphous Material Yields

### 4.2.1 Survey over Yield Data

To help the reader find his way through the multitude of existing experimental sputtering-yield data, we shall in the following sections briefly discuss systematics in the data with regard to the variation of projectile energy, atomic number, and angle of incidence as well as target atomic number. Wherever possible, comparisons to relevant theory will be attempted, particularly where such systematics may be helpful in providing extrapolations to unmeasured parameter combinations.

Before such a systematization is attempted, however, it is necessary to discuss the dependence of the yield on a number of target-characterizing parameters; hence such a discussion will be the subject of the following section.

### 4.2.2 Dependence of Yield on Target Properties and Dose Effects

The sputtering yield will depend on a number of target-characterizing parameters that are not directly functions of the target atomic number or sublimation energy. The dependence of parameters such as crystallinity, temperature, and surface topography will be discussed here while surface and bulk purity have been considered experimental problems and hence discussed in Sect. 4.1. A large group of dose changes of the yield will be explained as being caused by irradiation-induced changes in the above parameters.

In the title of the present chapter, sputtering yields of polycrystalline and amorphous targets are tentatively equated. If this is going to hold, single-crystal-yield mechanisms, in particular, (Chaps. 3 and 5) must not contribute strongly to the total yield. The characteristic feature of single-crystal sputtering is the preferred emission along close-packed directions. If the relative number of sputtered particles appearing above background in the spots is low, crystalline effects will play no large role for the total emission of material from polycrystalline targets. Actual quantities are approximately 20% [4.231, 232, 260, 262–267], and in the following, we shall thus equate polycrystalline and amorphous target materials, apart from their differences in surface-binding energy, see [4.268] and Chapter 2, 3 and 5.

Channeling of the incoming beam is known to cause a strong decrease in the sputtering yields [4.269] (see also Chaps. 3 and 5). If individual crystallites of a polycrystalline target are not randomly oriented but show a preferred orientation, the targets are said to be textured and channeling of the incoming beam may be important. Texture will probably appear in rolled and evaporated target materials. Possible texture may conveniently be detected and quantified when a light-ion beam is at hand [4.270], but the experimentalist should be aware that the ion-bombardment process may transform a non-textured into a textured target [4.271, 272].

Channeling of the incoming beam is responsible for the strong changes seen in the sputtering yield of a number of semiconductors when some critical temperature is passed. Below this temperature, which depends on dose rate, the target will amorphasize during ion bombardment; above it, it will remain crystalline and allow channeling if bombarded within the critical channeling angle. This effect was first demonstrated by *Anderson* [4.273, 274] for germanium and *Farren* and *Scaife* [4.275] for a nonelemental semiconductor (GaAs). Since then, the effect has been studied in detail by a large number of investigators, e.g., *Holmén* [4.15].

The temperature of the target has generally very little direct influence on the sputtering yield [4.276–278]. Very close to the melting point a yield increase was found by *Nelson* [4.278] and by *Vaulin* et al. [4.279], but apart from that, indirect effects may be seen through the temperature-dependent gas-sticking coefficients and diffusion coefficients for beam-induced impurities, as well as temperature-dependent surface topographies [4.145].

The phase of the target material is usually not thought to influence sputtering yields apart from changes in sublimation energy connected with phase transitions. For liquid-solid transitions, this was shown to be the case for Al, In, Ga, and Pb [4.280] and Sn [4.281]. Recent information that ferromagnetic transitions may lead to large changes in sputtering yield are not fully understood [4.282] and may have been an experimental artifact.

The surface topography may have a strong influence on the average sputtering yield. Roughness on a scale comparable to or larger than individual cascade sizes generally leads to higher yields than that of a flat surface [4.283], but opening angles from low-lying parts of the surface are probably important. Retrapping of sputtered material may thus perhaps lead to the effective yield being lower than the average yield.

In Industry, sandblasting of sputter cathodes is used to increase their roughness and hence their yield [4.284]. The sputtering-off of thin, quartz-crystal based films is particularly well suited to illustrate the influence of target roughness. As mentioned in the previous chapter, *Andersen* and *Bay* [4.12, 70–73] used unpolished quartz substrates, while *EerNisse* [4.28, 31, 75] used polished substrates, as did *Blank* [4.25] and *Blank* and *Wittmaack* [4.285]. Both for silicon ([4.72] vs. [4.25, 285]) and gold films ([4.73] vs. [4.31]), the unpolished substrates lead to substantially higher yields than the polished ones. For gold, large doses induce a common surface topography independent of the individual situation, resulting in reproducible yields (see also [4.33] and [4.286]). A similar situation does not hold for silicon. The rough-sample yields remain higher than the polished-sample yields. Experiments by *Hofmann* et al. [4.138] with Ni/Er sandwiches on rough substrates also showed higher yields than they did for smooth substrates.

The idea of retrapping behind the proposal of *Cramer* and *Oblow* [4.287] was to provide a thermonuclear reactor with a honeycomb-shaped first-wall surface. A tungsten surface covered with micro-dendrites actually had lower yields than smooth samples [4.288] as does copper sputtered by vanadium [4.33], where cone-covered surfaces [Ref. 4.57a, Chap. 6] are presumably created during irradiation. Direct evidence of the usefulness of honeycomb structures has been given recently [4.289]. Contrary to the above evidence, experiments by *Whitton* et al. [4.290] on pure copper irradiated with argon showed consistently higher yields for cone-covered surfaces than for smooth ones.

The extreme surface topographies induced by irradiation under blistering and exfoliation conditions will be discussed in [Ref. 4.57a, Chap. 7].

A number of the target properties mentioned in the present section may change during ion irradiation. Most notable are changes in surface topography (see, e.g., [4.3] and, for more recent work, [4.291–295]). Such changes are thought to be responsible for part of the strong dose effects seen, e.g., for gold [4.25, 73, 285, 296]. Changes in target-crystalline texture may also induce strong dose effects [4.271].

Remarkable dose effects may appear through beam-implantation alloy effects except for self-sputtering. Apart from the difficulties in determining yields by mass-change measurements (discussed in Sect. 4.1.3a), trapped projectiles may also cause changes in surface-binding energies (for a number of examples see *Almén* and *Bruce* [4.33]). A specific discussion of the bismuth-copper system was given by *Andersen* [4.74]. If heavy projectiles are implanted into a light substrate, they may cause strong dose effects because their presence changes the deposited-energy depth distribution. This effect is well documented for the case of xenon irradiation of silicon [4.26].

It may be concluded that for most systems, the sputtering yield is not only a function of projectile species and energy and of initial target material properties, but also of dose. The dose effects probably contribute heavily to the scatter of the data presented in the following section.

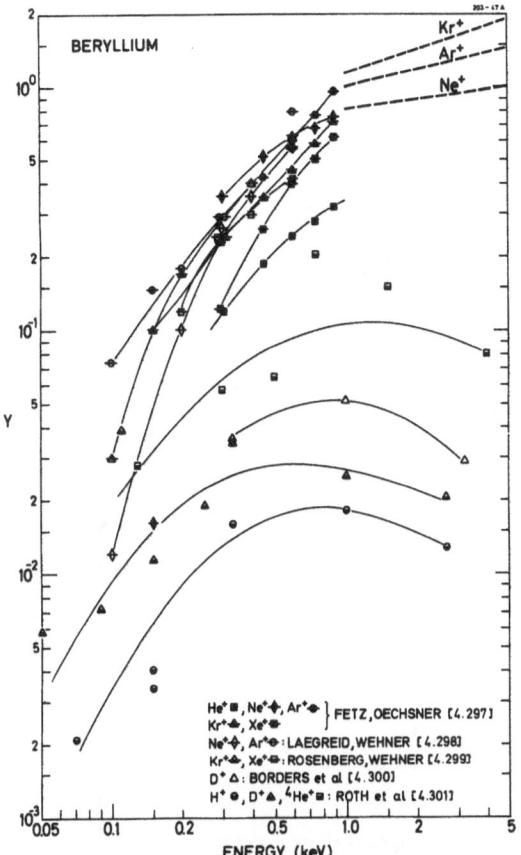

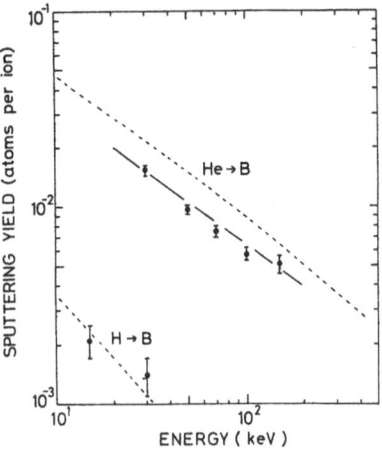

**Fig. 4.4.** Experimental sputtering-yield data for boron. (From *Miyagawa* et al. [4.302])

**Fig. 4.3.** Experimental sputtering-yield data for Be. As for all figures up to and including Fig. 4.29, heavy dashed lines indicate predictions according to (4.2.2) and thin full lines connect data points from individual publications

## 4.2.3 Dependence of Sputtering Yields on Projectile Species and Energy and on Target Materials

The present section attempts a presentation of a large fraction of the published data on sputtering yields at perpendicular incidence. The total material will allow a discussion of the agreement between the results of sputtering theory as presented in Chap. 2 and experimental data. Such a discussion will be necessary, apart from its inherent interest, to allow the reader to judge where it is possible to use theory as a guideline when extrapolating experimentally uninvestigated parameter combinations, and where not to use theory. For regions where the general theory does not apply, empirical rules to rationalize the experimental data will be discussed in some detail.

The total data set as presented in Figs. 4.3–29 is based on a systematic search of the literature up till July 1977. It has been possible to include a number of later references during revisions of the paper, but they are necessarily of a somewhat sporadic nature. The data are arranged according to target element. In each figure, it is possible to identify the projectile through the

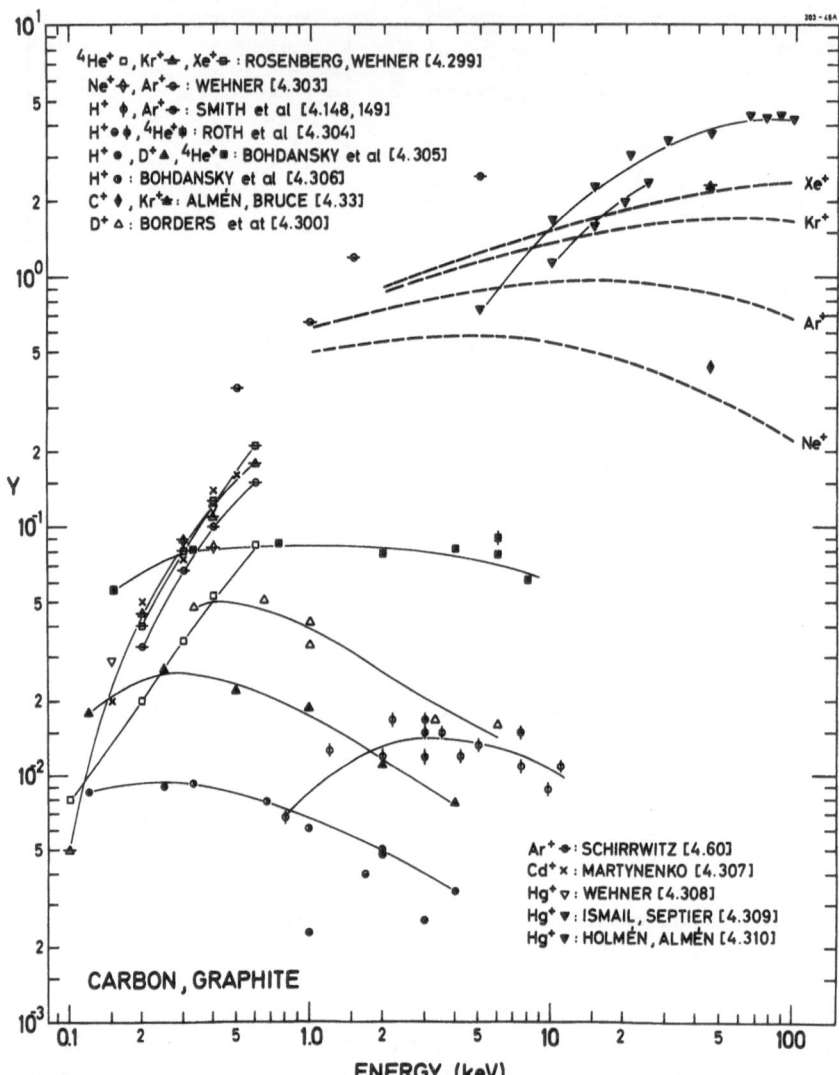

$^4$He$^+$ □ , Kr$^+$♣ , Xe$^+$♠ : ROSENBERG, WEHNER [4.299]
Ne$^+$♦ , Ar$^+$♠ : WEHNER [4.303]
H$^+$ ♦ , Ar$^+$♠ : SMITH et al [4.148, 149]
H$^+$♦♦ , $^4$He$^+$♦ : ROTH et al [4.304]
H$^+$ ● , D$^+$▲ , $^4$He$^+$■ : BOHDANSKY et al [4.305]
H$^+$ ● : BOHDANSKY et al [4.306]
C$^+$ ♦ , Kr$^+$♣ : ALMÉN, BRUCE [4.33]
D$^+$ △ : BORDERS et al [4.300]

Xe$^+$
Kr$^+$
Ar$^+$
Ne$^+$

Ar$^+$♦ : SCHIRRWITZ [4.60]
Cd$^+$× : MARTYNENKO [4.307]
Hg$^+$▽ : WEHNER [4.308]
Hg$^+$▼ : ISMAIL, SEPTIER [4.309]
Hg$^+$▼ : HOLMÉN, ALMÉN [4.310]

CARBON, GRAPHITE

Y

ENERGY (keV)

**Fig. 4.5.** Experimental sputtering-yield data for carbon. *Smith* et al. [4.147, 148] used evaporated targets, while *Holmén* and *Almén* [4.310], *Roth* et al. [4.304], and *Bohdansky* et al. [4.305, 306] used graphite from different suppliers, as did *Borders* et al. [4.300]. The remaining data have been obtained on unspecified "carbon"

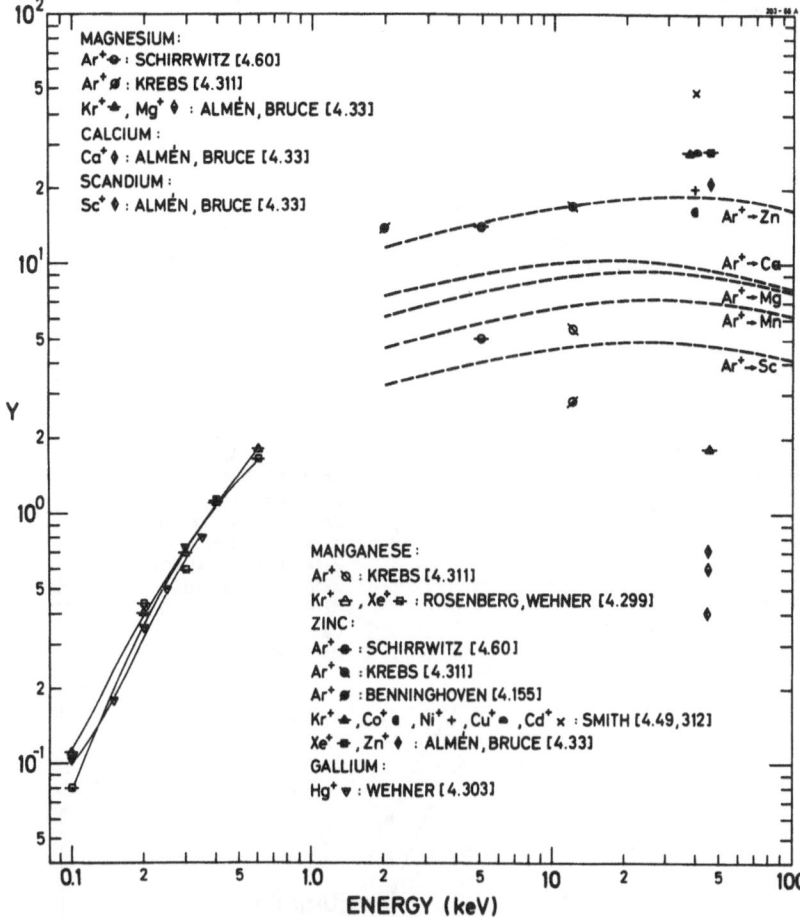

**Fig. 4.6.** Experimental sputtering-yield data for Mg, Ca, Sc, Mn, Zn, and Ga targets

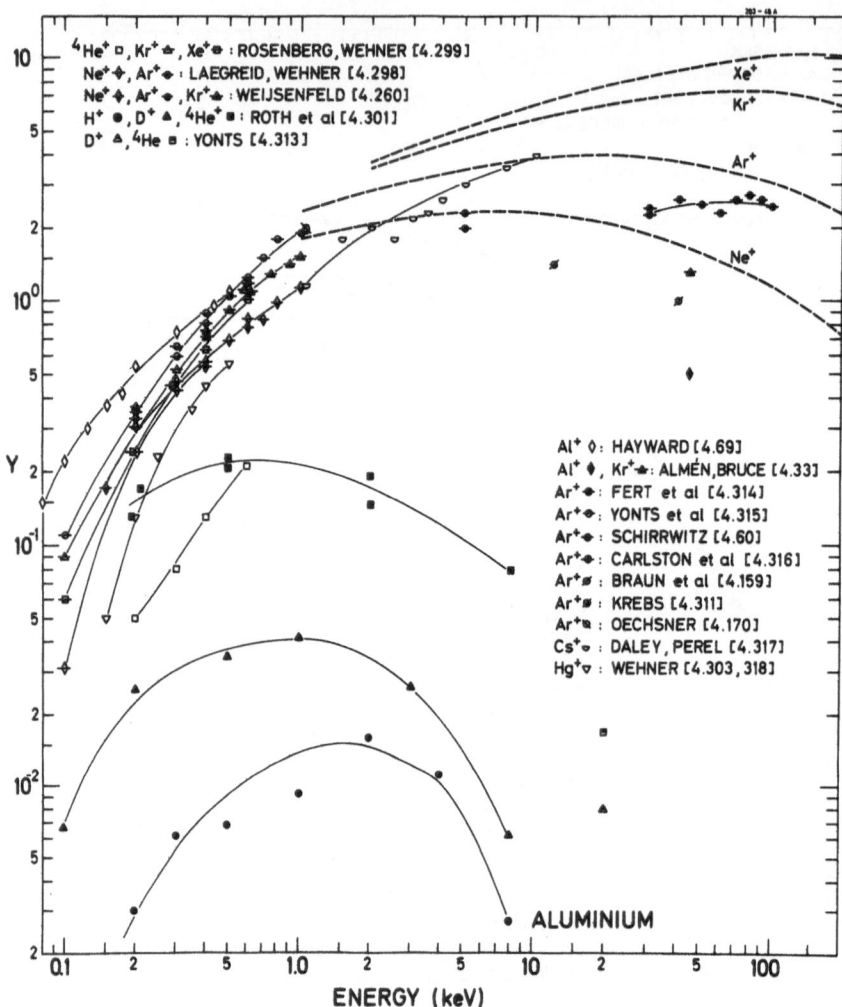

**Fig. 4.7.** Experimental sputtering-yield data for Al targets

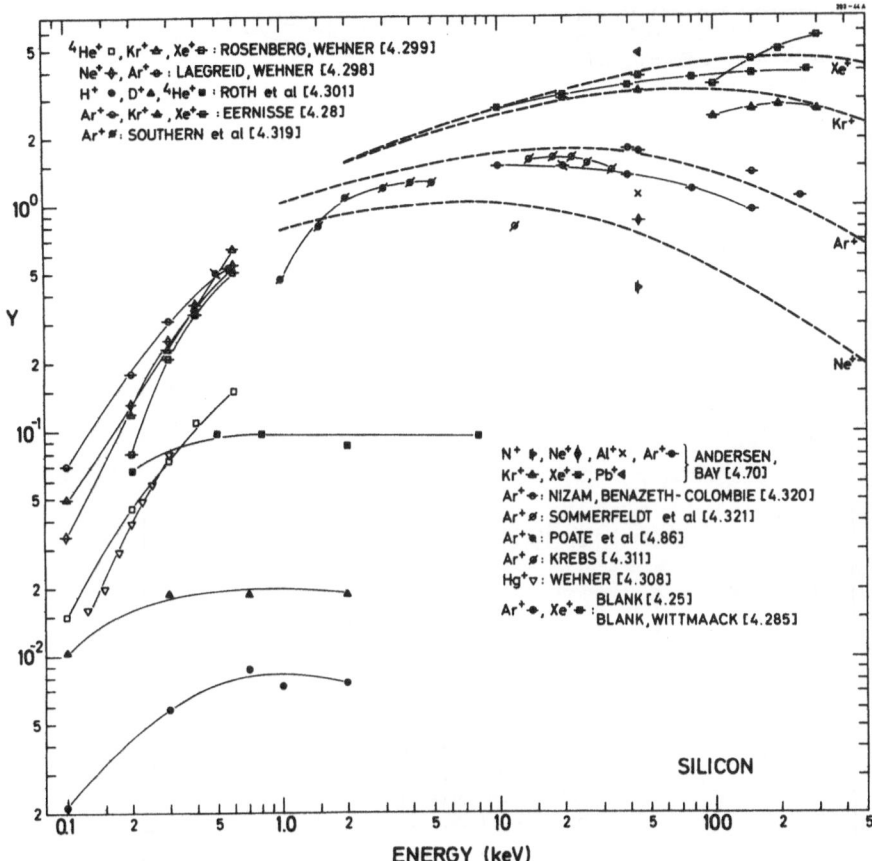

**Fig. 4.8.** Experimental sputtering-yield data for Si targets

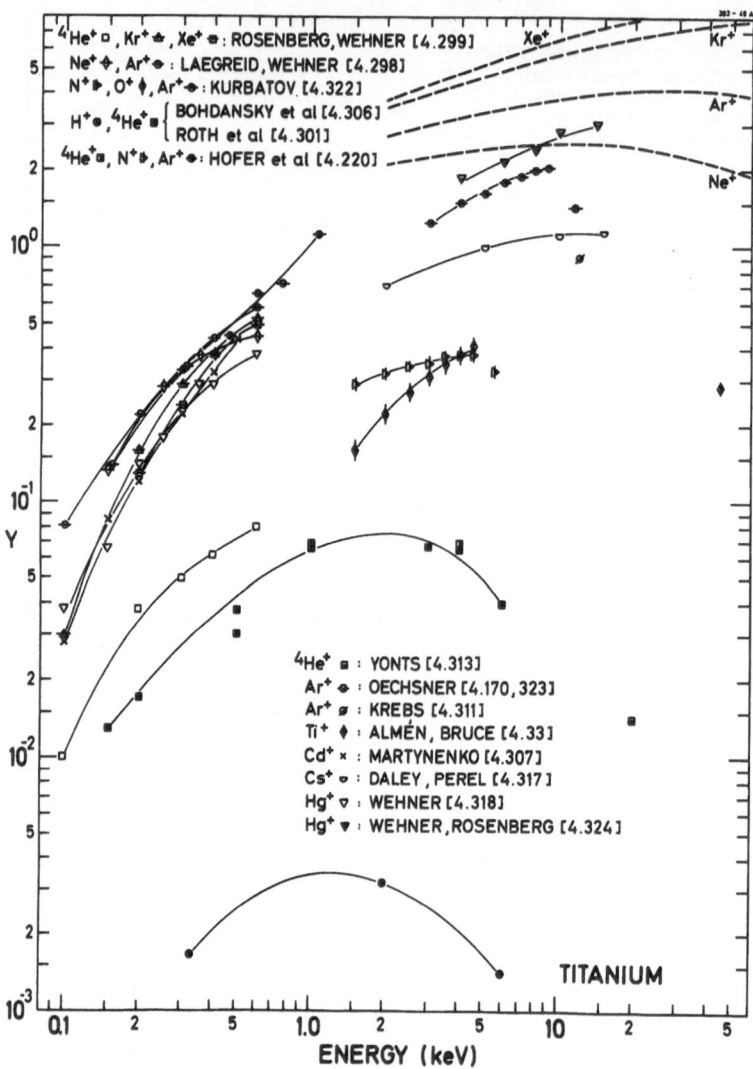

**Fig. 4.9.** Experimental sputtering-yield data for Ti targets

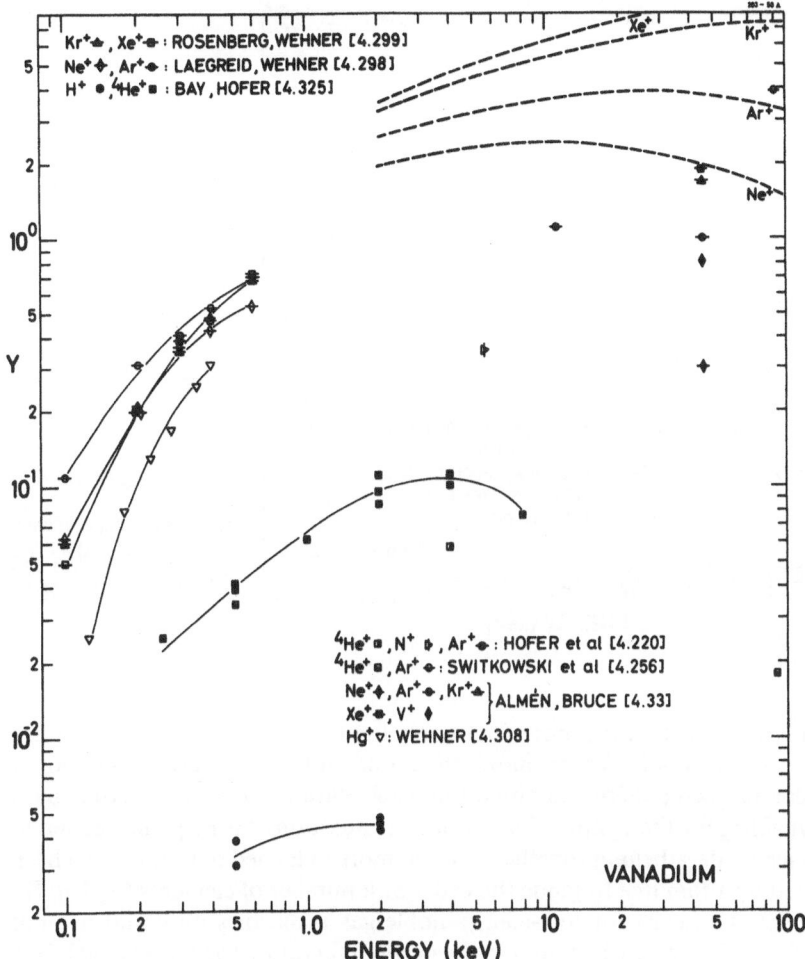

**Fig. 4.10.** Experimental sputtering-yield data for V targets

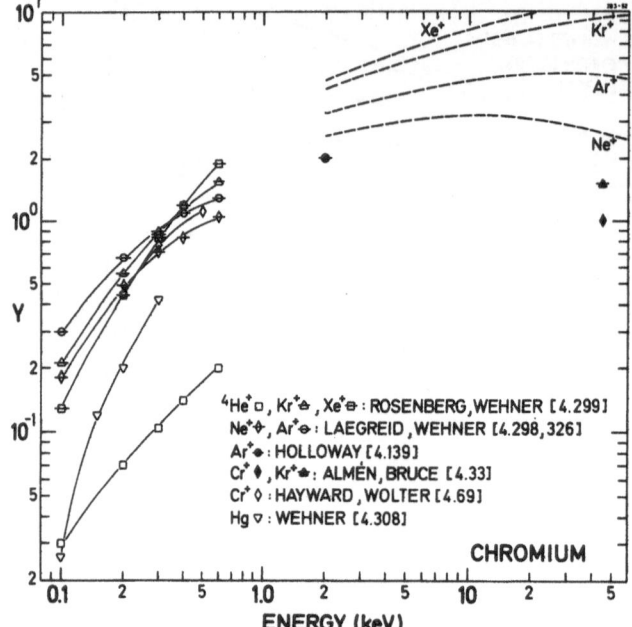

**Fig. 4.11.** Experimental sputtering-yield data for Cr targets

Legend within figure:

$^4$He$^+$ □, Kr$^+$ ⊕, Xe$^+$ ⊖ : ROSENBERG, WEHNER [4.299]
Ne$^+$ ✦, Ar$^+$ ⊝ : LAEGREID, WEHNER [4.298, 326]
Ar$^+$ ● : HOLLOWAY [4.139]
Cr$^+$ ◆, Kr$^+$ ▲ : ALMÉN, BRUCE [4.33]
Cr$^+$ ◊ : HAYWARD, WOLTER [4.69]
Hg ▽ : WEHNER [4.308]

CHROMIUM

symbol used. Note that the more commonly used ions are always characterized by the same symbol. Where more than one publication gives data for a particular projectile-target combination, each data set is characterized by a different filling of the symbol for the ion in question. To help the reader to discern which data belong together in a set, more extended data sets have been connected by a thin line to guide the eye. For a number of elements (Y, Ru, Hf, Re, Os, Ir), data exist for low-energy noble-gas projectiles only and are not depicted here. They may be found in the original literature [4.298, 299, 303, 308, 324]. For erbium, some data exist (not represented here) with the target in the form of erbium deuteride [4.219].

According to the theory presented in Chap. 2 (2.3.7, 11, 35), the sputtering yield (whether forward or backward) is given by the formula

$$Y(E, \vartheta, x) = \frac{0.042 F_D(E, \vartheta, x)}{N U_0}. \tag{4.2.1}$$

Here, $x$ is the depth of the sputtering surface from the entrance point of the projectile (i.e., $x = 0$ for backward sputtering), $E$ is the projectile energy, and $\vartheta$ is the angle of incidence to the surface. $F_D(E, \vartheta, x)$ is the deposited-energy depth distribution, and $U_0$ and $N$ are the average surface-binding energy and the density of the target material, respectively. A necessary condition for the yield to be proportional to the deposited energy at the surface is that $E \gg U_0$.

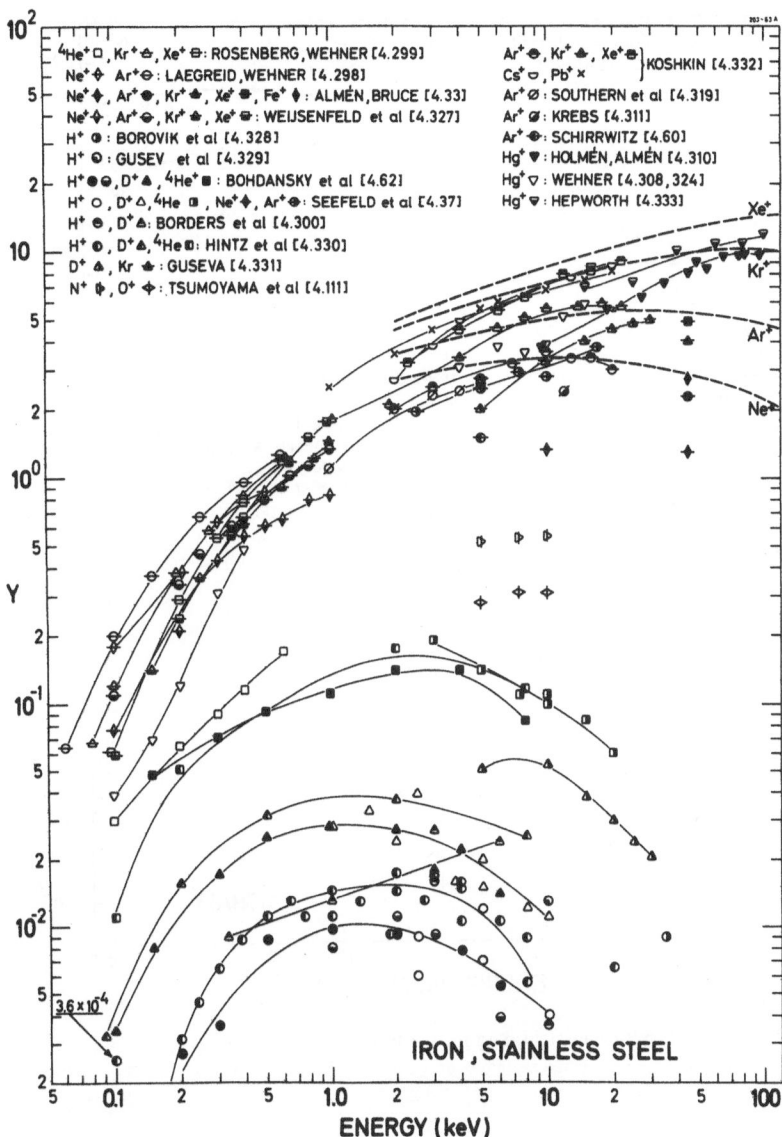

**Fig. 4.12.** Experimental sputtering-yield data for Fe and stainless-steel targets. *Borovik* et al. [4.328], *Guseva* [4.331], and *Gusev* et al. [4.329] used unspecified stainless steel; *Hepworth* [4.333] used EB 58 B, *Southern* et al. [4.319] and von *Seefeld* et al. [4.37] used SS316 stainless steel. Points by *Bohdansky* et al. [4.62] symbolized by ◐ were measured on SS304, while filled-point data were obtained on SS316. The remaining points were measured on Fe

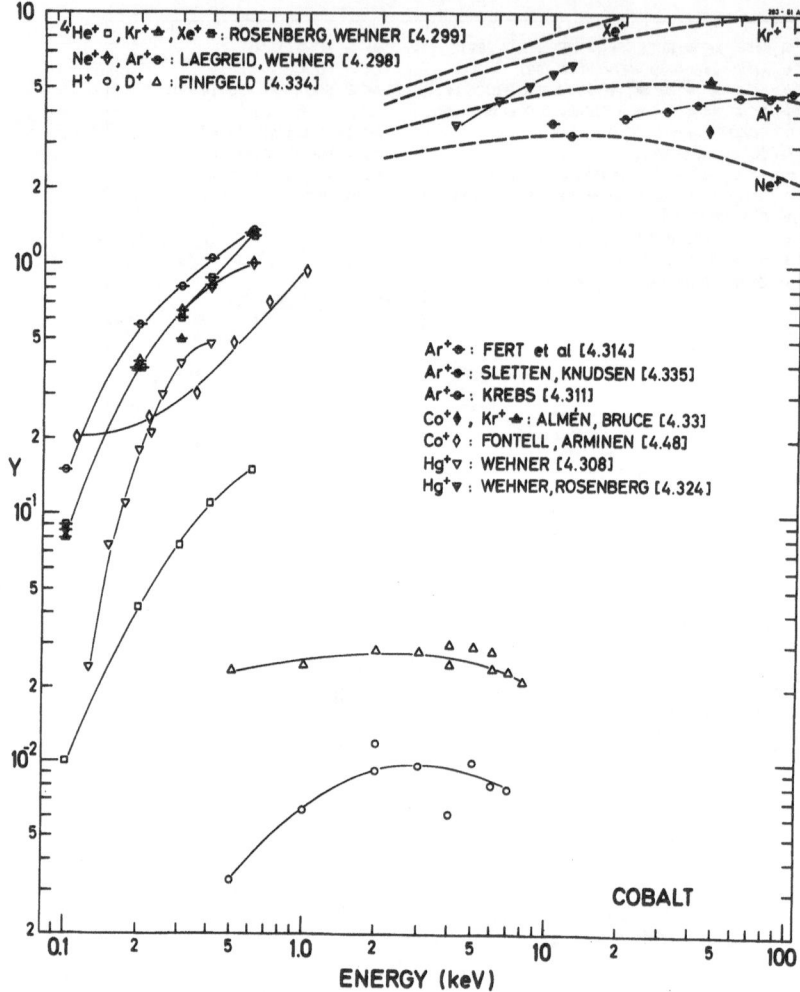

Fig. 4.13. Experimental sputtering-yield data for Co targets

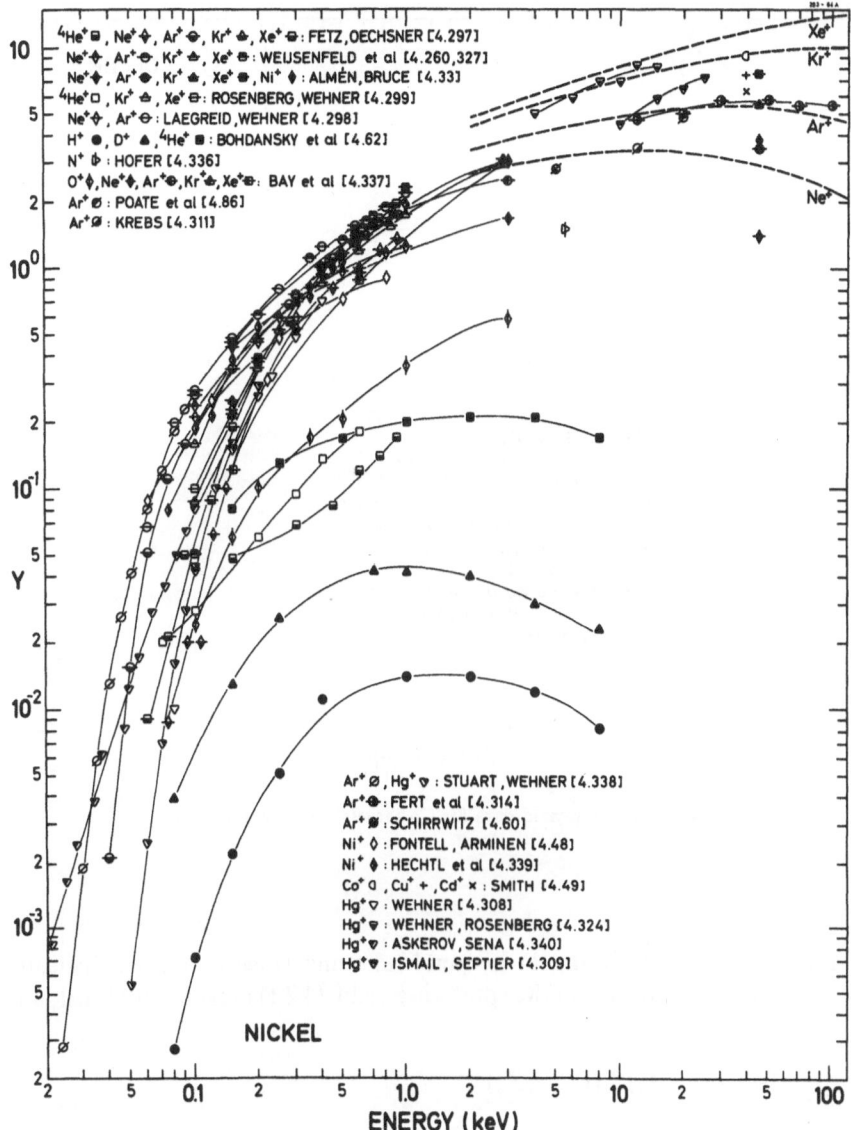

**Fig. 4.14.** Experimental sputtering-yield data for Ni targets

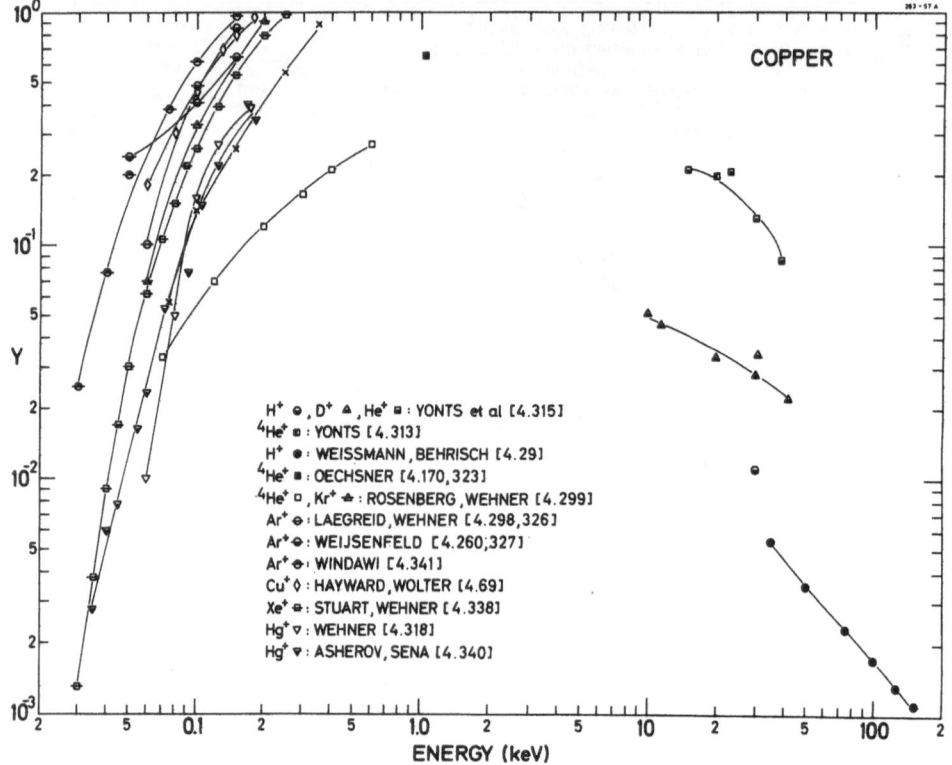

**Fig. 4.15.** Experimental sputtering-yield data for Cu targets. See also Fig. 4.16

For backward sputtering at perpendicular incidence and when inelastic effects have little influence on the sputtering yield, (4.2.1) may be simplified into

$$Y = \frac{0.042\alpha(M_2/M_1)S_n(E, Z_1, Z_2)}{NU_0}. \tag{4.2.2}$$

Here, $S_n$ is the nuclear-stopping power and $\alpha$ is an energy-independent function of the mass ratio between the target ($M_2$) and projectile ($M_1$) atoms. The $\alpha$ function is depicted below in Fig. 4.31.

Four dashed lines are shown in each data figure to indicate the theoretical yield for Ne, Ar, Kr, and Xe projectiles. For these projectiles, at least the chemical effects of implanted projectile species should be negligible and hence allow a somewhat better basis for comparisons with experiment than is the case for most other projectiles.

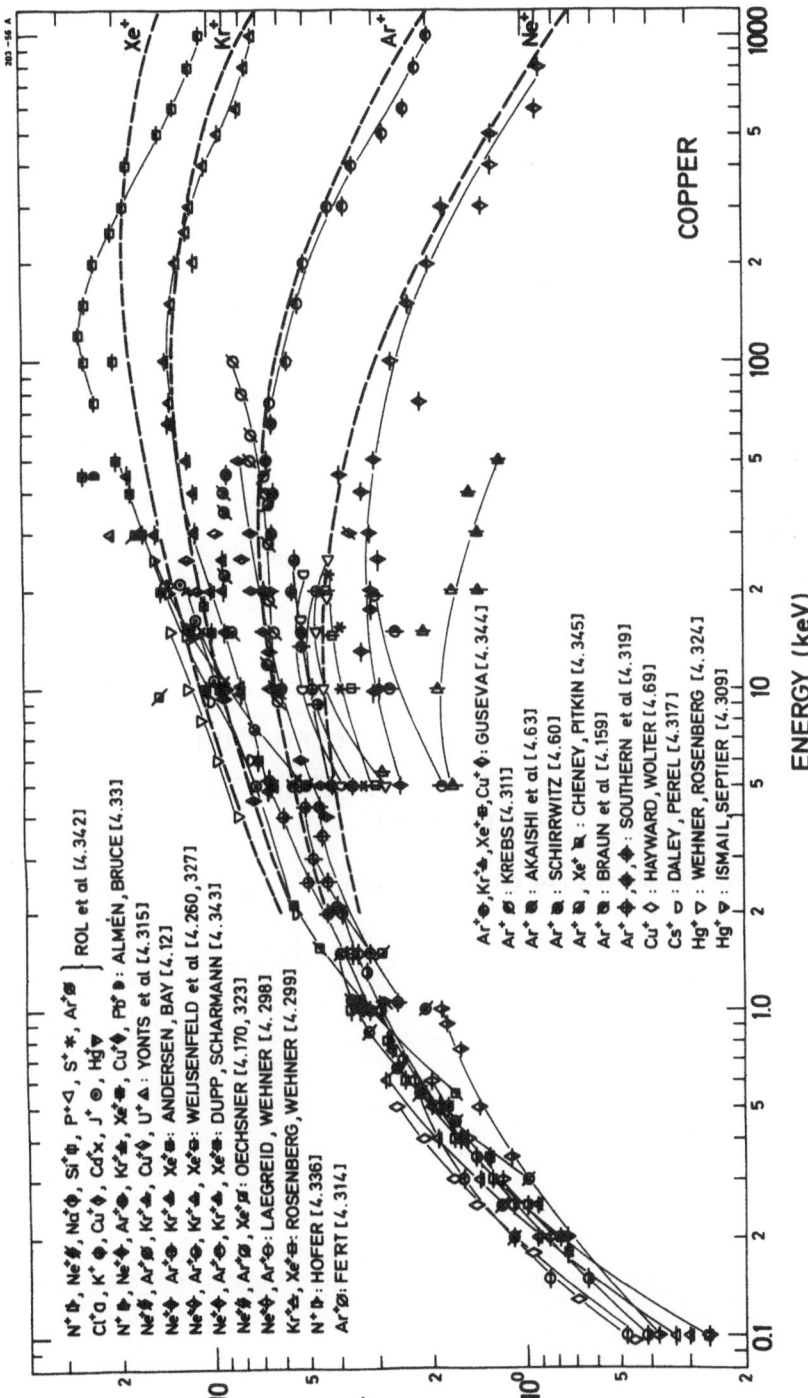

Fig. 4.16. Experimental sputtering-yield data for Cu targets. See also Fig. 4.15

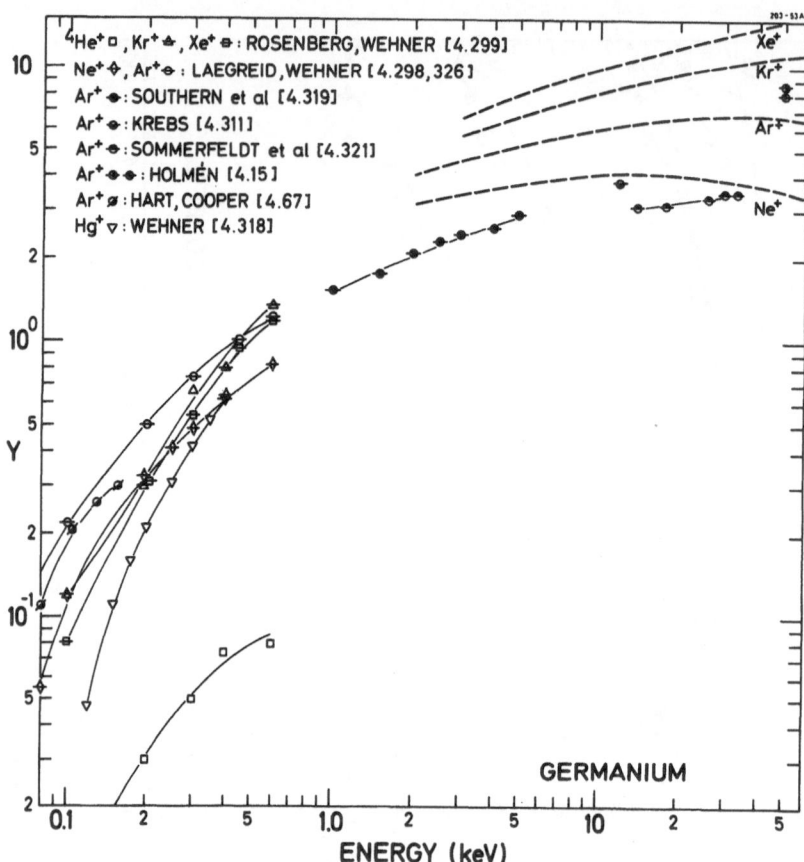

Fig. 4.17. Experimental sputtering-yield data for Ge targets

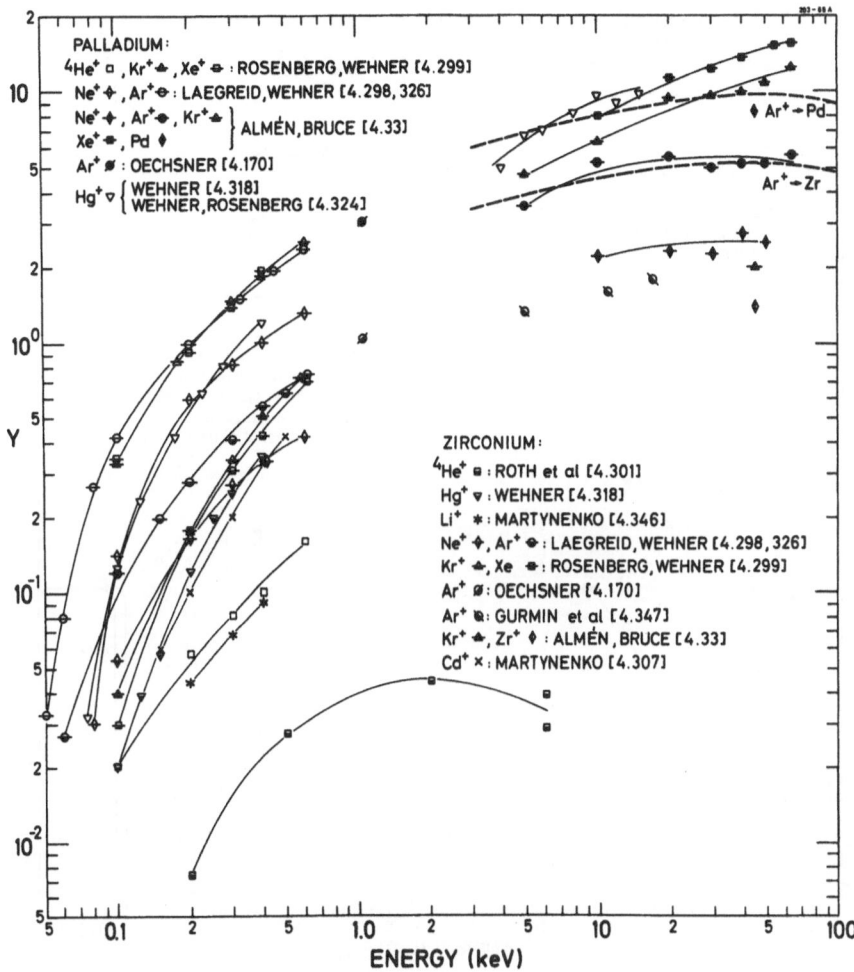

**Fig. 4.18.** Experimental sputtering-yield data for Zr and Pd targets

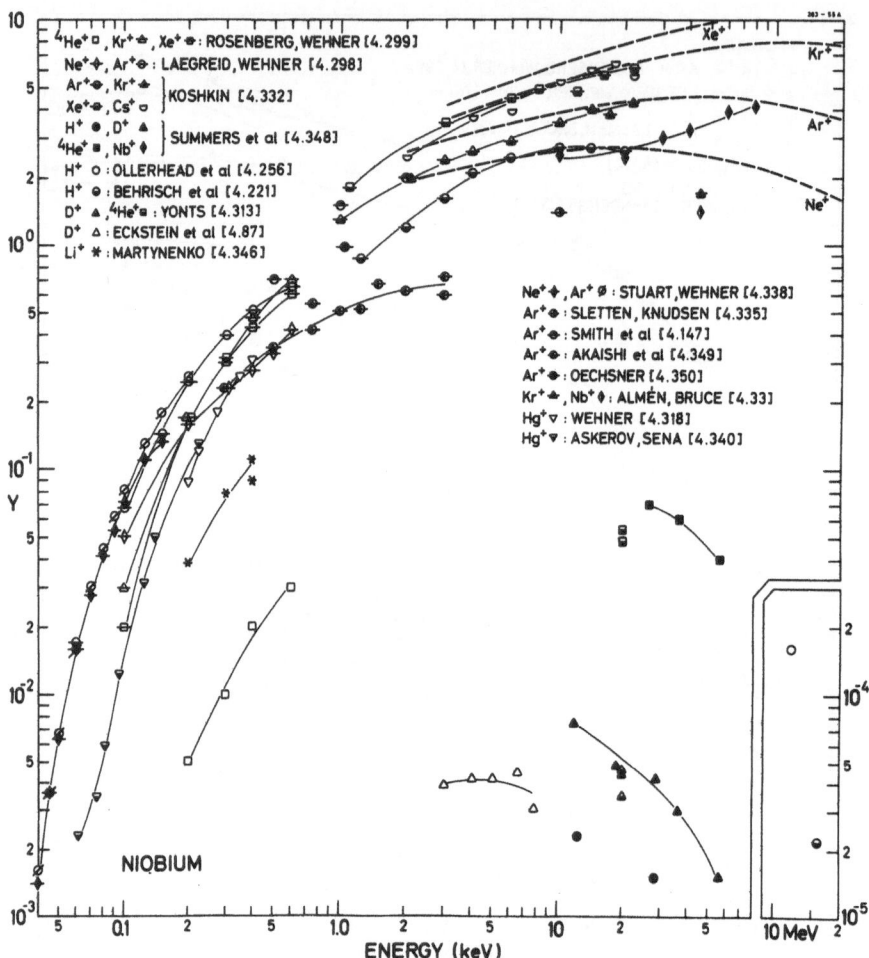

**Fig. 4.19.** Experimental sputtering-yield data for Nb targets

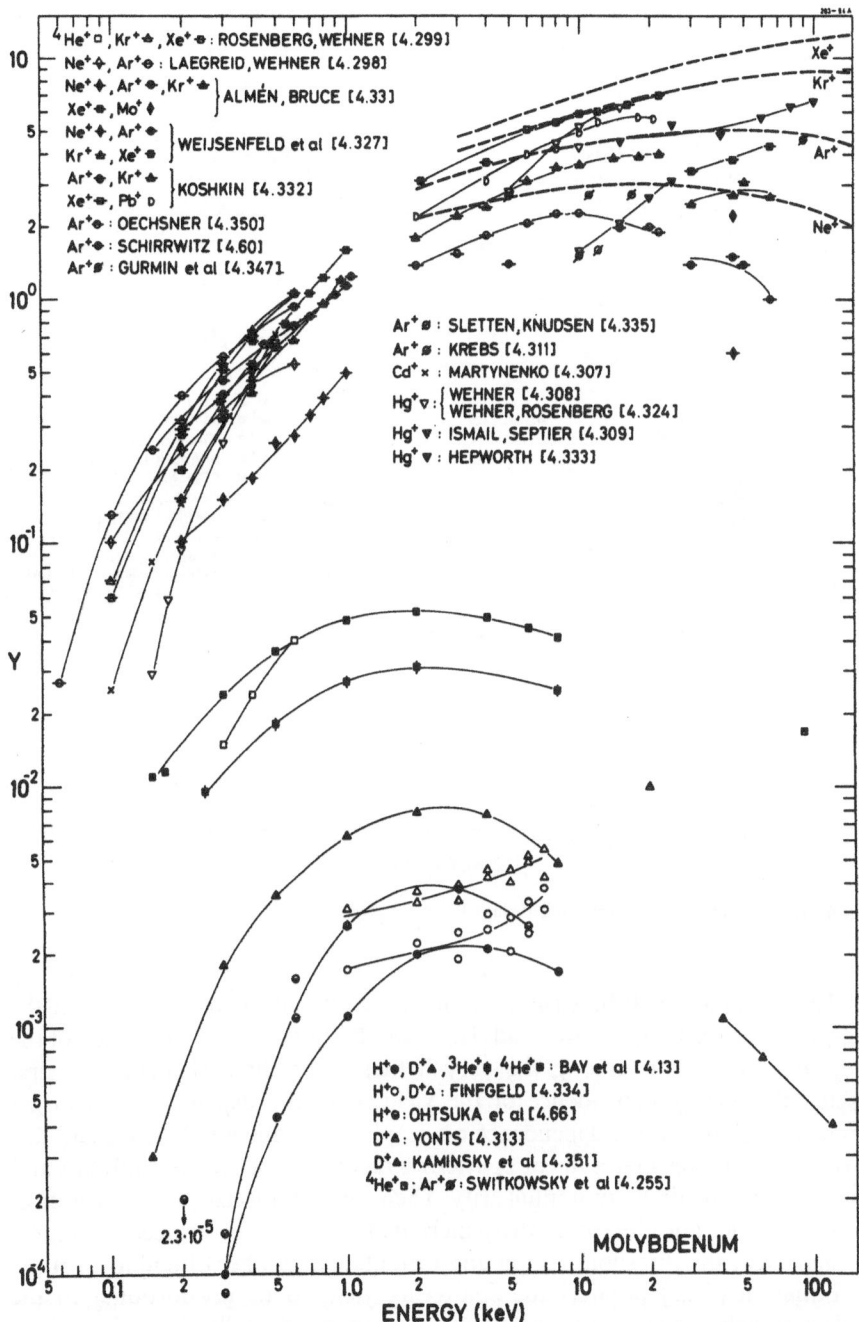

**Fig. 4.20.** Experimental sputtering-yield data for Mo targets

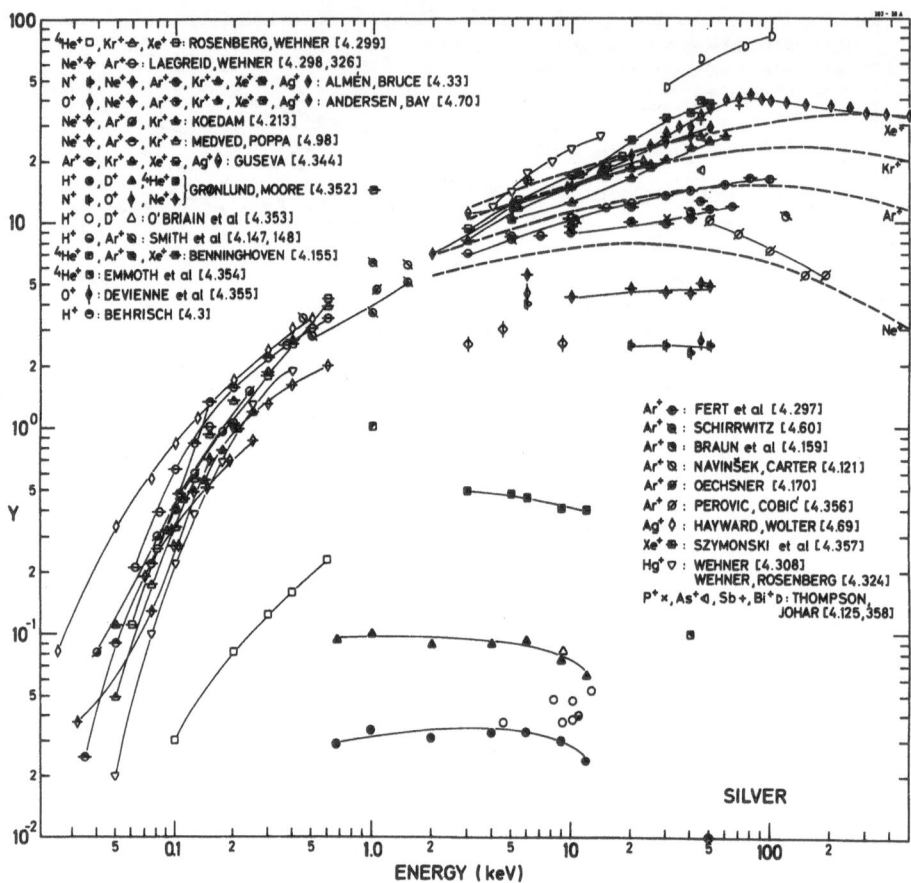

Fig. 4.21. Experimental sputtering-yield data for Ag targets

The discussion will be structured in such a manner that we first treat the energy region well above threshold. Here, we shall first discuss the yields of the projectiles not being among the very lightest. For these elements, we first discuss the energy dependence of the yield, then the dependence on target element, and, finally, the dependence on projectile atomic number. We end this part by concluding that a linear sputtering theory is not always sufficient and try to outline the limits to nonlinearity. Then we shall discuss the behaviour of light projectiles and the cases where inelasticity may not be neglected. Finally, we shall treat the threshold region and formulate a number of empirical rules.

Equation (4.2.2) predicts the sputtering yield to be proportional to the *nuclear-stopping power* and this dependence to be virtually the only one on *energy*. The data allow this prediction to be compared to theory. Take, for example, the copper data (Fig. 4.16). Neon and argon sputtering show a close agreement between the energy dependence of theoretical and experimental

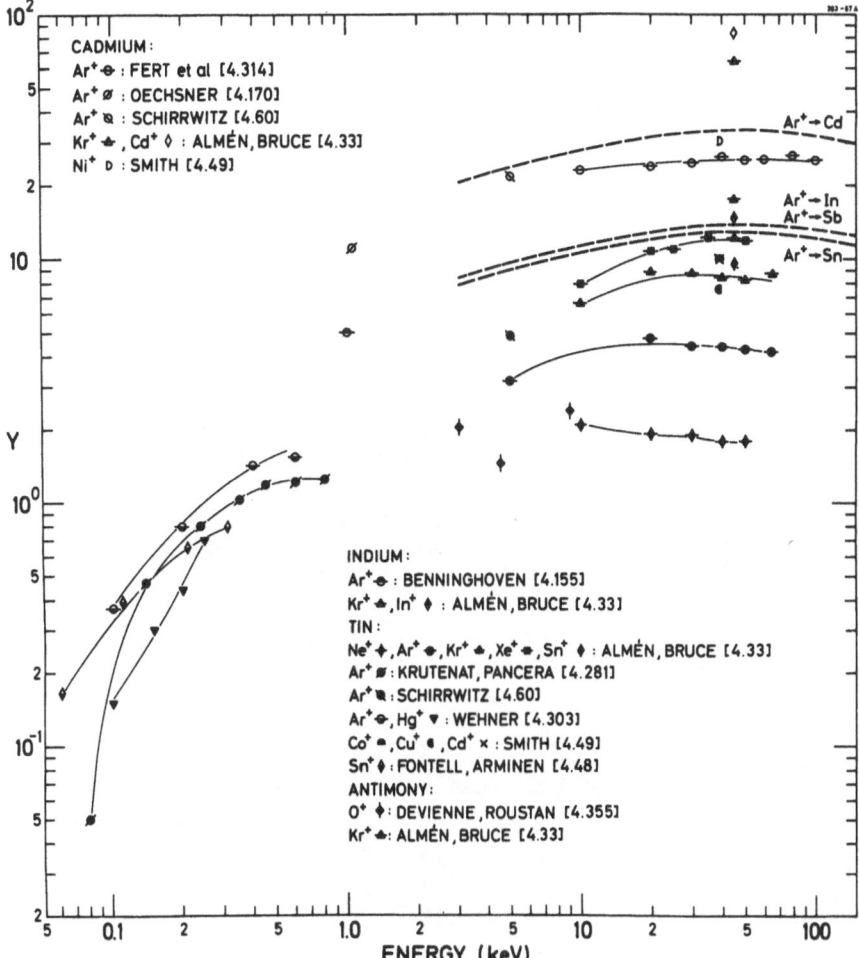

**Fig. 4.22.** Experimental sputtering-yield data for Cd, In, Sn, and Sb targets

results. For krypton, experimental data are slightly more peaked around the maximum of the nuclear-stopping power than predicted theoretically, and for xenon, this effect is rather pronounced. For silver self-sputtering (Fig. 4.21), a comparison between theory (Xe→Ag) and experiment (Ag→Ag) also shows a stronger peak in the experimental data. The most pronounced effect is probably seen for gold self-sputtering (Fig. 4.27).

The set of figures (Fig. 4.3–29) for all target elements allows a direct comparison with theoretical predictions concerning target-material dependence of sputtering yields. In spite of difficulties when comparing absolute yields, it is evident that yields of high-yield materials are predicted reasonably

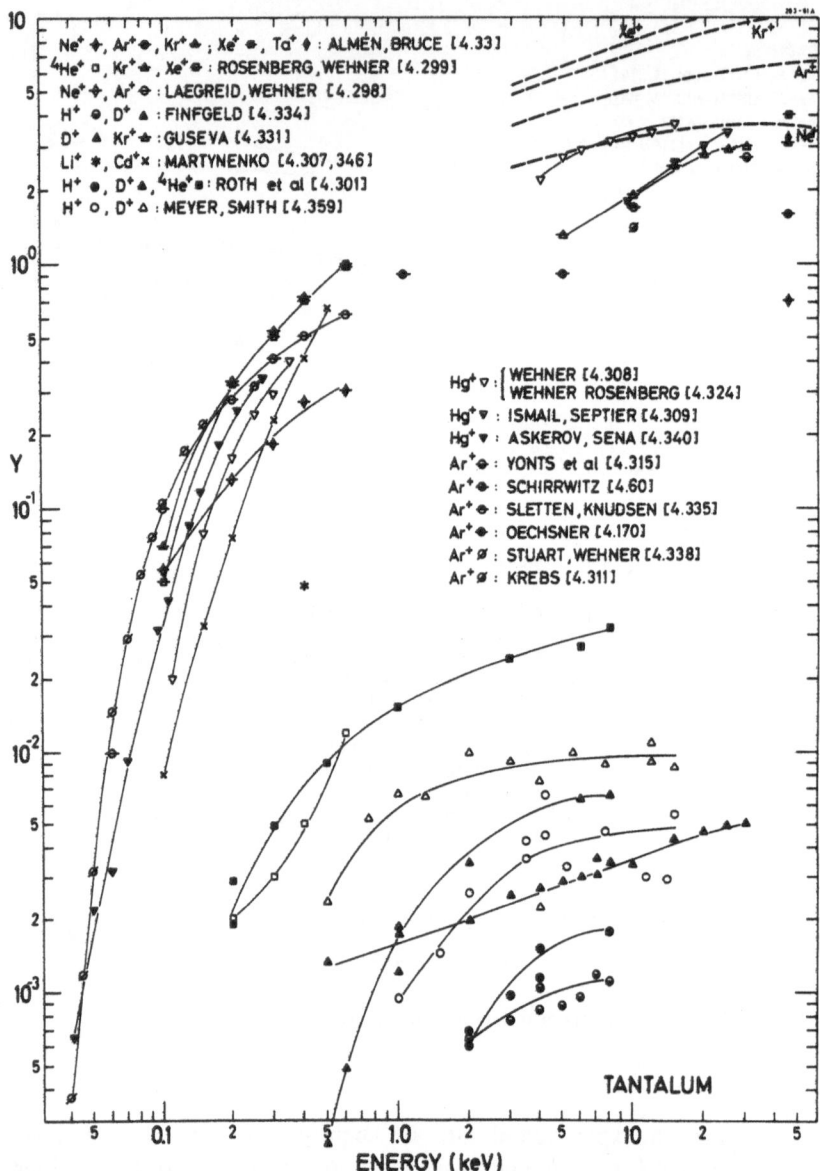

Fig. 4.23. Experimental sputtering-yield data for Ta targets

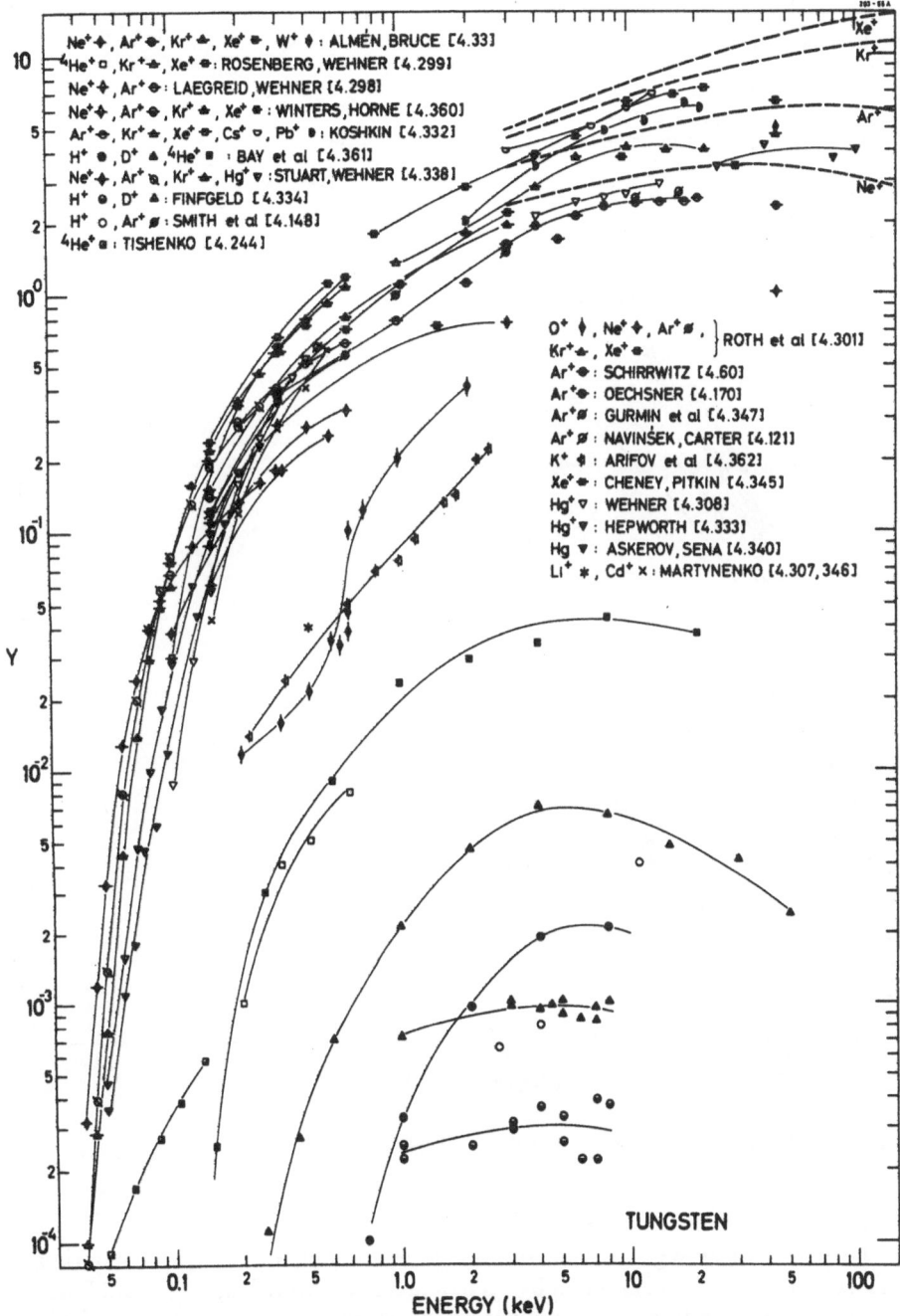

Fig. 4.24. Experimental sputtering-yield data for W targets

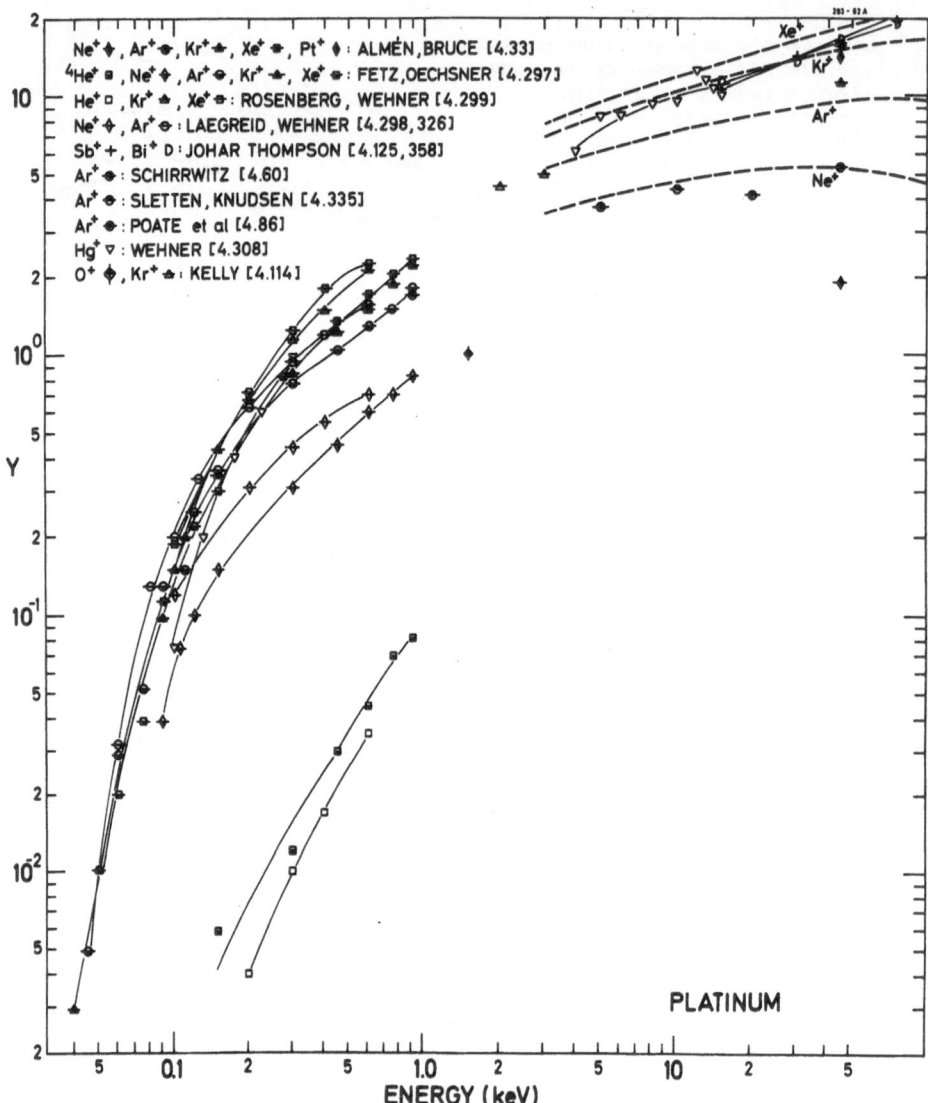

**Fig. 4.25.** Experimental sputtering-yield data for Pt targets

well by theory (see, e.g., Cu, Zn, Ag, Au), while the yield for low-yield materials (Ti, Nb, W, Ta) is overestimated by up to a factor of three. This statement holds rather independently of projectile species or energy.

The dependence on *projectile species* for a given target material, like the energy dependence, does not require a determination of absolute yields and may thus be performed in considerably greater detail than the dependence on

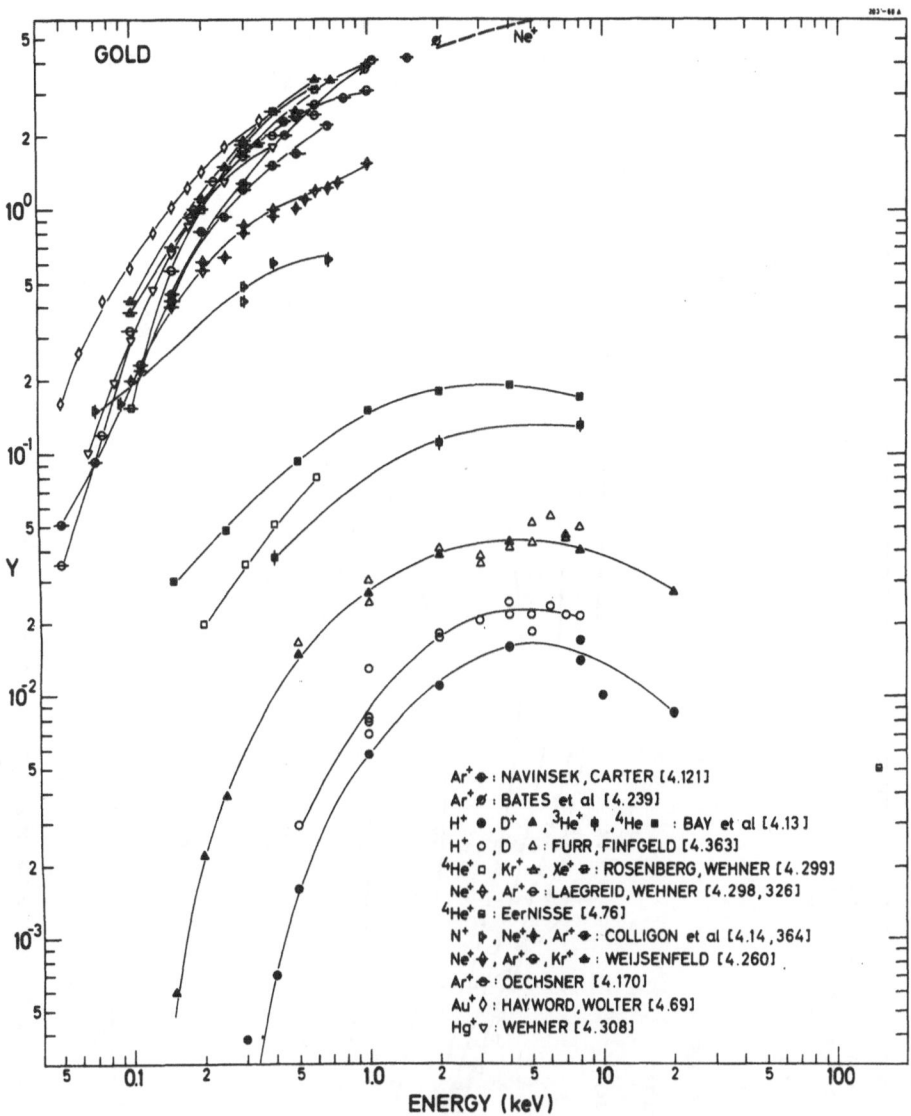

**Fig. 4.26.** Experimental sputtering-yield data for Au targets. See also Fig. 4.27

target material. Figure 4.30 shows relative sputtering yields at 45 keV for Si, Cu, Ag, and Au targets [4.12, 70, 73]. Again we see that the predictions of theory are followed very well for light targets, but for heavier targets, in particular in combination with heavier projectiles, the yield increases substantially faster with $Z_1$ than predicted by theory. However, we can not immediately decide

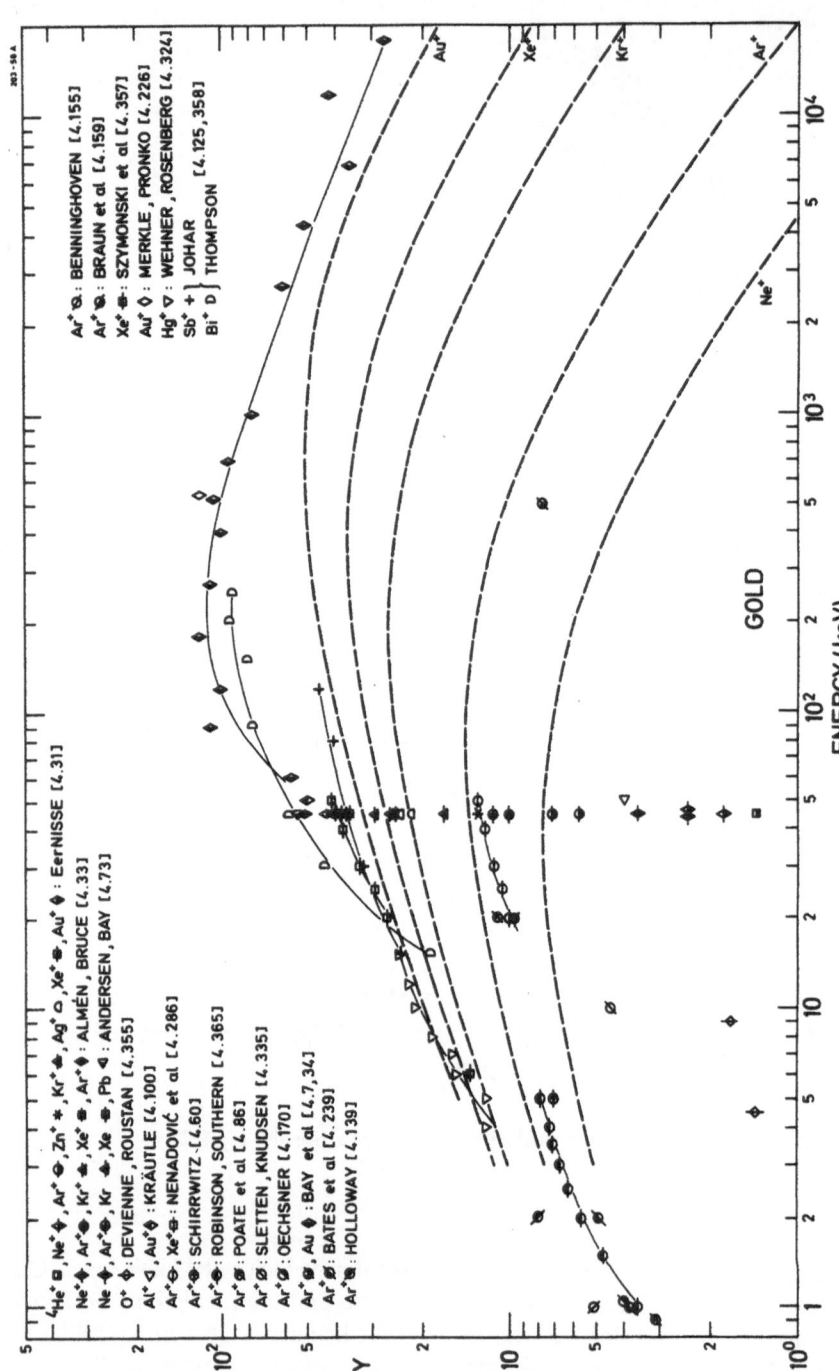

Fig. 4.27. Experimental sputtering-yield data for Au targets. See also Fig. 4.26

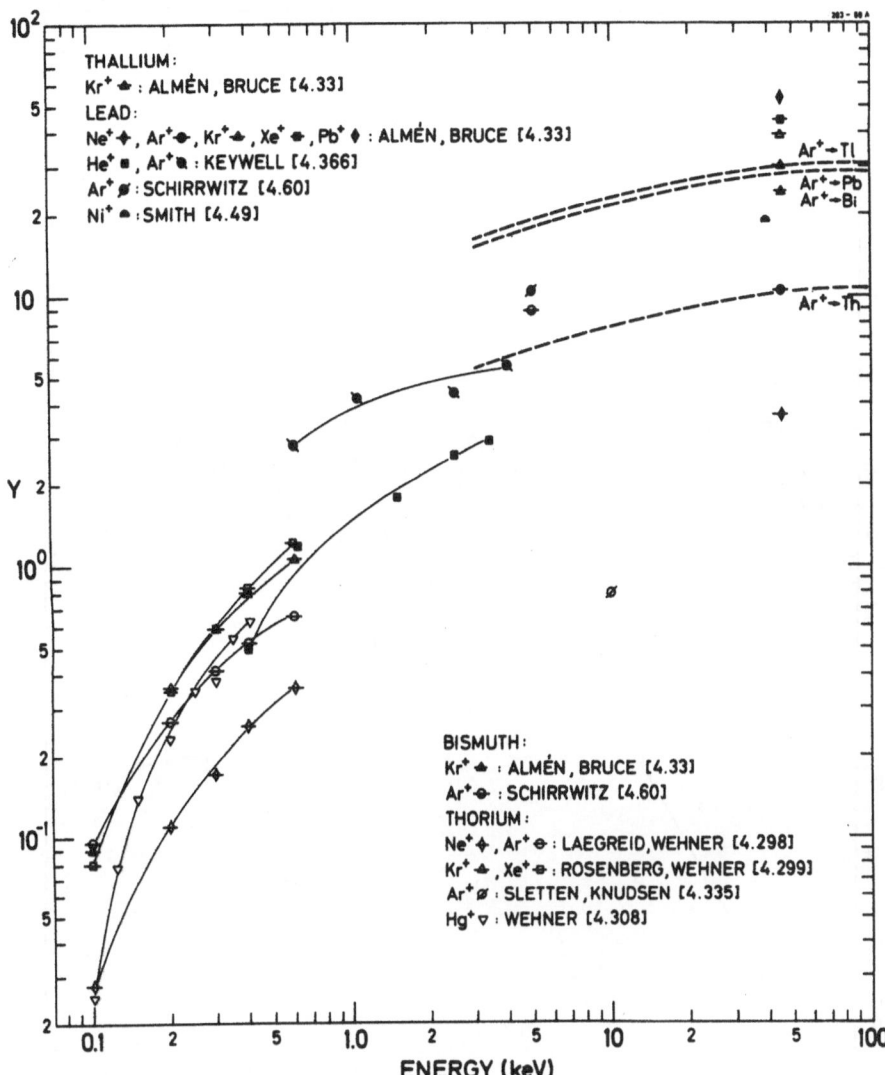

**Fig. 4.28.** Experimental sputtering-yield data for Tl, Pb, Bi, and Th targets

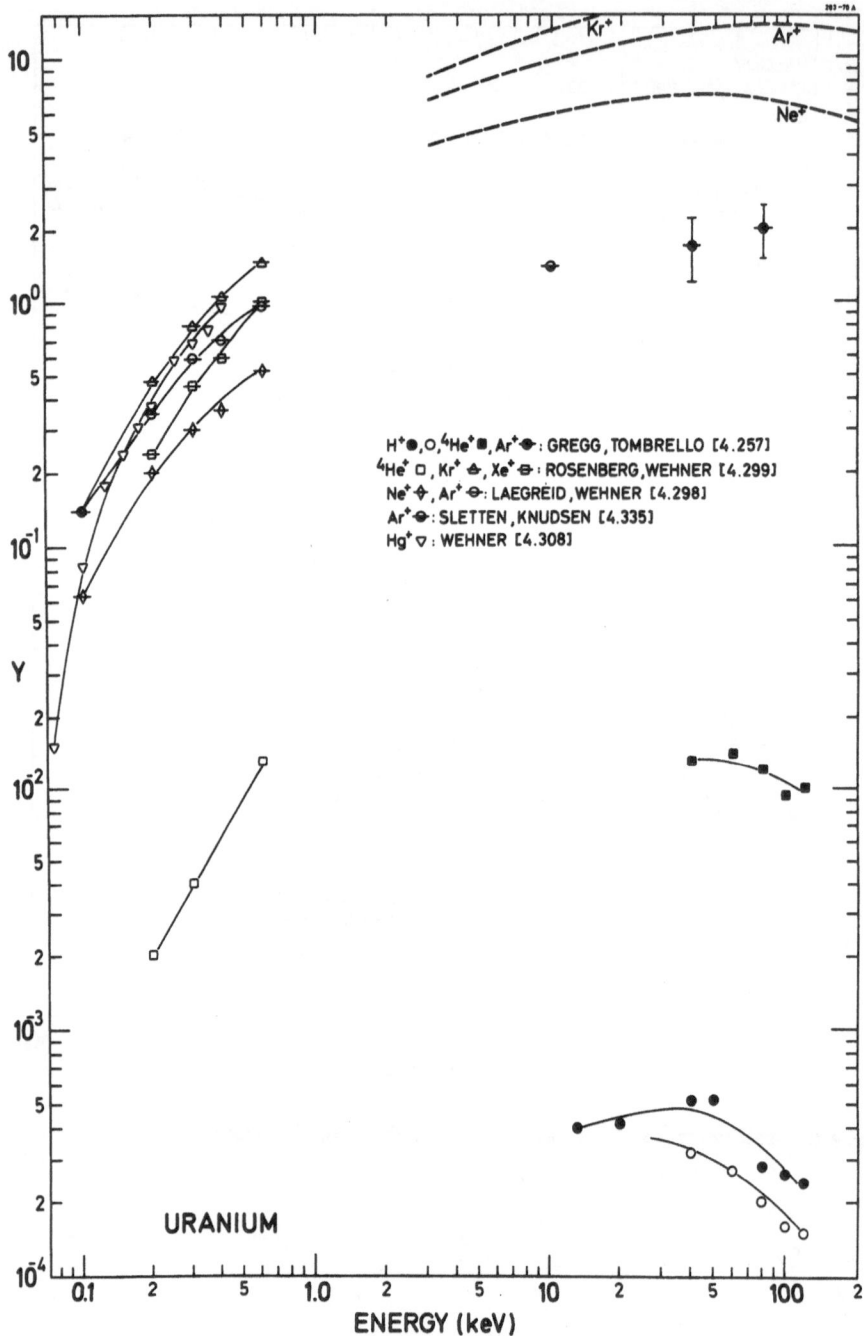

**Fig. 4.29.** Experimental sputtering-yield data for U targets

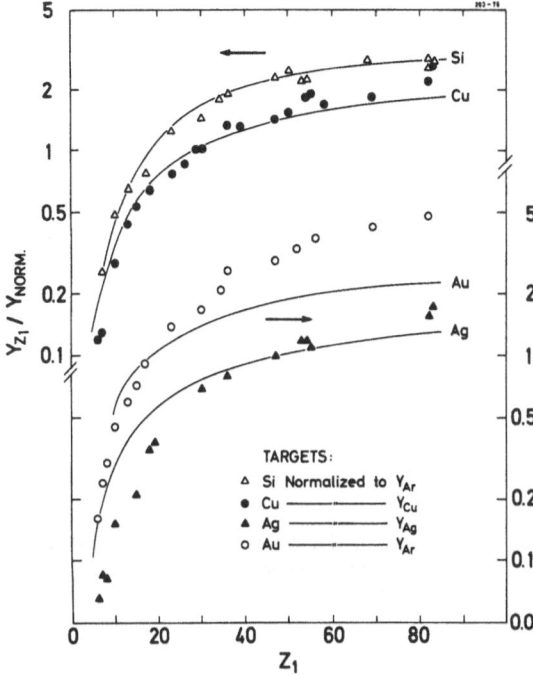

**Fig. 4.30.** Relative sputtering yields for 22 different ions at 45 keV for Si, Cu, Ag, and Au targets. To avoid influence of dose effects, the data have been obtained relative to selfsputtering (Ag, Cu) and Ar sputtering (Si, Au). Solid curves represent theoretical values according to [(4.2.2)]. (Data from *Andersen* and *Bay* [4.12, 70, 73])

whether the above effect has solely the same cause as does the more peaked energy maximum seen for heavy projectiles and targets. Hence, we depict the function $\alpha(M_2/M_1)$, (4.2.2), in Fig. 4.31. Apart from the upward bend for low mass ratios, particularly for gold targets, we see that the predicted existence of a single mass-ratio dependent $\alpha$ function is well fulfilled for the data of *Andersen* and *Bay* [4.12, 70, 73]. However, the deduced function does not agree too well with theoretical predictions for large mass ratios (light projectiles), as discussed in detail below.

The discrepancies between theoretical and experimental results for *heavy ions and/or targets* were interpreted as being caused by *nonlinear effects* in the collision cascade [4.70, 71]. Figure 4.32 presents direct evidence for the existence of such nonlinear effects. A gold target is irradiated with alternating beams of atomic and molecular ions at the same velocity. At both energies, selenium atoms arriving via $Se_2^+$ ions are found to be considerably more effective in causing sputtering than those arriving as $Se^+$ ions. Similar experiments have been performed recently by *Johar* and *Thompson* [4.125], *Thompson* and *Johar* [4.358, 369] on platinum and gold. The nonlinear effects, particularly seen with $Sb_2^+$ and $Sb_3^+$ ions, are much larger for gold than for platinum targets. The latter material has a substantially higher surface-binding energy (sublimation energy) than gold. The energy dependence of the nonlinear effect has not been investigated in any great detail, but Fig. 4.33

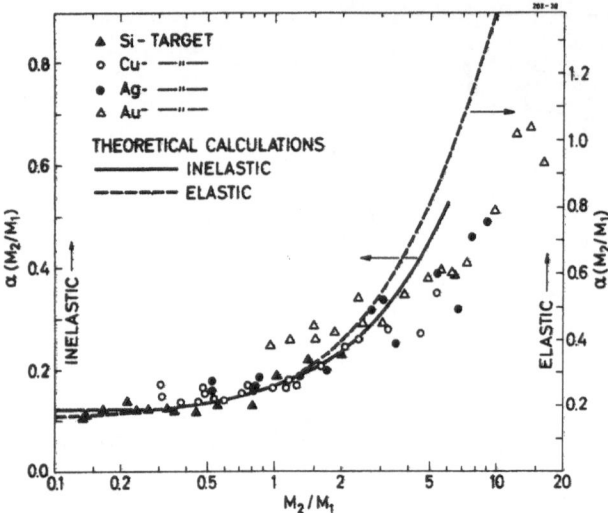

**Fig. 4.31.** Experimental and theoretical values of the function $\alpha(M_2/M_1)$ (4.2.2). Experimental data from Fig. 4.30 [4.73]. Curve marked "elastic" from *Sigmund* [4.89] curve marked "inelastic" from *Winterbon* [4.368]

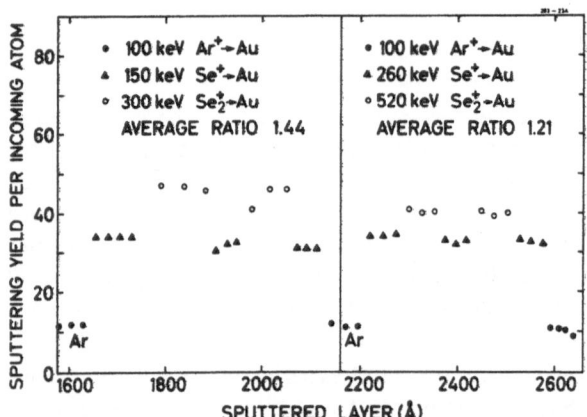

**Fig. 4.32.** Sputtering yields per atom of Au irradiated with 150-keV $Se^+$ and 300-keV $Se_2^+$ ions and 260-keV $Se^+$ and 520-keV $Se_2^+$ ions. (From [4.73])

shows that the excess yield is a slowly decreasing function of energy. Energy dependence of the atomic sputtering yield (e.g., Au→Au, Fig. 4.27) indicates that the effect will also drop off at lower energies.

The trend as a function of $Z_1$ and $Z_2$ may be found from the Table 4.1 giving the relative increase in yields per atom at energies close to the maximum in nuclear-stopping power [4.73, 358].

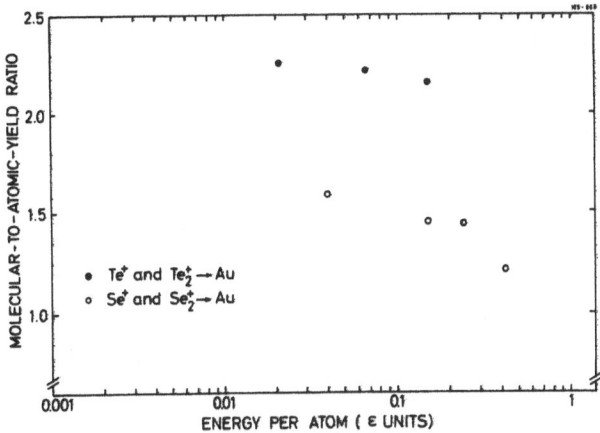

**Fig. 4.33.** Molecular-to-atomic yield ratios of Au for Se and Te irradiations as a function of Thomas-Fermi reduced energy ε. (From [4.73])

**Table 4.1.** Ratios between the sputtering yield *per atom* for irradiation with molecular and atomic ions for different ion-target combinations at energies close to the maximum in nuclear-stopping power

| Projectiles | Targets | | |
| --- | --- | --- | --- |
| | Si | Ag | Au |
| $Cl-Cl_2$ | — | 1.09 | — |
| $Se-Se_2$ | 1.15 | 1.44 | 1.44 |
| $Sb-Sb_2$ | — | 2.48 | 2.59 |
| $Te-Te_2$ | 1.30 | 1.67 | 2.15 |
| $Bi-Bi_2$ | — | 3.97 | 3.92 |

Detailed theoretical discussions of the mechanisms behind the nonlinear effects will be found in Chap. 2.

Hence, for energies well above threshold and for projectile masses larger than ~10 amu, where inelastic effects to a large extent may be neglected, it is concluded that the predicted proportionality to nuclear stopping holds except for heavy projectiles and/or targets.

The discrepancies between *Sigmund*'s [4.89] theoretical α function and the experimentally deduced values for large-mass ratios (Fig. 4.31) is considered to be due to the neglect of electronic stopping and the application of an imaginary surface plane in an infinite target in the calculations [4.73]. For *light projectiles*, the electronic stopping may be comparable to or even larger than the nuclear-stopping power and was included in more recent calculations of $F_D(E, \vartheta, x)$ by *Winterbon* [4.368], by which he obtained a better agreement with the experimental data (Fig. 4.31). The assumption of an infinite target also leads to an overestimation of the sputtering yield since this allows projectiles to be reflected

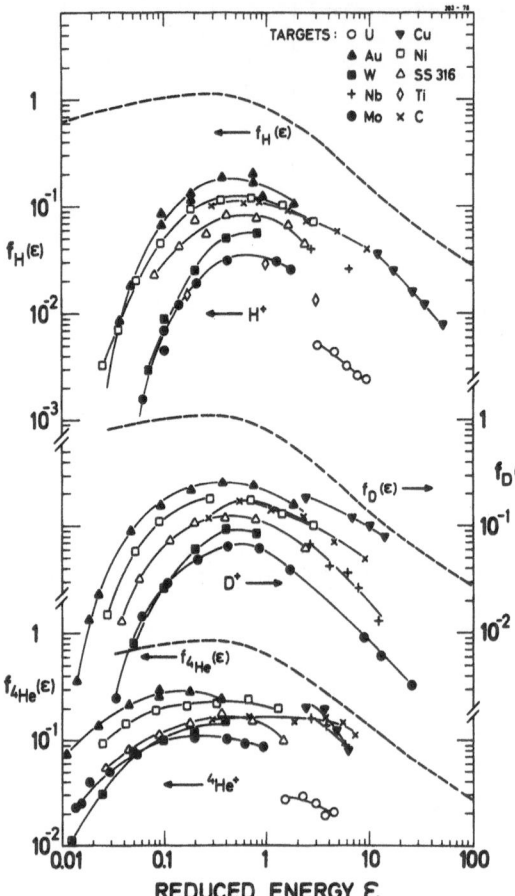

**Fig. 4.34.** Reduced deposited-energy function $f(\varepsilon, 0)$ at the target surface (4.2.3). The dashed lines represent theoretical calculations of *Littmark* and *Maderlechner* [4.370]. Experimental points are obtained by use of (4.2.1, 3) from experimental yields of Cu [4.29, 315], Nb [4.348], Ta and Ti [4.306], Ni and SS316 [4.62], Mo and Au [4.13], W [4.361], C [4.305], Mo [4.351], and U [4.258]

several times across the imaginary target surface. Hence this surface effect must increase with increasing reflection coefficient, i.e., with increasing mass ratio [4.45, 367] in accordance with the experimental findings.

In the extreme case of *proton sputtering*, the discrepancy increases to an order of magnitude or more, as shown in Fig. 4.34. To facilitate comparisons of experimental data from different target materials, the reduced deposited-energy function $f(0, \varepsilon)$ is depicted here instead of the sputtering yield, where $f(0, \varepsilon)$ is defined by

$$f(\varepsilon, 0) = (\varepsilon/E)(x/\varrho)F_D(E, 0, 0), \tag{4.2.3}$$

and $F_D(E, \vartheta, x)$ is the deposited-energy distribution, see (4.2.1), with $\varepsilon$ and $\varrho$ being Lindhard's reduced energy and range parameters [4.371] given by $\varepsilon = EaM_2/Z_1Z_2e^2(M_1 + M_2)$ and $\varrho = xNM_24\pi a^2M_1/(M_1 + M_2)^2$ with $a = 0.8853\, a_0(Z_1^{2/3} + Z_2^{2/3})^{-1/2}$ and the Bohr radius $a_0 = 0.529$ Å. The calculated

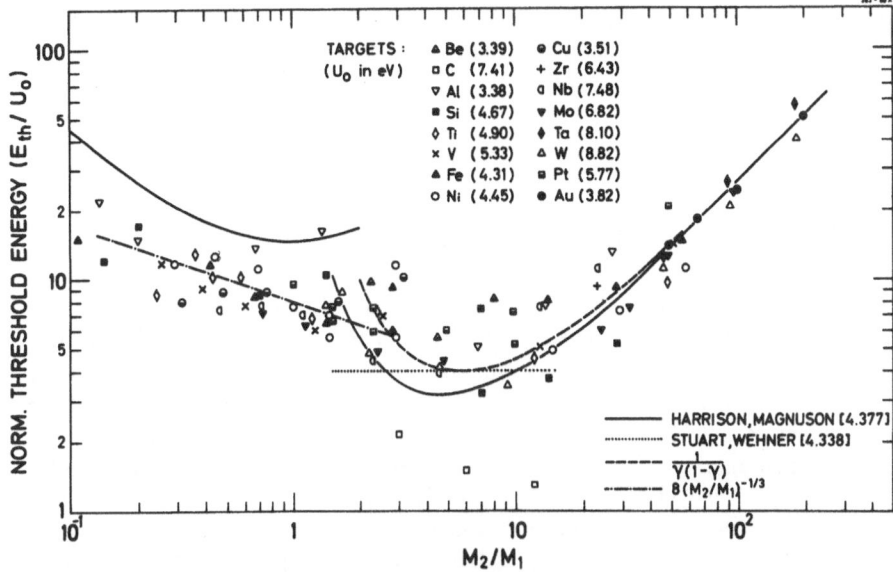

**Fig. 4.35.** Normalized threshold energies $E_{th}/U_0$ for 16 elements obtained through extrapolation of sputtering-yield data by use of the normalized sputtering-yield function [Fig. 4.36 and (4.2.6)]. All data shown in Figs. 4.3–29 for the 16 different target materials have been analyzed, but only those that approximately fit the energy dependence given by the normalization function have been included. The data for Be and Al are probably influenced by oxide layers on the surface. The carbon points are too low, indicating that the actual surface-binding energy is lower than the sublimation energy. For surface-binding energies, the sublimation energies $U_0$ have been used and are depicted in the figure (eV/atom) as found from *Smithells* [4.375] and *Stull* and *Prophet* [4.376]

$f(\varepsilon, 0)$ depends only very slightly on the type of projectile and target material. The experimental values are deduced by combining (4.2.1) and (4.2.3).

Although the energy dependence of $f(\varepsilon)$ seems reasonable at high energies $(\varepsilon \gtrsim 2)$, it is obvious from Fig. 4.34 that (4.2.1) is not suitable for a description of light-ion sputtering yields. Besides the overestimate due to the absence of a surface-correction term in (4.2.1), the experimentally deduced $f(\varepsilon)$ values show a strong dependence on target material, contrary to the theoretical prediction. The rapid increase in the discrepancy with decreasing energy is inherent with the assumption in the theory of an isotropic cascade, which is no longer fulfilled at low energies. A similar tendency is seen for heavier ions (see Figs. 4.5–29). It should be noted, however, that for light ions, the problematic energy region may extend to energies of the order of 10 keV due to the small energy-transfer factor between projectile and target atoms.

The collision-cascade theory of sputtering resulting in (4.2.1) demands the projectile energy to be substantially larger (i.e., two or three orders of magnitude) than the surface-binding energy. Hence, the *threshold region* is rather broad. The experimental threshold energy, i.e., the projectile energy below which no sputtering occurs, is sometimes obtained from experimental data by linear extrapolation of the yield-versus-energy curve to zero yield [4.372]. As the energy dependence near the threshold is largely unknown, the yield must be measured as near as possible to the threshold energy to obtain a reliable value. This requires a measuring technique which works satisfactorily for sputtering yields varying by several orders of magnitude and clearly calls for particular caution during the measurements.

Unfortunately, most investigations of the threshold energy have been performed in a plasma discharge [4.308, 318, 338, 340]. The measurements, where a mass and energy-analyzed beam was used, were performed by an insufficiently sensitive method or were not specifically devoted to the investigation of threshold energies [4.13, 334, 363, 364]. With retarded ion beams, neutral particles in the beam may constitute a serious problem [4.13, 68, 174] which may be overcome by electrostatic deflection of the beam immediately in front of the target [4.339, 373].

The most extensive investigations of threshold energies have been made by *Stuart* and *Wehner* [4.338], *Askerov* and *Sena* [4.340], and *Oechsner* [4.323]. *Stuart* and *Wehner* concluded that the threshold energy is independent of projectile mass with $E_{th} \simeq 4 \times U_0$, whereas *Oechsner* [4.323] found a weak variation for $M_1 > 20$ and a much stronger dependence for light projectiles. This latter result was qualitatively confirmed by measurements by *Bay* et al. [4.13].

In recent work, *Bohdansky* et al. [4.374] extracted threshold energies from low-energy sputtering-yield data by fitting the data to a universal sputtering-yield curve, see (4.2.5) below, instead of using a linear extrapolation. Figure 4.35 shows their threshold energies together with values extracted from Figs. 4.3–29. For large mass ratios $(M_2/M_1 \gtrsim 5)$, the threshold energy for metal targets is roughly given by

$$E_{th} = U_0/(1-\gamma)\gamma, \tag{4.2.4}$$

where $\gamma$ is the energy transfer factor $\gamma = 4M_1 M_2/(M_1 + M_2)^2$. This result can be interpreted to the effect that the sputtering near the threshold is caused by reflected projectiles on their way *out* through the surface. This mechanism has previously been discussed for hydrogen and helium sputtering by *Behrisch* and *Weissmann* [4.84], *Bay* et al. [4.13], and in most detail by *Behrisch* et al. [4.378].

For $M_2/M_1 \sim 1$, the threshold energy reaches a minimum about $4U_0$. For small mass ratios $(M_2/M_1 \lesssim 1)$, the projectile cannot be reflected in a single collision. Hence, the sputtering is mainly determined by collisions between target atoms, which makes the threshold energy nearly independent of the projectile mass in spite of the maximum in $\gamma$ at $M_2/M_1 = 1$.

The lack of a suitable analytical theory for light-ion sputtering calls for other methods to predict sputtering yields, particularly in the threshold region, which might be of special importance for fusion-reactor applications. *Behrisch* et al. [4.378] and *Haggmark* and *Wilson* [4.379] have been able to reproduce the energy dependence of the experimental data by computer simulation in the

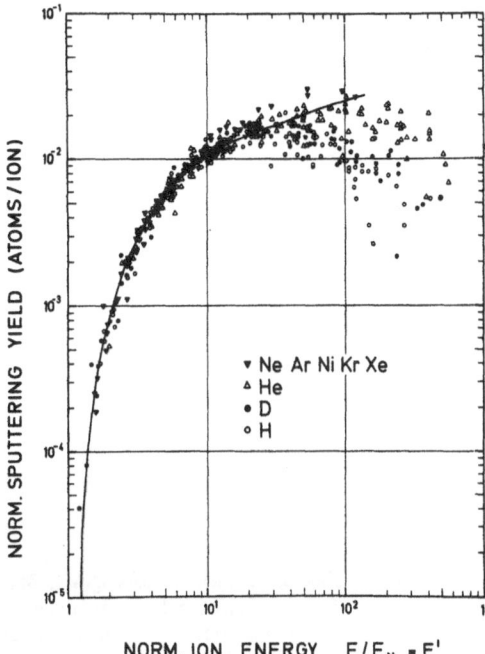

NORM. SPUTTERING YIELD (ATOMS / ION)

▼ Ne Ar Ni Kr Xe
△ He
• D
○ H

NORM. ION ENERGY  $E / E_{th} = E'$

**Fig. 4.36.** Normalized sputtering yields for low-energy sputtering as a function of $E' = E/E_{th}$. The yield factors $Q(M_1, M_2, U_0)$ are given in Fig. 4.37 and $E_{th}$ in Fig. 4.35. (From *Bohdansky* et al. [4.374])

energy region from the threshold to about 10 keV and even obtain reasonable absolute values. The fit of *Haggmark* and *Wilson* is, however, based on the use of unphysical values for a volume-binding energy. A *scaling law for low-energy, light-ion sputtering* was proposed by *Bay* et al. [4.13] with an energy parameter $E'$ defined by $E' = E/E_{th}$ and found to work successfully up to about 20 times the threshold energy for the target materials investigated. The scaling law was further developed by *Bohdansky* et al. [4.374]. By also considering the threshold energy as an adjustable parameter, they found the scaling law to work even for heavy-ion sputtering. The threshold energies for metals for $M_2/M_1 \gtrsim 5$ are related to each other through $E_{th} = E_B/(1 - \gamma)\gamma$. $E_B$ may be interpreted as an average surface-binding energy, which is always found to be very close to the sublimation energy $U_0$, see Fig. 4.35 and (4.2.4). Figure 4.36 shows sputtering yields for a large number of projectile-target combinations as a function of $E'$. The yields are normalized by means of experimentally determined yield factors $Q(M_1, M_2, E_B)$ given by

$$Y(E') = Q(M_1, M_2, E_B) f(E'), \tag{4.2.5}$$

where $f(E')$ is a universal yield-energy curve given approximately by

$$f(E') \cong 8.5 \times 10^{-3} E'^{1/4} (1 - 1/E')^{7/2}. \tag{4.2.6}$$

The yield factor $Q$ depends on a projectile-target combination through $M_1, M_2$, and $E_B$. So far, the yield factors must be considered fitting parameters only.

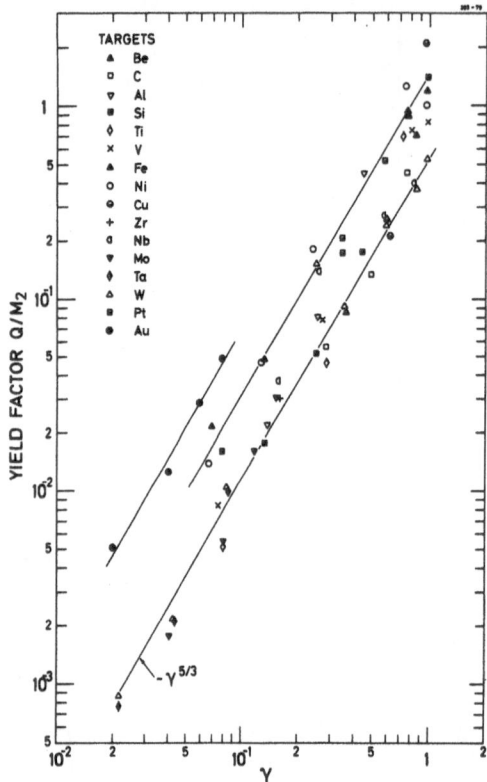

Fig. 4.37. Yield factors $Q(M_1, M_2, U_0)$ for $M_2/M_1 > 1$ for 16 different target materials as a function of $\gamma = 4M_1M_2/(M_1 + M_2)^2$. The straight lines are proportional to $\gamma^{5/3}$. (For further details, see caption to Fig. 4.35)

They are depicted in Fig. 4.37 as a function of $\gamma$. It is seen that

$$Q(M_1, M_2, E_B) \cong a \cdot M_2 \gamma^{5/3}, \tag{4.2.7}$$

where $a$ depends on the target material only, mainly through a weak dependence on the surface-binding energy (Fig. 4.37). As an approximation accurate within a factor of two, the value $a = 0.75$ may be used except for gold. By means of (4.2.6, 7), low-energy sputtering yields can be calculated if the threshold energy $E_{th}$ is known. For light projectiles, (4.2.4) may be used to find $E_{th}$.

The dependence of the sputtering yield on projectile species at low energies has been investigated by *Fetz* and *Oechsner* [4.297]. Their results for beryllium, nickel, and platinum targets show a dependence on projectile mass which varies with energy. For low energies they found a maximum at a target-projectile mass ratio of $M_2/M_1 \simeq 1$. Recently, the measurements on nickel were repeated by *Bay* et al. [4.337], who used a mass-analyzed beam. At 150 eV projectile energy, they found the maximum in the yield at $M_2/M_1 \simeq 2$ in better agreement with the minimum in the threshold energy at mass ratio 5 (Fig. 4.35). Both results are shown in Fig. 4.38 together with data from *Laegreid* and

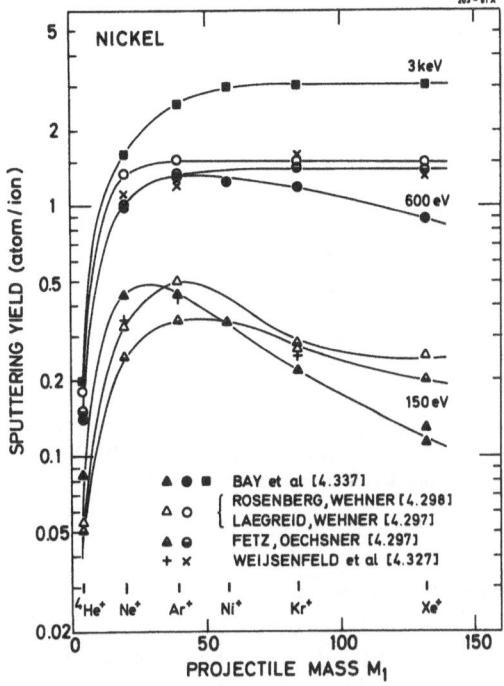

**Fig. 4.38.** The $M_1$ dependence of the sputtering yield of nickel at normal incidence for bombarding energies 150 eV, 600 eV, and 3 keV

*Wehner* [4.298], *Rosenberg* and *Wehner* [4.299], and *Weijsenfeld* et al. [4.327]. The disagreement in the $M_1$ dependence between the older results and those of *Bay* et al. [4.337] is thought to be due to the influence of secondary-electrons and impurities in the plasma in the former experiments, which were all performed in a plasma discharge. At higher energies, the dependence on the type of projectile changes to a monotonic increase in the yield with increasing projectile mass (or projectile atomic number, see Fig. 4.30). This tendency can already be observed in the nickel data in Fig. 4.38.

In the light of the above discussion of the total body of experimental data, the region of validity of the linear, analytic sputtering theory may be summarized by means of Fig. 4.39. The region is a simple-connected subspace of the three-dimensional $(Z_1, Z_2, E)$ space. Figure 4.39 shows two cuts through such a space. The upper figure is a $(Z_1, Z_2)$ plane for $E \simeq 50$ keV, the lower a $(E, Z_1)$ plane for $Z_2 \simeq 50$. Within each shaded region, the main effect contributing to the breakdown of the theory is given. Boundaries are of course not sharp. They have tentatively been drawn to indicate deviations from theory of a similar magnitude for each effect. The region marked "inelasticity" indicates a breakdown of (4.2.2) only, not of (4.2.1). Monte-Carlo simulations may be used to extend the theory into all the shaded regions except that marked "spikes" as simulation models are, at least in the currently used forms, also linear.

The use of the logarithmic energy scale in Fig. 4.39 camouflages the fact that (4.2.2) is an amazingly good prediction for a very large number of $(Z_1, Z_2, E)$ combinations of practical interest.

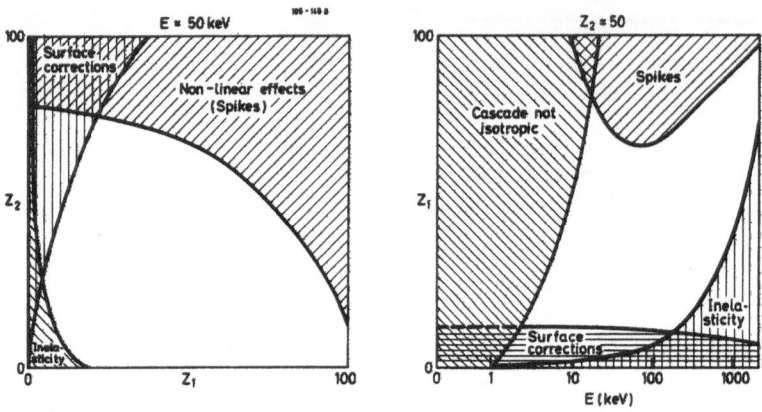

**Fig. 4.39.** The region of validity of the linear, analytic sputtering theory illustrated by two planar cuts through the $(Z_1, Z_2, E)$ space. Left figure shows the $(Z_1, Z_2)$ plane for $E = 50$ keV, while right figure shows an $(E, Z_1)$ plane for $Z_2 = 50$. Boundaries are approximate only

### 4.2.4 Dependence on Angle of Incidence

The first qualitative information on the angular dependence of the sputtering yield was obtained by *Fetz* [4.118] and *Wehner* [4.380, 381], while more quantitative results were awaiting a development of mass-analyzed and well-collimated ion beams. For heavy-ion bombardment in the keV-energy region *Molchanov* and *Tel'kovskii* [4.382], *Dushikov* et al. [4.383], and *Mashkova* and *Molchanov* [4.384] found a dependence on the angle of incidence given by $\psi(\vartheta)/\psi(0^\circ) = 1/\cos\vartheta$, where $\vartheta$ is measured from the surface normal (Fig. 4.40). Such a dependence was predicted by *Goldman* and *Simon* [4.385] and *Pease* [4.386] and may also be obtained from (4.2.1) by simple arguments. *Sigmund* [4.89], however, showed that the $(\cos\vartheta)^{-1}$ dependence should be expected only for large $M_2/M_1$. For smaller mass ratios, a faster variation with angle should be expected. A faster dependence was in fact found by *Cheney* and *Pitkin* [4.345] for Xe→Mo, while a $(\cos\vartheta)^{-1}$ trend fitted their Xe→Cu data well. Hence their results displayed a mass-ratio trend opposite to the one predicted by *Sigmund* [4.89].

The results of *Rol* et al. [4.342], *Almén* and *Bruce* [4.33], *Dupp* and *Scharmann* [4.387], and *Ismail* and *Septier* [4.309] all showed the somewhat stronger dependence on $\vartheta$ than given by $1/\cos\vartheta$ and could be reasonably well described by the formula given by *Sigmund* [4.89] (Fig. 4.40). An even stronger variation with $\vartheta$ was found by *Ramer* et al. [4.388] for argon sputtering of copper and silver and by *Holmén* and *Almén* [4.310] for mercury-ion sputtering of iron and graphite. Finally, *Edwin* [4.42] investigated the angular dependence of the sputtering yield for argon bombardment of fused silica and found a very steep increase with increasing angle of incidence, which does not follow any of the theoretical predictions. The fine structure in the angular dependence found by *Evdokimov* and *Molchanov* [4.389, 390] must be ascribed to crystal effects,

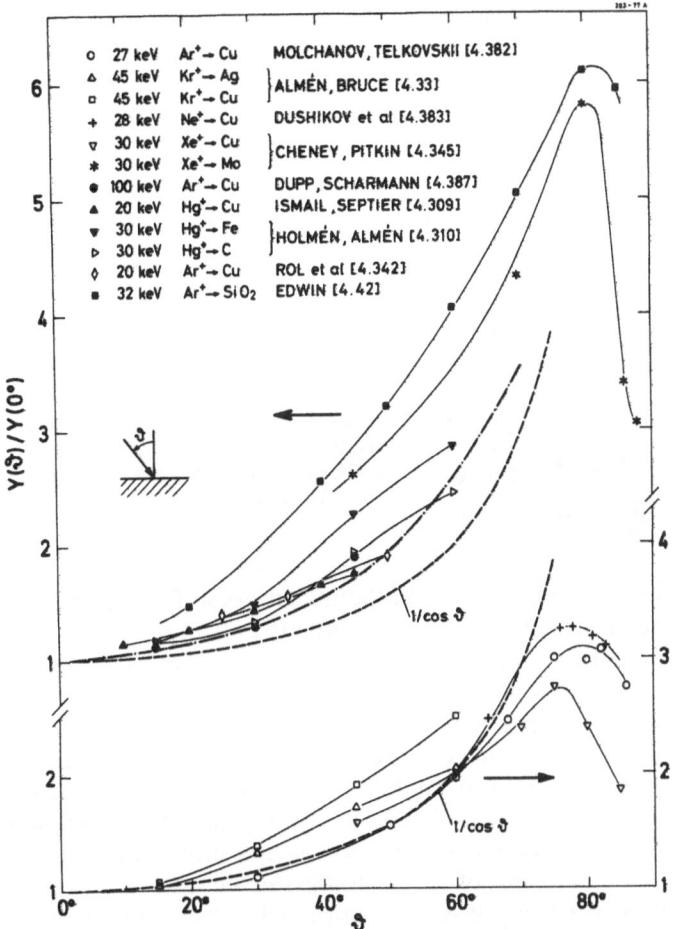

**Fig. 4.40.** Heavy-ion sputtering yields in the keV energy region as a function of angle of incidence. Dash-and-dot curve represents the prediction of *Sigmund* [4.89]. The thin curves are drawn only to guide the eye

and also the results for niobium selfsputtering by *Summers* et al. [4.349] show a distinct influence of the texture of their targets.

In all cases investigated, the yield reaches a maximum at angles between 60° and 80°, while it decreases rapidly for larger angles. Such behaviour is thought to be caused by a rapid increase in the reflection coefficient as the direction of incidence approaches the glancing one.

No particular trend may be discerned in the experimental data, which often disagree with one another (e.g. *Cheney* and *Pitkin* [4.345] found a much stronger dependence on $\vartheta$ for molybdenum than for copper targets, while *Mashkova* and *Molchanov* [4.384] came to the opposite result). One reason for the lack of

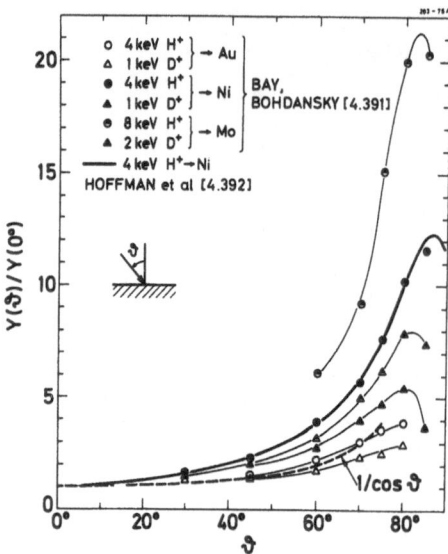

Fig. 4.41. Light-ion sputtering yields as a function of angle of incidence for Ni, Mo, and Au targets. Thin solid lines are drawn merely to guide the eye. The thick solid line represents the computation of *Hoffmann* et al. [4.392], experimental data are from *Bay* and *Bohdansky* [4.391]

reproducibility of the data of Fig. 4.40 may be the influence of surface topography, to which most of the investigators have paid only little attention.

At energies around 1 keV, the angular dependence has been investigated systematically for noble gases for a number of target materials by *Oechsner* [4.170]. For not too large angles of incidence, the normalized increase in the yield defined by $\Delta Y(\vartheta)/Y(\vartheta_{opt})$, where $\Delta Y(\vartheta) = Y(\vartheta) - Y(0^0)$ and $\vartheta_{opt}$ are the angle of incidence at maximum yield, could be described by a single function $F(\vartheta') = 1.2 \vartheta'^2$, where $\vartheta' = \vartheta/\vartheta_{opt}$. This change in the angular dependence was shown by *Oechsner* [4.170] to be in agreement with simple theoretical estimates and is caused by the anisotropy of the collision cascade at low projectile energies (Fig. 4.39). The results have been reviewed recently by *Oechsner* [4.171].

The dependence on angle of incidence for light-ion sputtering in the low-keV range has most recently been investigated by *Bay* and *Bohdansky* [4.391]. The most interesting features of their results is the very rapid variation of the yield with angle of incidence. Further, the maximum is not reached before angles of $\vartheta_{opt} \gtrsim 80°$ (Fig. 4.41). The maximum in the normalized yield $Y(\vartheta_{opt})/Y(0°)$ was found to be higher the lighter the projectiles, and to increase with increasing projectile energy and increasing surface-binding energy of the target material. By means of simple estimates using power-law potentials, the latter result could be related to the contribution to the sputtering from direct knock-off surface atoms.

The results of *Bay* and *Bohdansky* [4.391] do not agree with previously published data for 1.05-keV $He^+ \rightarrow Cu$ by *Oechsner* [4.170] and for 12.2-keV $D^+ \rightarrow Nb$ by *Summers* et al. [4.348] as these publications showed a much weaker dependence on $\vartheta$. However, they do agree with the result of *Dushikov* et al. [4.383] for 29-keV $He \rightarrow Cu$. The latter authors also found $\vartheta_{opt} > 80°$.

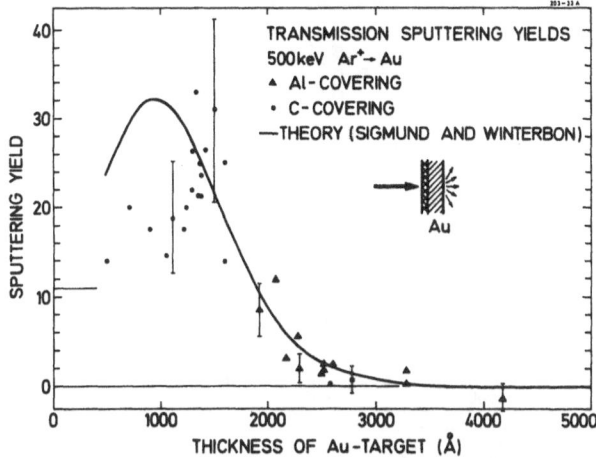

**Fig. 4.42.** Transmission-sputtering yield of Au irradiated with 500-keV Ar ions as a function of Au-film thickness [4.34]. Solid curve represents the deposited-energy function normalized to the measured backsputtering yield

Recently, *Hoffman* et al. [4.392] computed the angular dependence of proton-sputtering yields of nickel numerically and found excellent agreement with the experimental data of *Bay* and *Bohdansky* [4.391]. In their computation, the optimum angle $\vartheta_{opt}$ could be investigated in more detail and was found to increase from 85° at 1 keV to about 88° at 8 keV.

### 4.2.5 Transmission Sputtering

Transmission-sputtering measurements present rather unique possibilities for checking one of the basic results of collision-cascade theory, but experiments up till now have been few and scattered in their aim. This trend may be due to the fact that transmission-sputtering experiments are even more difficult to perform in a reproducible way than backscattering ones.

Transmission-sputtering experiments may be divided into three groups according to the main objective of the measurements, which are: (i) studies of deposited-energy depth distributions, (ii) investigations of the mechanisms responsible for very small yields far beyond the depth covered by the deposited-energy function, and (iii) neutron-sputtering simulations.

According to current sputtering theory, the sputtering yield from a free surface is proportional to the intensity of the deposited-energy function at that same surface. Provided the target foil is thick compared to recoil ranges in the target, the existence of a target-front surface will not influence the distribution function at the back surface. Hence, measurements of transmission sputtering may be used to investigate the deep tail of deposited-energy distribution functions. This tail is difficult to investigate by means of more conventional methods used to measure depth distributions of radiation damage.

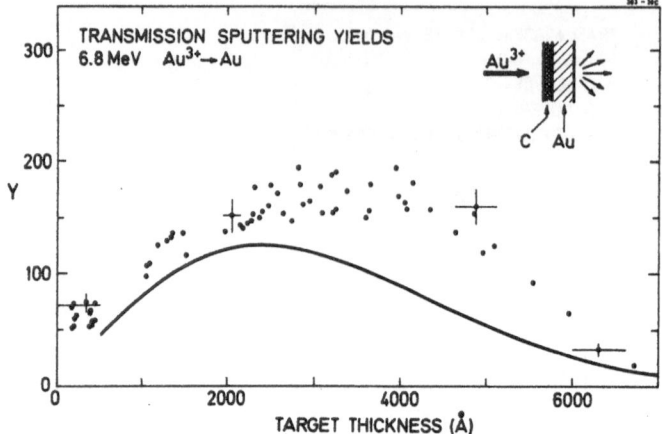

**Fig. 4.43.** Transmission-sputtering yield of Au irradiated with 6.8-MeV Au ions as a function of Au-film thickness [4.35]. Solid curve represents the deposited-energy function normalized to the measured backsputtering yield

Measurements have been performed by *Robison* [4.243] (15-keV $H^+$, $D^+ \rightarrow Au$), *Mertens* [4.97, 192] (250-keV Ar→Al, Cu), *Bay* et al. [4.34] (500-keV Ar→Au), *Bay* et al. [4.35] (6.8-MeV Au→Au), *Marwick* [4.393] (46.7-MeV Ni→Au), and *Ecker* and *Merkle* [4.124] (560-keV Bi, 400-keV Ag, 400-keV Ar, 50-keV He→Au). The data of *Mertens* are not presented in a form where they are easy to interpret, and all other data have been obtained on gold targets. Figure 4.42 shows the data of *Bay* et al. [4.34] for 500-keV Ar on gold. In spite of the large scatter in the data, the depth distribution is in good agreement with theory. For heavier projectiles (6.8-MeV Au→Au), Fig. 4.43 shows the maximum at a larger depth and with a higher value than predicted. This is explained as being due to the nonlinear effects discussed in Sect. 4.2.3. For much higher energies and, again, for somewhat lighter projectiles, *Marwick* [4.393] found the position of the maximum to correspond with the theoretical prediction (46.7-MeV Ni→Au), but the distribution was narrower than expected. At such high energies, the position of the theoretically calculated maximum depends critically on the electronic-stopping power used and does not carry much information on other aspects of cascade theory. For argon irradiations, *Ecker* and *Merkle* [4.124] found, as did *Bay* et al. [4.34], that the yield in the maximum was lower than expected, while 560-keV Bi on gold showed good agreement (see the figures in the following paragraph).

Figure 4.42 shows that no substantial contribution to the transmission-sputtering yields stems from mechanisms other than random-collision cascades (less than a 10% contribution from mechanisms with ranges larger than 100 Å [4.34]). Measurements where the sputtered material is collected are much better suited for a study of the penetrating tail of the distribution. As possible contributions from long-range focussons (see Chaps. 3 and 5) have been in the

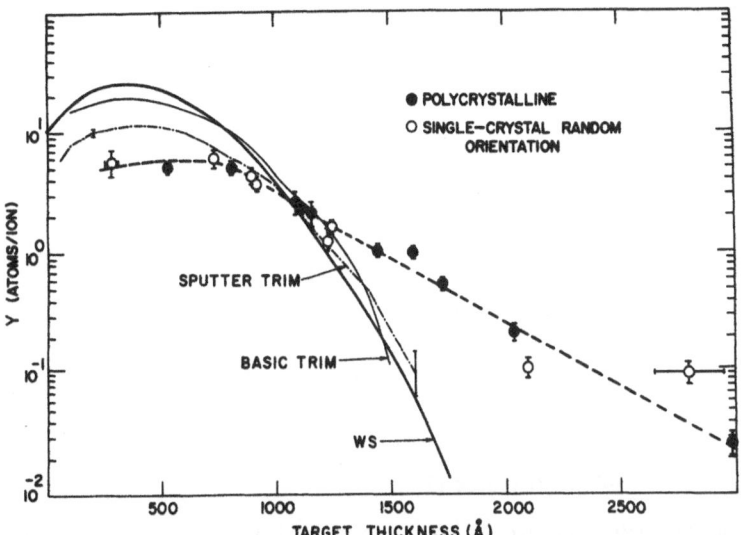

**Fig. 4.44.** Transmission-sputtering yield of Au irradiated with 200-keV Ar ions as a function of Au-film thickness. Note the logarithmic scale to accentuate the weak but penetrating tail. (From *Ecker* and *Merkle* [4.124])

center of interest for a number of years, a substantial part of transmission-sputter measurements have been devoted to a study of those, among them some early studies of proton-transmission sputtering [4.251, 394, 395].

Systematic studies of the penetrating tails were taken up by *Ecker* [4.229] and continued by *Ecker* and *Merkle* [4.124]. *Ecker* used 0.28–8-keV noble-gas ions, while his investigations with *Merkle* utilized much more energetic ions, as listed above. Figures 4.44, 4.45 show results for 200-keV argon and 560-keV bismuth irradiations. The most notable aspect is the exponential tail (note the logarithmic plot in contrast to Figs. 4.42 and 4.43). The authors concluded that the tail is caused by dechanneling of channeled projectiles in the target and that focussons with a range longer than 50 Å do not contribute to transmission sputtering of gold. This conclusion is based on both energy, and, in particular, projectile-species dependence of the intensity in the tail. Similar tails were found by *Robison* [4.243] and *Ayrault* et al. [4.396, 397]. The latter authors interpreted their data in terms of long-range focussed collision sequences.

The last group of measurements utilizes the fact that high-energy protons, which may be available in large fluences from tandem accelerators, may be used to simulate the sputtering of 14-MeV neutrons where intense beams are very difficult to create. Measurements with approximately 16-MeV protons on niobium and gold have been performed [4.8, 221, 398, 399]. A discussion of these results belongs together with a general discussion of neutron sputtering, but it may be noted that the ratio between backward and transmission-

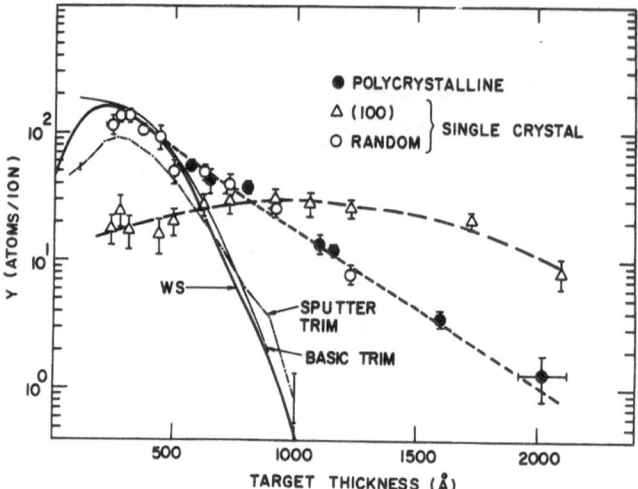

**Fig. 4.45.** Transmission-sputtering yield of Au irradiated with 560-keV Bi ions as a function of Au-film thickness. (From *Ecker* and *Merkle* [4.124])

sputtering yields in such experiments is often at variance with conclusions drawn from the results of energy-reflection measurements [4.400].

## 4.2.6 Energy-Reflection Coefficients (Sputtering Efficiency)

While transmission sputtering carries information on the deep tail of the deposited-energy function, the part of this function, which in an infinite medium lies in front of the projectile starting point, may be investigated by means of the energy-reflection coefficient. This coefficient, which naturally lends itself to investigation by calorimetric methods, was originally termed sputtering efficiency [4.401].

For mass ratios $M_2/M_1$ not substantially larger than one, the main part of the reflected energy is carried by the sputtered species. This will be the region discussed here. In the opposite region, where reflected projectiles dominate, substantial amounts of data exist [4.402, 403, 404]. They are of particular interest for the plasma-wall interaction problem but only weakly related to sputtering and will not be further discussed here.

The calorimetric-collection method was introduced by *Almén* and *Bruce* [4.405] by *Andersen* [4.406, 407] and later used by *Hildebrandt* and *Manns* [4.408, 409], *Düsterhöft* et al. [4.410], and *Hildebrandt* et al. [4.411]. The light-ion measurements mentioned above show that the method yields results that are accurate to within 10%. Discrepancies between *Andersen*'s results and those of *Hildebrandt*'s group for a few target materials (Cu, Si) are not understood as the results of the two groups agree nicely for other target materials.

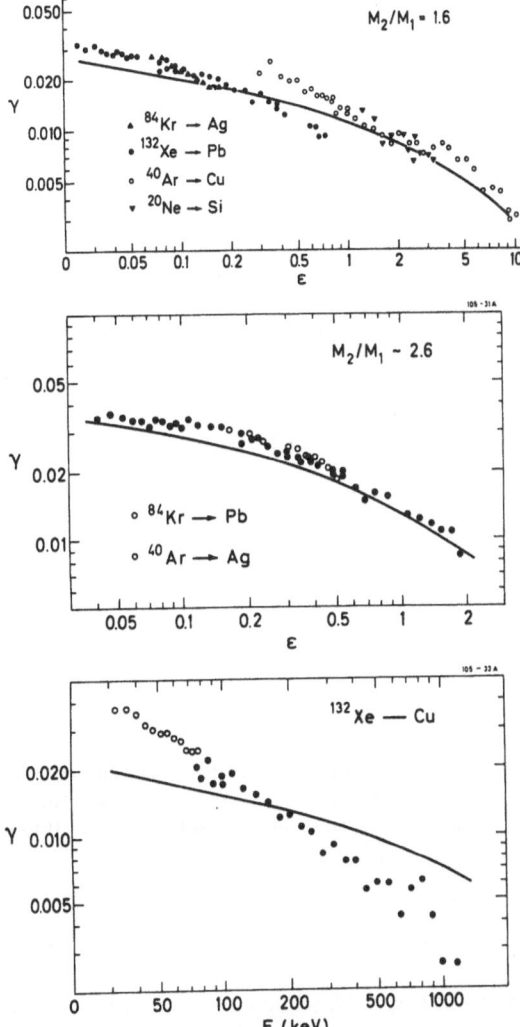

**Fig. 4.46.** Energy-reflection coefficient for mass ratio 1.6 as a function of Thomas-Fermi reduced energy. Solid curve is the theoretical result calculated from *Winterbon* [4.368], while experimental data stem from *Andersen* [4.407]

**Fig. 4.47.** Energy-reflection coefficient for mass ratio 2.6 as a function of Thomas-Fermi reduced energy. Solid curve is calculated from *Winterbon* [4.368], while experimental data stem from *Andersen* [4.407]

**Fig. 4.48.** Energy-reflection coefficient for mass ratio 0.5 as a function of Thomas-Fermi reduced energy. Solid curve is the theoretical result calculated from *Winterbon* [4.368], while experimental data stem from *Andersen* [4.407]

*Andersen*'s [4.406, 407] results cover a broad energy region. Generally, it is found that the main parameters are mass ratio $M_2/M_1$ and Thomas-Fermi reduced energy $\varepsilon$. Figures 4.46, 4.47 show results (energy region roughly 30–1200 keV) for mass ratios 1.6 and 2.6 as compared to theoretical results extracted from *Winterbon*'s [4.368] tables. For mass ratio 0.5, the agreement is much less convincing, as shown in Fig. 4.48.

The energy-reflection coefficient has also been measured as a function of angle of incidence. Empirically, it was found that

$$\gamma(\vartheta) = \gamma(0°) + [1/2 - \gamma(0°)](1 - \cos\vartheta)^2 \qquad (4.2.8)$$

up to at least $\vartheta = 45°$.

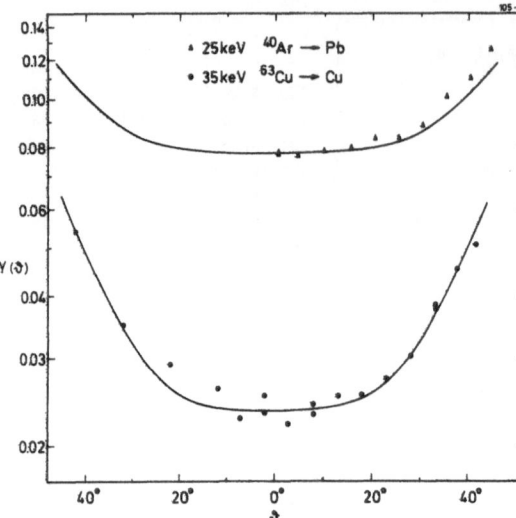

**Fig. 4.49.** Dependence of energy-reflection ratio $\gamma$ for Ar and Cu on Pb and Cu as a function of angle of incidence. The curves represent the functions $\gamma(\vartheta) = \gamma(0°) + [1/2 - \gamma(0°)] (1 - \cos\vartheta)^2$ [(4.2.8)]. (From *Andersen* [4.407])

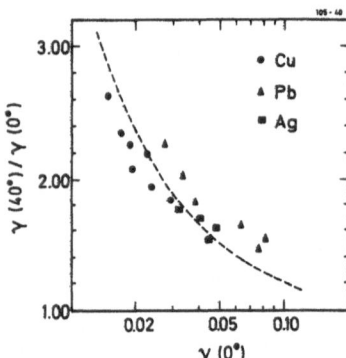

**Fig. 4.50.** $\gamma(40°)/\gamma(0°)$ for 17 different combinations of target materials, ion species, and energy. Broken curve represents (4.2.8) with $\vartheta = 40°$. (From *Andersen* [4.407])

The quality of the fit is illustrated in Fig. 4.49, while a number of projectile-target combinations are represented in Fig. 4.50, which gives the ratio $\gamma(40°)/\gamma(0°)$ as a function of $\gamma(0°)$ and compared to (4.2.8). This fit also looks reasonable.

It is concluded that for mass ratios between 1 and 5, $\gamma(0°)$ may safely be extracted from *Winterbon's* [4.368] tables. For $M_2/M_1$ smaller than 1, the experimental data show a considerably stronger energy dependence than predicted [4.407, 411] and experimental data must be used. The dependence on angle of incidence is satisfactorily represented by (4.2.8).

*Acknowledgements.* The careful technical assistance of Tove Asmussen in preparing the figures and that of Alice Grandjean in preparing the manuscript are gratefully acknowledged.

# References

4.1     W.R.Grove: Philos. Trans. R. Soc. London, (1852) p. 87
4.1a    R.Behrisch (ed.): *Sputtering by Particle Bombardment*, III, Topics in Applied Physics
        (Springer, Berlin, Heidelberg, New York 1981) to be published
4.2     J.Plücker: Ann. Phys. Chem. (Leipzig) **103**, 88 (1858)
4.3     R.Behrisch: Ergeb. Exakten Naturwiss. **35**, 295 (1964)
4.4     O.Christensen: Thin Solid Films **15**, 1 (1973)
4.5     G.Carter, W.A.Grant: *Ion Implantation of Semiconductors* (Halsted, New York 1976)
4.6     G.Dearnaley, J.H.Freeman, R.S.Nelson, J.Stephen (eds.): *Ion Implantation* (North-Holland,
        Amsterdam 1973)
4.7     H.L.Bay, H.H.Andersen, W.O.Hofer, O.Nielsen: Nucl. Instrum. Methods **132**, 301 (1976)
4.8     J.E.Robinson, D.A.Thompson: Phys. Rev. Lett. **33**, 1569 (1974)
4.9     J.H.Freeman: In *Ion Implantation* ed. by G.Dearnaley, J.H.Freeman, R.S.Nelson, J.Stephen
        (North-Holland, Amsterdam 1973) p. 255
4.10    A.Weiss, L.Heldt, W.J.Moore: J. Chem. Phys. **29**, 7 (1958)
4.11    R.S.Bhattacharya, C.B.W.Kerdijk, D.Hoonhout, F.W.Saris: In *Proc. 7th Int. Vacuum
        Congress and 3rd Int. Conf. on Solid Surfaces*, ed. by R.Dobrozemsky, F.Rüdenauer,
        F.P.Viehböck, A.Breth (private publisher, Wien 1977) p. 2193
4.12    H.H.Andersen, H.Bay: Radiat. Eff. **13**, 67 (1972)
4.13    H.L.Bay, J.Roth, J.Bohdansky: J. Appl. Phys. **48**, 4722 (1977)
4.14    J.S.Colligon, C.M.Hicks, A.P.Neokleous: Radiat. Eff. **18**, 119 (1973)
4.15    G.Holmén: Radiat. Eff. **24**, 7 (1975)
4.16    G.Holmén, O.Kugler, O.Almén: Nucl. Instrum. Methods **105**, 545 (1972)
4.17    O.C.Yonts, D.E.Harrison, Jr.: J. Appl. Phys. **31**, 1583 (1960)
4.18    E.Taglauer, W.Heiland, J.Onsgaard: Nucl. Instrum. Method **168**, 571 (1980)
4.19    E.Taglauer, U.Beitat, G.Marin, W.Heiland: J. Nucl. Mater. **63**, 193 (1976)
4.20    H.F.Winters, P.Sigmund: J. Appl. Phys. **45**, 4760 (1974)
4.21    T.Hirao, K.Inoue, S.Takayanagi, Y.Yaegashi: Appl. Phys. Lett. **31**, 505 (1977)
4.22    R.A.Moline, A.G.Cullins: Appl. Phys. Lett. **26**, 551 (1975)
4.23    P.Sigmund: J. Appl. Phys. **50**, 7261 (1979)
4.24    R.Kelly, J.B.Sanders: Surf. Sci. **57**, 143 (1976)
4.25    P.Blank: Dissertation, München (1977)
4.26    P.Blank, K.Wittmaack: Radiat. Eff. **27**, 29 (1975)
4.27    P.Blank, K.Wittmaack, F.Schulz: Nucl. Instrum. Methods **132**, 387 (1976)
4.28    E.P.EerNisse: J. Appl. Phys. **42**, 480 (1971)
4.29    R.Weissmann, R.Behrisch: Radiat. Eff. **19**, 69 (1973)
4.30    R.J.MacDonald, D.Haneman: J. Appl. Phys. **37**, 1609 (1966)
4.31    E.P.EerNisse: Appl. Phys. Lett. **29**, 14 (1976)
4.32    G.D.Peterson, E.P.EerNisse: Rev. Sci. Instrum. **47**, 1153 (1976)
4.33    O.Almén, G.Bruce: Nucl. Instrum. Methods **11**, 257 and 279 (1961)
4.34    H.L.Bay, H.H.Andersen, W.O.Hofer: Radiat. Eff. **28**, 87 (1976)
4.35    H.L.Bay, H.H.Andersen, W.O.Hofer, O.Nielsen: Appl. Phys. **11**, 289 (1976)
4.36    R.Behrisch, B.M.U.Scherzer: Thin Solid Films **19**, 247 (1973)
4.37    H. von Seefeld, H.Schmidl, R.Behrisch, B.M.U.Scherzer: J. Nucl. Mater. **63**, 215 (1976)
4.38    K.Pearmain, B.A.Unvala: Vacuum **25**, 3 (1975)
4.39    F.Schulz, K.Wittmaack, J.Maul: Radiat. Eff. **18**, 211 (1973)
4.40    R.P.Gittings, D.V.Morgan, G.Dearnaley: J. Phys. D**5**, 1654 (1972)
4.41    G.S.Anderson, W.N.Mayer, G.K.Wehner: J. Appl. Phys. **33**, 2991 (1962)
4.42    R.P.Edwin: J. Phys. D**6**, 833 (1973)
4.43    B.Navinšek: J. Appl. Phys. **36**, 1678 (1965)
4.44    J.Bøttiger, J.A.Davies: Radiat. Eff. **11**, 61 (1971)
4.45    J.Bøttiger, J.A.Davies, P.Sigmund, K.B.Winterbon: Radiat. Eff. **11**, 69 (1971)

4.46   E. Arminen, A. Fontell: Ann. Acad. Sci. Fenn. Ser. A No. 357 (1971)
4.47   E. Arminen, A. Fontell, V. K. Lindroos: Phys. Status Solidi (a)4, 663 (1971)
4.48   A. Fontell, E. Arminen: Can. J. Phys. 47, 2405 (1969)
4.49   H. J. Smith: Radiat. Eff. 18, 55 and 65 (1973)
4.50   R. Behrisch, J. Bøttiger, W. Eckstein, J. Roth, B. M. U. Scherzer: J. Nucl. Mater. 56, 365 (1975)
4.51   J. Bøttiger, S. T. Picraux, N. Rud, T. Laursen: J. Appl. Phys. 48, 920 (1977)
4.52   J. Roth, S. T. Picraux, W. Eckstein, J. Bøttiger: J. Nucl. Mater. 63, 120 (1976)
4.53   A. W. Tinsley, W. A. Grant, G. Carter, M. J. Nobes: In *Ion Implantation in Semiconductors*, ed.
       by I. Ruge and J. Graul (Springer, Berlin, Heidelberg, New York 1971) p. 199
4.54   J. L. Whitton, G. Carter, J. N. Baruah, W. A. Grant: Radiat. Eff. 16, 101 (1972)
4.55   S. K. Erents, G. M. McCracken: J. Phys. D2, 1397 (1969)
4.56   G. M. McCracken, D. K. Jeffries, P. Goldsmith: Inst. Phys. Conf. Ser. 5, 149 (1968)
4.57   G. M. McCracken: Rep. Prog. Phys. 38, 241 (1975)
4.57a  R. Behrisch (ed.): *Sputtering by Particle Bombardment*, II, Topics in Applied Physics
       (Springer, Berlin, Heidelberg, New York 1981) to be published
4.58   B. M. U. Scherzer, R. Behrisch, W. Eckstein, U. Littmark, J. Roth, M. K. Sinha: J. Nucl. Mater.
       63, 100 (1976)
4.59   J. M. Fluit, P. K. Rol: Physica 30, 857 (1964)
4.60   H. Schirrwitz: Beitr. Plasmaphys. 2, 188 (1962)
4.61   R. Behrisch, J. Bohdansky, G. H. Oetjen, J. Roth, G. Schilling, H. Verbeek: J. Nucl. Mater. 60,
       321 (1976)
4.62   J. Bohdansky, H. L. Bay, J. Roth: In *Proc. 7th Int. Vacuum Congress and 3rd Int. Conf. on Solid
       Surfaces*, ed. by R. Dobrozemsky, F. Rüdenauer, F. P. Viehböck, A. Breth (private publisher,
       Wien 1977) p. 1509
4.63   K. Akaishi, A. Kiyahara, Z. Kabeya, M. Komizo, T. Gotoh: In *Proc. 7th Int. Vacuum Congress
       and 3rd Int. Conf. on Solid Surfaces*, ed. by R. Dobrozemsky, F. Rüdenauer, F. P. Viehböck,
       A. Breth (private publisher, Wien 1977) p. 1477
4.64   A. I. Aizentson, V. S. Karpukhin: Zh. Tekh. Fiz. 42, 2607 (1972); Engl. transl.: Sov. Phys.-
       Tech. Phys. 17, 2025 (1973)
4.65   T. Narusawa, O. Tsukakoshi, T. Satake, H. Mizuno, H. Ohtsuka, K. Sone, S. Komiya: In *Proc.
       7th Int. Vacuum Congress and 3rd Int. Conf. on Solid Surfaces*, ed. by R. Dobrozemsky,
       F. Rüdenauer, F. P. Viehböck, A. Breth (private publisher, Wien 1977) p. 371
4.66   H. Ohtsuka, R. Yamada, K. Sone, M. Saidoh, T. Abe: J. Nucl. Mater. 76, 188 (1978)
4.67   R. G. Hart, C. B. Cooper: J. Vac. Sci. Technol. 13, 553 (1976)
4.68   D. McKeown: Rev. Sci. Instrum. 32, 133 (1961)
4.69   W. H. Hayward, A. R. Wolter: J. Appl. Phys. 40, 2911 (1969)
4.70   H. H. Andersen, H. L. Bay: Radiat. Eff. 19, 139 (1973)
4.71   H. H. Andersen, H. L. Bay: J. Appl. Phys. 45, 953 (1974)
4.72   H. H. Andersen, H. L. Bay: J. Appl. Phys. 46, 1919 (1975)
4.73   H. H. Andersen, H. L. Bay: J. Appl. Phys. 46, 2416 (1976)
4.74   H. H. Andersen: Radiat. Eff. 19, 257 (1973)
4.75   E. P. EerNisse: J. Appl. Phys. 43, 1330 (1972)
4.76   E. P. EerNisse: J. Nucl. Mater. 53, 226 (1974)
4.77   E. P. EerNisse: J. Vac. Sci. Technol. 11, 408 (1974)
4.78   E. P. EerNisse: J. Vac. Sci. Technol. 12, 564 (1975)
4.79   H. H. Andersen, J. Bøttiger, H. Knudsen (eds.): *Proc. 4th Int. Conf. on Ion-Beam Analysis*
       (North-Holland, Amsterdam 1980); Nucl. Instrum. Methods 168 (1980)
4.80   J. W. Mayer, J. F. Ziegler (eds.): *Ion Beam Surface Layer Analysis* (Elsevier, Lausanne 1974)
4.81   O. Meyer, G. Linker, F. Käppeler (eds.): *Ion Beam Surface Layer Analysis* (Plenum, New
       York 1975)
4.82   E. A. Wolicki, J. W. Butler, P. A. Treado (eds.): *Proc. 3rd Intern. Conf. Ion Beam Analysis*
       (North-Holland, Amsterdam 1978); Nucl. Instrum. Methods 149 (1978)
4.83   H. H. Andersen, H. Knudsen, P. Møller Petersen: J. Appl. Phys. 49, 5638 (1978)
4.84   R. Behrisch, R. Weissmann: Phys. Lett. 30A, 506 (1970)
4.85   Z. L. Liau, W. L. Brown, R. Homer, J. M. Poate: Appl. Phys. Lett. 30, 626 (1977)

4.86   J.M.Poate, W.L.Brown, R.Homer, W.M.Augustyniak: Nucl. Instrum. Methods **132**, 345 (1976)

4.87   W.Eckstein, B.M.U.Scherzer, H.Verbeek: Radiat. Eff. **18**, 135 (1973)

4.88   H.H.Andersen: Unpublished (1976)

4.89   P.Sigmund: Phys. Rev. **184**, 383 (1969)

4.90   F.Folkmann: In *Ion Beam Surface Layer Analysis* ed. by O.Meyer, G.Linker, F.Käppeler (Plenum, New York 1975) p. 695

4.91   J.Bøttiger: J. Nucl. Mater. **78**, 161 (1978)

4.92   M.A.Kirk, R.A.Conner, D.G.Wozniak, L.R.Greenwood, R.L.Malewicki, R.R.Heinrich: Phys. Rev. B**19**, 87 (1979)

4.93   G.S.Anderson: J. Appl. Phys. **40**, 2884 (1969)

4.94   C.R.Fritzsche, W.Rothemund: Appl. Phys. **7**, 39 (1975)

4.95   J.Kirschner, H.W.Etzkorn: Appl. Surf. Sci. **3**, 251 (1979)

4.96   J.Tardy, J.Pivot, J.P.Dupin, A.Cachard: In *Proc. 7th Int. Vacuum Congress and 3rd Int. Conf. on Solid Surfaces*, ed. by R.Dobrozemsky, F.Rüdenauer, F.P.Viehböck, A.Breth (private publisher, Wien 1977) p. 1481

4.97   P.Mertens: Nucl. Instrum. Methods **132**, 307 (1976)

4.98   D.B.Medved, H.Poppa: J. Appl. Phys. **33**, 1759 (1962)

4.99   D.Fuller, J.S.Colligon, J.S.Williams: Surf. Sci. **54**, 647 (1976)

4.100  H.Kräutle: Nucl. Instrum. Methods **137**, 553 (1976)

4.101  C.R.Fritzsche, W.Rothemund: J. Electrochem. Soc. **119**, 1243 (1972)

4.102  D.W.Ormond, E.A.Irene, J.E.E.Baglin, B.L.Crowder: In *Ion Implantation in Semiconductors*, ed. by F.Chernow, J.A.Borders, D.K.Brice (Plenum, New York 1977) p. 305

4.103  E.P.EerNisse, C.B.Norris: J. Appl. Phys. **45**, 5196 (1975)

4.104  W.Primak: *The Compacted State of Vitreous Silica* (Gordon and Breach, New York 1975)

4.105  G.J.Ogilvie, M.J.Ridge: J. Phys. Chem. Solids **10**, 217 (1959)

4.106  R.L.Hines, R.Waller: J. Appl. Phys. **32**, 202 (1961)

4.107  H.Bach: Nucl. Instr. Methods **84**, 4 (1970)

4.108  H.Bach: Z. Naturforsch. **27a**, 333 (1972)

4.109  H.Bach: Int. J. Mass Spectrom. and Ion Phys. **9**, 247 (1972)

4.110  H.Bach, J.Kitzmann, H.Schröder: Radiat. Eff. **21**, 31 (1974)

4.111  K.Tsunoyama, Y.Ohashi, T.Suzuki, K.Tsuruoka: Jpn. J. Appl. Phys. **13**, 1683 (1974)

4.112  K.Tsunoyama, T.Suzuki, Y.Ohashi: Jpn. J. Appl. Phys. **15**, 349 (1976)

4.113  T.Kimura, J.Kobayashi, S.Okuda, H.Akimune: Jpn. J. Appl. Phys. **15**, 2479 (1976)

4.114  R.Kelly: J. Appl. Phys. **39**, 5298 (1968)

4.115  L.Q.Nghi, R.Kelly: Can. J. Phys. **48**, 137 (1970)

4.116  J.W.Guthrie, R.S.Blewer: Rev. Sci. Instrum. **43**, 654 (1972)

4.117  E.H.Sondheimer: Adv. Phys. **1**, 1 (1952)

4.118  H.Fetz: Z. Phys. **119**, 590 (1942)

4.119  H.G.Scott: J. Appl. Phys. **30**, 2011 (1962)

4.120  V.Teodosić: Appl. Phys. Lett. **9**, 209 (1966)

4.121  B.Navinšek, G.Carter: In *Physics of Ionized Gases 1970. Contributed Papers*, ed. by B.Navinšek (Institute of Physics, Ljubljana 1970) p. 65

4.122  D.K.Murti, R.Kelly: Thin Solid Films **33**, 149 (1976)

4.123  T.E.Parker, R.Kelly: J. Phys. Chem. Solids **36**, 377 (1975)

4.124  K.H.Ecker, K.L.Merkle: Phys. Rev. B**18**, 1020 (1978)

4.125  S.S.Johar, D.A.Thompson: Surf. Sci. **90**, 310 (1979)

4.126  J.M.Walls, R.M.Boothby, H.N.Southworth: Surf. Sci. **61**, 419 (1976)

4.127  J.M.Walls, A.D.Martin, H.N.Southworth: Surf. Sci. **50**, 360 (1975)

4.128  J.M.Walls, H.N.Southworth, E.Bram: Vacuum **24**, 471 (1974)

4.129  R.Dobrozemsky, F.Rüdenauer, F.P.Viehböck, A.Breth (eds.): *Proc. 7th Int. Vacuum Congress and 3rd Int. Conf. on Solid Surfaces* (private publishers, Wien 1977)

4.130  A.Benninghoven, C.A.Evans, Jr., R.A.Powell, R.Shimizu, H.A.Storms (eds.): *Secondary Ion Mass Spectrometry (SIMS) II*, Springer Ser. Chem. Phys., Vol. 9 (Springer, Berlin, Heidelberg, New York 1980)

4.131  G.C.Nelson: J. Vac. Sci. Technol. **13**, 974 (1976)
4.132  W.L.Harrington, R.E.Honig, A.M.Goodman, R.Williams: Appl. Phys. Lett. **24**, 644 (1975)
4.133  R.Behrisch, B.M.U.Scherzer, P.Staib: Thin Solid Films **19**, 57 (1973)
4.134  W.K.Chu, J.K.Howard, R.F.Lever: J. Appl. Phys. **47**, 4500 (1976)
4.135  C.M.Garner, Y.D.Shen, J.S.Kim, G.L.Pearson, W.E.Spicer, J.S.Harris, Jr., D.D.Edwall: J. Appl. Phys. **48**, 3147 (1977)
4.136  H.Goretzki, A.Mühlratzer, J.Nickl: In *Proc. 7th Int. Vacuum Congress and 3rd Int. Conf. on Solid Surfaces*, ed. by R.Dobrozemsky, F.Rüdenauer, F.P.Viehböck, A.Breth (private publisher, Wien 1977) p. 2387
4.137  K.Goto, T.Koshikawa, K.Ishikawa, R.Shimuzu: In *Proc. 7th Intern. Vacuum Congress and 3rd Int. Conf. on Solid Surfaces*, ed. by R.Dobrozemsky, F.Rüdenauer, F.P.Viehböck, A.Breth (private publisher, Wien 1977) p. 1493
4.138  S.Hofmann, J.Erlewein, A.Zalar: Thin Solid Films **43**, 275 (1977)
4.139  P.H.Holloway: Surf. Sci. **66**, 479 (1977)
4.140  T.Koshikava, K.Goto, N.Saeki, R.Shimizu: In *Proc. 7th Int. Vacuum Congress and 3rd Int. Conf. on Solid Surfaces*, ed. by R.Dobrozemsky, F.Rüdenauer, F.P.Viehböck, A.Breth (private publisher, Wien 1977) p. 1489
4.141  M.Ono, Y.Takasu, K.Nakayama, T.Yamashina: Surf. Sci. **26**, 313 (1971)
4.142  A. van Oostrom: J. Vac. Sci. Technol. **13**, 225 (1976)
4.143  D.T.Quinto, V.S.Sundaram, W.D.Robertson: Surf. Sci. **28**, 504 (1971)
4.144  L.E.Rehn, P.R.Okamoto, D.I.Potter, H.Wiedersich: J. Nucl. Mater. **74**, 242 (1978)
4.145  J.Roth, J.Bohdansky, W.O.Hofer, J.Kirschner: In *Plasma Wall Interactions* (Pergamon, London 1977) p. 308
4.146  H.Shimizu, M.Ono, K.Nakayama: Surf. Sci. **36**, 817 (1973)
4.147  J.N.Smith, Jr., C.H.Meyer, Jr., J.K.Layton: J. Appl. Phys. **46**, 4291 (1975)
4.148  J.N.Smith, Jr., C.H.Meyer, Jr., J.K.Layton: Nucl. Technol. **29**, 318 (1976)
4.149  J.N.Smith, Jr., C.H.Meyer, Jr., J.K.Layton: J. Nucl. Mater. **67**, 234 (1977)
4.150  M.L.Tarng, G.K.Wehner: J. Vac. Sci. Technol. **8**, 23 (1971)
4.151  M.L.Tarng, G.K.Wehner: J. Appl. Phys. **42**, 2449 (1971)
4.152  M.L.Tarng, G.K.Wehner: J. Appl. Phys. **43**, 2268 (1972)
4.153  L.A.West: J. Vac. Sci. Technol. **13**, 198 (1976)
4.154  C.E.KenKnight, G.K.Wehner: J. Appl. Phys. **35**, 322 (1964)
4.155  A.Benninghoven: Z. Angew. Phys. **27**, 51 (1969)
4.156  W.O.Hofer, H.Liebl: Appl. Phys. **8**, 359 (1975); In *Ion Beam Surface Layer Analysis*, ed. by O.Meyer, G.Linker, F.Käppeler (Plenum, New York 1976) p. 659
4.157  W.O.Hofer, P.J.Martin: Appl. Phys. **16**, 271 (1978)
4.158  H.Oechsner, H.Schoof, E.Stumpe: In *Proc. 7th Int. Vacuum Congress and 3rd Int. Conf. on Solid Surfaces*, ed. by R.Dobrozemsky, F.Rüdenauer, F.P.Viehböck, A.Breth (private publisher, Wien 1977) p. 1497; Surf. Sci. **76**, 343 (1978)
4.159  M.Braun, B.Emmoth, R.Buchta: Radiat. Eff. **28**, 77 (1976)
4.160  H.M.Windawi: Surf. Sci. **55**, 573 (1976)
4.161  H.H.Andersen: Appl. Phys. **18**, 131 (1979)
4.162  K.Wittmaack: In *Inelastic Ion-Surface Collisions*, ed. by N.H.Tolk, J.C.Tully, W.Heiland, C.W.White (Academic, New York 1977) p. 153
4.163  R.P.Stein, F.C.Hurlbut: Phys. Rev. **123**, 790 (1961)
4.164  J.M.Fluit, L.Friedman, H.J.M.Boerbom, J.Kistemaker: J. Chem. Phys. **35**, 1143 (1961)
4.165  C.P.Können, J.Grosser, A.Haring, A.E. de Vries, J.Kistemaker: Radiat. Eff. **21**, 171 (1974)
4.166  R.E.Honig: J. Appl. Phys. **29**, 549 (1958)
4.167  J.Comas, C.B.Cooper: J. Appl. Phys. **38**, 2956 (1967)
4.168  H.Oechsner: Phys. Rev. Lett. **24**, 583 (1970)
4.169  H.Oechsner: Z. Phys. **238**, 433 (1970)
4.170  H.Oechsner: Z. Phys. **261**, 37 (1973)
4.171  H.Oechsner, W.R.Gesang: Phys. Lett. **39**A, 236 (1971)
4.172  H.Oechsner, L.Reichert: Phys. Lett. **23**, 90 (1966)
4.173  R.V.Stuart, G.K.Wehner: J. Appl. Phys. **35**, 1819 (1964)

4.174  J.R.Woodyard, C.B.Cooper: J. Appl. Phys. **35**, 1107 (1964)
4.175  F.Bernhardt, H.Oechsner, E.Stumpe: Nucl. Instrum. Methods **132**, 329 (1976)
4.176  J.W.Coburn, E.Kay: Appl. Phys. Lett. **19**, 350 (1971)
4.177  J.W.Coburn, E.Taglauer, E.Kay: J. Appl. Phys. **45**, 1779 (1974)
4.178  W.Gerhard, H.Oechsner: Z. Phys. B**22**, 41 (1975)
4.179  H.Oechsner, W.Gerhard: Surf. Sci. **44**, 480 (1974)
4.180  H.Oechsner, E.Stumpe: Appl. Phys. **14**, 43 (1977)
4.181  H.Oechsner, K.Rühe, E.Stumpe: Surf. Sci. **85**, 289 (1979)
4.182  R.Bruckmüller, W.Husinsky, P.Blum: In *Proc. 7th Int. Vacuum Congress and 3rd Int. Conf.
        on Solid Surfaces*, ed. by R.Dobrozemsky, F.Rüdenauer, F.P.Viehböck, A.Breth (private
        publisher, Wien 1977) p. 1469
4.183  D.Hammer, E.Benes, P.Blum, W.Husinsky: Rev. Sci. Instrum. **47**, 1178 (1976)
4.184  W.Husinsky, R.Bruckmüller, P.Blum: Nucl. Instrum. Methods **170**, 287 (1980)
4.185  A.Elbern: In *Plasma Wall Interactions* (Pergamon, London 1977) p. 489
4.186  A.Elbern, P.Mioduszewski, E.Hintz: In *Proc. 7th Int. Vacuum Congress and 3rd Int. Conf. on
        Solid Surfaces*, ed. by R.Dobrozemsky, F.Rüdenauer, F.P.Viehböck, A.Breth (private
        publisher, Wien 1977) p. 1473
4.187  B.Schweer, H.L.Bay: In *Proc. 4th Int. Conf. Solid Surfaces and 3rd European Conf. Surface
        Science*, Societé Francaise du Vide Paris (1980) p. 1349
4.188  P.Bogen, E.Hintz: Comm. Plasma Phys. Cont. Fusion **4**, 115 (1978)
4.189  K.Wittmaack: Nucl. Instrum. Methods **168**, 343 (1980)
4.190  T.Ishitani, R.Shimizu: Appl. Phys. **6**, 277 (1975)
4.191  A.R.Krauss, D.M.Gruen: J. Nucl. Mater. **63**, 380 (1976)
4.192  P.Mertens: Thesis. Berlin (1975)
4.193  H.H.Andersen, P.Tykesson: IEEE Trans. Nucl. Sci. NS**22**, 1632 (1975)
4.194  V.E.Krohn, Jr.: J. Appl. Phys. **33**, 3523 (1962)
4.195  R.Middleton, C.T.Adams: Nucl. Instrum. Methods **118**, 329 (1974)
4.196  H.Sporn: Z. Phys. **112**, 279 (1939)
4.197  N.H.Tolk, J.C.Tully, W.Heiland, C.W.White (eds.): *Inelastic Ion-Surface Collisions*
        (Academic, New York 1977)
4.198  N.H.Tolk, I.S.T.Tsong, C.W.White: Anal. Chem. **49**, 16A (1977)
4.199  S.S.Pop, I.P.Zaposochnyi, A.I.Imre, S.A.Evdokimov, A.I.Daschenko: Zh. Eksp. Teor. Fiz.
        **73**, 90 (1977); Engl. transl.: Sov. Phys.-JETP **46**, 46 (1977)
4.200  S.K.Erents, G.M.McCracken: J. Appl. Phys. **44**, 3139 (1973)
4.201  S.K.Erents, G.M.McCracken: Culham Laboratory Report CLM-P360 (1973)
4.202  M.W.Thompson, B.W.Farmery, P.A.Newson: Philos. Mag. **18**, 361 (1968)
4.203  M.R.Weller, T.A.Tombrello: Radiat. Eff. **49**, 239 (1980)
4.204  W.Krüger, K.Rödelsperger, A.Scharmann: Z. Phys. **262**, 315 (1973)
4.205  J.J.Cuomo, R.J.Gambino, J.M.E.Harper, J.D.Kuptsis, J.C.Webber: J. Vac. Sci. Technol.
        **15**, 281 (1978)
4.206  H.L.Bay, J.Bohdansky, W.O.Hofer, J.Roth: Appl. Phys. **21**, 327 (1980)
4.207  V.E.Dubinskii, S.Ya.Lebedev: Fiz. Tverd. Tela **12**, 1906 (1970); Engl. transl.: Sov. Phys.-
        Solid State **12**, 1516 (1971)
4.208  S.Ya.Lebedev, G.V.Lysova: Fiz. Tverd. Tela **17**, 3068 (1975); Engl. transl.: Sov. Phys.-Solid
        State **17**, 2014 (1976)
4.209  R.J.MacDonald, J.Reid: J. Phys. F**4**, 1832 (1974)
4.210  L.T.Chadderton, A.Johansen, L.Sarholt-Kristensen, H.Schumacher: Phys. Lett. **40**A, 231
        (1972)
4.211  L.T.Chadderton, A.Johansen, S.Steenstrup, L.Sarholt-Kristensen: Radiat. Eff. **17**, 281
        (1973)
4.212  W.O.Hofer: Radiat. Eff. **21**, 141 (1974)
4.213  M.Koedam: Thesis, Utrecht (1961) and Physica **24**, 692 (1962)
4.214  G.E.Chapman, J.C.Kelly: J. Sci. Instrum. **44**, 261 (1967)
4.215  K.Rödelsperger, W.Krüger, A.Scharmann: Z. Phys. **269**, 83 (1974)

4.216 K. Takatsu, T. Toda: In *Proc. Vht Intern. Conf. Ionization Phenomena in Gases*, ed. by Maecker (North-Holland, Amsterdam 1962) p. 96
4.217 L. L. Tongson, C. B. Cooper: Radiat. Eff. **24**, 187 (1975)
4.218 J. Bohdansky, J. Roth, M. K. Sinha, W. Ottenberger: J. Nucl. Mater. **63**, 115 (1976)
4.219 E. P. EerNisse, D. K. Brice: Nucl. Instrum. Methods **132**, 363 (1976)
4.220 W. O. Hofer, H. L. Bay, P. J. Martin: J. Nucl. Mater. **76**, 156 (1978)
4.221 R. Behrisch, O. K. Harling, M. T. Thomas, R. L. Brodzinski, L. H. Jenkins, G. J. Smith, J. F. Wendelkeng, M. J. Saltmarsh, M. Kaminsky, S. K. Das, C. M. Logan, R. Meisenheimer, J. E. Robinson, M. Shimotomai, D. A. Thompson: J. Appl. Phys. **48**, 3914 (1977)
4.222 O. K. Harling, M. T. Thomas, R. L. Brodzinski, L. A. Ranticelli: Phys. Rev. Lett. **34**, 1340 (1975)
4.223 O. K. Harling, M. T. Thomas, R. L. Brodzinski, L. A. Ranticelli: J. Nucl. Mater. **63**, 422 (1976)
4.224 L. H. Jenkins, T. S. Noggle, R. E. Reed, M. J. Saltmarsh, G. S. Smith: Appl. Phys. Lett. **26**, 426 (1975)
4.225 M. Kaminsky, S. K. Das: J. Nucl. Mater. **60**, 111 (1976)
4.226 K. L. Merkle, P. P. Pronko: J. Nucl. Mater. **53**, 231 (1974)
4.227 H. H. Andersen, F. Besenbacher, P. Goddiksen: In *Symposium on Sputtering*, Institute für Allgemeine Physik, ed. by P. Varga, G. Betz, F. P. Viehböck (Technical University, Vienna 1980) p. 446
4.228 K. Rödelsperger, A. Scharmann: Nucl. Instrum. Methods **132**, 355 (1976)
4.229 K. H. Ecker: Radiat. Eff. **23**, 171 (1974)
4.230 E. H. Hasseltine, N. T. Olson, H. P. Smith: J. Appl. Phys. **39**, 1417 (1968)
4.231 W. O. Hofer: Dissertation, München (1972)
4.232 W. O. Hofer: Radiat. Eff. **19**, 263 (1973)
4.233 W. O. Hofer: Mikrochim. Acta Suppl. **7**, 185 (1977)
4.234 M. Kaminsky, S. K. Das: J. Nucl. Mater. **53**, 162 (1974)
4.235 M. Kaminsky, J. H. Peavey, S. K. Das: Phys. Rev. Lett. **32**, 599 (1974)
4.236 G. K. Wehner: Appl. Phys. Lett. **30**, 185 (1977)
4.237 M. Braun, B. Emmoth: Appl. Phys. Lett. **29**, 545 (1976)
4.238 D. M. Gruen, S. L. Gaudisio, R. L. McBeth, J. L. Lerner: J. Chem. Phys. **60**, 89 (1974)
4.239 J. Bates, D. M. Gruen, R. Varma: Rev. Sci. Instrum. **48**, 1506 (1976)
4.240 G. E. Thomas, E. E. de Kluizenaar, W. J. van Kollenburg, L. C. Bastings: Anal. Chem. **47**, 2357 (1975)
4.241 C. Cassignol, G. Rang: C. R. Acad. Sci. **248**, 1988 (1959)
4.242 W. J. Moore, C. D. O'Brian, A. Lindner: Ann. N. Y. Acad. Sci. **67**, 600 (1957)
4.243 C. H. Robison: J. Appl. Phys. **39**, 3441 (1968)
4.244 V. D. Tischenko: Radiotekh. Elektron. [Engl. transl.: Rad. Eng. Electron. Phys. (USSR) **13**, 1431 (1968)]
4.245 V. M. Agranovich, O. I. Kapusta, S. Ya. Lebedev, L. P. Semenov: Fiz. Tverd. Tela **11**, 2816 (1969); Engl. transl.: Sov. Phys.-Solid State **11**, 2280 (1970)
4.246 H. Patterson, D. H. Tomlin: Proc. Roy. Soc. (London) A**265**, 474 (1962)
4.247 I. Reid, B. W. Farmery, M. W. Thompson: Nucl. Instrum. Methods **132**, 317 (1976)
4.248 M. W. Thompson: Phys. Lett. **6**, 24 (1963)
4.249 J. F. Cuderman, J. J. Brady: Surf. Sci. **10**, 410 (1968)
4.250 R. S. Nelson: Radiat. Eff. **7**, 263 (1971)
4.251 R. S. Nelson, M. W. Thompson: Proc. Roy. Soc. (London) A**259**, 458 (1961)
4.252 R. S. Nelson, M. W. Thompson, H. Montgomery: Philos. Mag. **7**, 1385 (1962)
4.253 R. S. Nelson, M. W. Thompson: Phys. Lett. **2**, 124 (1962)
4.254 R. S. Nelson: Philos. Mag. **8**, 693 (1963)
4.255 Z. E. Switkowski, F. M. Mann, D. W. Kneff, R. W. Ollerhead, T. A. Tombrello: Radiat. Eff. **29**, 65 (1976)
4.256 R. W. Ollerhead, F. M. Mann, D. W. Kuff, Z. E. Switkowski, T. A. Tombrello: Phys. Rev. Lett. **36**, 439 (1976)
4.257 J. P. Biersack, D. Fink, P. Mertens: J. Nucl. Mater. **53**, 194 (1974)
4.258 R. Gregg, T. A. Tombrello: Radiat. Eff. **35**, 243 (1978)

4.259  C.H.Weijsenfeld: In *Proc. 6ieme Intern. Conf. Phénomènes d'Ionization dans les Gaz*, Vol. II, Paris (1963) p. 43

4.260  C.H.Weijsenfeld: Philips Res. Rep. Suppl. No. 2 (1967)

4.261  S.A.Drentje: Phys. Lett. **25A**, 433 (1967) and Thesis, Groningen (1968)

4.262  V.M.Bukhanov, V.G.Morozov, V.E.Yurasova: Radiat. Eff. **19**, 215 (1973)

4.263  V.M.Bukhanov, D.D.Odintsov, V.E.Yurasova: Fiz. Tverd. Tela **12**, 2425 (1970); Engl. transl.: Sov. Phys.-Solid State **12**, 1937 (1971)

4.264  R.G.Musket, H.G.Smith, Jr.: J. Appl. Phys. **39**, 3579 (1968)

4.265  N.T.Olson, H.P.Smith, Jr.: Phys. Rev. **157**, 241 (1967)

4.266  H.P.Smith, Jr., R.G.Musket: J. Appl. Phys. **40**, 3859 (1969)

4.267  V.E.Yurasova: In *Physics of Ionized Gases 1976*, ed. by B.Navinšek (Institute of Physics, Ljubljana 1976) p. 493

4.268  C.B.Cooper, R.G.Hart, J.C.Riley: J. Appl. Phys. **44**, 5183 (1973)

4.269  D.Onderdelinden: Can. J. Phys. **46**, 739 (1968) and Thesis, Amsterdam (1968)

4.270  H.H.Andersen, K.N.Tu, J.F.Ziegler: Nucl. Instrum. Methods **149**, 247 (1978)

4.271  H.J.Smith, G. van Wyk: Phys. Lett. **64A**, 327 (1977)

4.272  G. van Wyk, H.J.Smith: Radiat. Eff. **38**, 245 (1978)

4.273  G.S.Anderson: J. Appl. Phys. **37**, 2838 (1966)

4.274  G.S.Anderson: J. Appl. Phys. **38**, 1607 (1967)

4.275  J.Farren, W.J.Scaife: AERE-R5717, AERE Harwell Didcot (1968)

4.276  J.J.Ph.Elich, H.E.Roosendaal, D.Onderdelinden: Radiat. Eff. **14**, 93 (1972)

4.277  J.M.Fluit, C.Snoek, J.Kistemaker: Physica **30**, 144 (1964)

4.278  R.S.Nelson: Philos. Mag. **7**, 515 (1962)

4.279  E.P.Vaulin, N.E.Georgieva, T.D.Martynenko: Fiz. Tverd. Tela **19**, 1423 (1977); Engl. transl.: Sov. Phys.-Solid State **19**, 826 (1977)

4.280  H.L.Garvin: NASA CR-54678 (1968)

4.281  R.C.Krutenat, C.Panzera: J. Appl. Phys. **41**, 4953 (1970)

4.282  V.E.Yurasova, U.S.Chernysh, M.U.Kuvakin, L.B.Shelyakin: Zh. Eksp. Teor. Fiz. Pisma Red. **21**, 197 (1975); Engl. transl.: JETP Lett. **21**, 88 (1975)

4.283  U.Littmark, W.Hofer: J. Mater. Sci. **13**, 2577 (1978)

4.284  P.P.Davidse, L.T.Maissel: J. Appl. Phys. **37**, 574 (1966)

4.285  P.Blank, K.Wittmaack: J. Appl. Phys. **50**, 1519 (1979)

4.286  T.M.Nenadović, Z.B.Fotrić, T.S.Dimitrijević: Surf. Sci. **33**, 607 (1972)

4.287  S.N.Cramer, E.M.Oblow: Nucl. Fusion **15**, 339 (1975)

4.288  J.F.Ziegler, J.J.Cuomo, J.Roth: Appl. Phys. Lett. **30**, 268 (1977)

4.289  K.Ohasa, H.Maeda, S.Yamamoto, M.Nagami, H.Ohtsuka, S.Kasai, K.Odajima, H.Kimura, S.Sengoku, Y.Shimomura: J. Nucl. Mater. **76**, 489 (1978)

4.290  J.L.Whitton, W.O.Hofer, U.Littmark, M.Braun, B.Emmoth: Appl. Phys. Lett. **36**, 531 (1980)

4.291  R.J.Berg, G.J.Kominiak: J. Vac. Sci. Technol. **13**, 403 (1976)

4.292  R.S.Gvosdover, V.M.Efremenkova, L.B.Shelyakin, V.E.Yurasova: Radiat. Eff. **27**, 237 (1976)

4.293  W.Hauffe: Phys. Status Solidi (a) **4**, 111 (1971)

4.294  J.L.Whitton, G.Carter, M.J.Nobes, J.S.Williams: In *Physics of Ionized Gases 1976, Contributed Papers*, ed. by B.Navinšek (Institute of Physics, Ljubljana 1976) p. 246

4.295  J.L.Whitton, L.Tanović, J.S.Williams: Appl. Surf. Sci. **1**, 408 (1978)

4.296  H.H.Andersen: In *Physics of Ionized Gases 1974*, ed. by V.Vujnović (Institute of Physics, Zagreb 1974) p. 361

4.297  H.Fetz, H.Oechsner: In *Proc. 6ieme Int. Conf. Phénomènes d'Ionisations dans les Gaz*, Paris (1963) p. 39

4.298  N.Laegreid, G.K.Wehner: J. Appl. Phys. **32**, 365 (1961)

4.299  D.Rosenberg, G.K.Wehner: J. Appl. Phys. **33**, 1842 (1962)

4.300  J.A.Borders, R.A.Langley, K.L.Wilson: J. Nucl. Mater. **76**, 168 (1978)

4.301  J.Roth, J.Bohdansky, W.Ottenberger: IPP Report 9/26 (1979)

4.302  S.Miyagawa, Y.Ato, Y.Morita: J. Appl. Phys. **49**, 6164 (1978)

4.303  G.K.Wehner: General Mills Report No. 2309 (1962)

4.304  J.Roth, J.Bohdansky, W.Poschenrieder, M.K.Sinha: J. Nucl. Mater. **63**, 222 (1976)
4.305  J.Bohdansky, H.L.Bay, W.Ottenberger: J. Nucl. Mater. **76**, 163 (1978)
4.306  J.Bohdansky, J.Roth, M.K.Sinha: In *Proc. 9th Symp. on Fusion Technology* (Pergamon, London 1976) p. 541
4.307  T.P.Martynenko: Fiz. Tverd. Tela **9**, 2839 (1967); Engl. transl.: Sov. Phys.-Solid State **9**, 2232 (1968)
4.308  G.K.Wehner: Phys. Rev. **108**, 35 (1957)
4.309  H.Ismail, A.Septier: In *Proc. 3rd Int. Symp. on Discharges and Electrical Insulation in Vacuum* (1968) p. 95
4.310  G.Holmén, O.Almén: Ark. Fys. **40**, 429 (1970)
4.311  K.H.Krebs: In *Atomic and Molecular Data for Fusion* (IAEA, Wien 1977) p. 185
4.312  H.J.Smith: Phys. Lett. **37**A, 289 (1971)
4.313  O.C.Yonts: In *Proc. Nuclear Fusion Reactor Conference* (British Nuclear Energy Society, Culham 1969) p. 424
4.314  C.Fert, N.Colombie, B.Fagot, Pham van Chuong: In *Le Bombardement Ionique* (CNRS, Paris 1961) p. 67
4.315  O.C.Yonts, C.E.Normand, D.E.Harrison: J. Appl. Phys. **31**, 447 (1960)
4.316  C.E.Carlston, G.D.Magnuson, A.Comeaux, P.Mahavedan: Phys. Rev. **138**, 759 (1965)
4.317  H.Daley, J.Perel: AIAA J. **5**, 113 (1967)
4.318  G.K.Wehner: Phys. Rev. **112**, 1120 (1958)
4.319  A.Southern, W.R.Willis, M.T.Robinson: J. Appl. Phys. **34**, 153 (1963)
4.320  J.Nizam, N.Benazeth-Colombie: Rev. Phys. Appl. **10**, 183 (1975)
4.321  H.Sommerfeldt, E.S.Mashkova, V.A.Molchanov: Phys. Lett. **38**A, 237 (1972)
4.322  O.V.Kurbatov: Zh. Tekh. Fiz. **37**, 1814 (1967); Engl. transl.: Sov. Phys.-Tech. Phys. **12**, 1328 (1968)
4.323  H.Oechsner: Dissertation, Würzburg (1963)
4.324  G.K.Wehner, D.Rosenberg: J. Appl. Phys. **32**, 887 (1961)
4.325  H.L.Bay, W.O.Hofer: Unpublished (1977)
4.326  N.Laegreid, G.K.Wehner: In *Trans. 7th National Vacuum Symp.* (1960) p. 286
4.327  C.H.Weijsenfeld, A.Hoogendoorn, M.Koedam: Physica **27**, 763 (1961)
4.328  E.S.Borovik, N.P.Katrich, G.T.Nikolaev: At. Energ. U.S.S.R. **21**, 339 (1966)
4.329  W.M.Gusev, M.I.Guseva, Y.L.Krassoklin, S.U.Mirnov, A.V.Nedopassov, U.N.Stepanov: Fiz. Khim. Obrab. Mater. **1**, 15 (1976)
4.330  E.Hintz, D.Rusbüldt, B.Schweer, J.Bohdansky, J.Roth, P.A.Martinelli: J. Nucl. Mater. **93 and 94** (1980)
4.331  M.I.Guseva: Radiotekh. Elektron. **7**, 1680 (1962); Engl. transl.: Radio Eng. Electron. Phys. (USSR) **7**, 1563 (1962)
4.332  V.K.Koshkin: "Sergev Ordzhonikidze" Aeronauti Institut, Report (1975)
4.333  J.K.Hepworth: J. Phys. D**3**, 1475 (1970)
4.334  C.R.Finfgeld: ORO-3557-15, Oak Ridge Operations Offices, Oak Ridgé, Tennessee (1970)
4.335  G.Sletten, P.Knudsen: Nucl. Instrum. Methods **102**, 459 (1972)
4.336  W.O.Hofer: Private communication (1977)
4.337  H.L.Bay, J.Bohdansky, E.Hechtl: Radiat. Eff. **41**, 77 (1979)
4.338  R.V.Stuart, G.K.Wehner: J. Appl. Phys. **33**, 2345 (1962)
4.339  E.Hechtl, H.L.Bay, J.Bohdansky: Appl. Phys. **16**, 147 (1978)
4.340  Sh.G.Askerov, L.A.Sena: Fiz. Tverd. Tela **11**, 1591 (1969); Engl. transl.: Sov. Phys.-Solid State **11**, 1288 (1969)
4.341  H.M.Windawi, J.R.Katzer, C.B.Cooper: Phys. Lett. **59**A, 62 (1976)
4.342  P.K.Rol, J.M.Fluit, J.Kistemaker: Physica **26**, 1000 (1960)
4.343  G.Dupp, A.Scharmann: Z. Phys. **192**, 284 (1966)
4.344  M.I.Guseva: Fiz. Tverd. Tela **1**, 1540 (1959); Engl. transl.: Sov. Phys.-Solid State **1**, 1410 (1960)
4.345  K.B.Cheney, E.T.Pitkin: J. Appl. Phys. **36**, 3542 (1965)
4.346  T.P.Martynenko: Fiz. Tverd. Tela **10**, 2876 (1968) [Engl. Transl.: Sov. Phys.-Solid State **10**, 2274 (1969)

4.347 B.M.Gurmin, T.P.Martynenko, Ya.A.Ryzkov: Fiz. Tverd. Tela **10**, 411 (1968); Engl. transl.: Sov. Phys.-Solid State **10**, 324 (1968)

4.348 A.J.Summers, N.J.Freeman, N.R.Daly: J. Appl. Phys. **42**, 4774 (1971)

4.349 K.Akaishi: Private communication (1977)

4.350 H.Oechsner: Appl. Phys. **8**, 185 (1975)

4.351 M.Kaminsky, S.K.Das, P.Dusza, J.Cecchi: In *Int. Symp. Fusion Technology* (Euroatom, Padua 1978) p. 112

4.352 F.Grønlund, W.J.Moore: J. Chem. Phys. **32**, 1540 (1960)

4.353 C.D.O'Brian, A.Lindner, W.J.Moore: J. Chem. Phys. **29**, 3 (1958)

4.354 B.Emmoth, M.Braun, H.P.Palenius: J. Nucl. Mater. **63**, 482 (1976)

4.355 F.M.Devienne, A.Roustan: C. R. Acad. Sci. **268**, 1362 (1969)

4.356 B.Perović, B.Čobić: In *Proc. 5th Int. Conf. Ionization Phenomena in Gases*, ed. by H.Maecher (North-Holland, Amsterdam 1962) p. 1155

4.357 M.Szymoński, R.S.Bhattacharya, H.Overeijnder, A.E.deVries: J. Phys. D. **11**, 751 (1978) 751 (1978)

4.358 D.A.Thompson, S.S.Johar: Appl. Phys. Lett. **34**, 342 (1979)

4.359 C.H.Meyer, Jr., J.N.Smith: GA-A15134, General Atomics Inc., San Diego (1978); J. Vac. Sci. Technol. **16**, 248 (1979)

4.360 H.F.Winters, D.Horne: Phys. Rev. B**10**, 55 (1974)

4.361 H.L.Bay, J.Bohdansky, J.Roth: Unpublished (1977)

4.362 U.A.Arifov, A.Kh.Ayukhanov, V.A.Shustrov, R.M.Khasanov, V.I.Poltoratskii: Doklady Akad. Nauk SSSR **155**, 306 (1964); Engl. transl.: Sov. Phys. Doklady **9**, 214 (1964)

4.363 A.K.Furr, C.R.Finfgeld: J. Appl. Phys. **41**, 1739 (1970)

4.364 J.S.Colligon, R.W.Bramham: In *Atomic Collisions in Solids*, ed. by D.W.Palmer, M.W.Thompson, P.D.Townsend (North-Holland, Amsterdam 1970) p. 258

4.365 M.T.Robinson, A.L.Southern: J. Appl. Phys. **38**, 2969 (1967)

4.366 F.Keywell: Phys. Rev. **97**, 1611 (1955)

4.367 J.Bøttiger, H.Wolder Jørgensen, K.B.Winterbon: Radiat. Eff. **11**, 133 (1971)

4.368 K.B.Winterbon: *Ion Implantation Range and Energy Deposition Distributions*, Vol. 2, Low Incident Ion Energies (Plenum, New York 1975)

4.369 D.A.Thompson, S.S.Johar: Nucl. Instrum. Methods **170**, 281 (1980)

4.370 U.Littmark, G.Maderlechner: In *Physics of Ionized Gases 1976, Contributed Papers*, ed. by B.Navinšek (Institute of Physics, Ljubljana 1976) p. 139

4.371 J.Lindhard, M.Scharff, H.E.Schiøtt: K. Dan. Vidensk. Selsk. Mat. Fys. Medd. **33**, No. 14 (1963)

4.372 E.Hotston: Nucl. Fusion **15**, 544 (1975)

4.373 E.Hechtl: Nucl. Instrum. Methods **139**, 79 (1976)

4.374 J.Bohdansky, J.Roth, H.L.Bay: J. Appl. Phys. **51**, 2861 (1980)

4.375 C.J.Smithells (ed.): *Metals Reference Book* (Butterworths, London 1975)

4.376 D.P.Stull, H.Prophet (eds.): *JANAF Thermochemical Tables*, 2nd ed., NSRDS-NBS No. 37, National Bureau of Standards, Washington, D.C. (1971)

4.377 D.E.Harrison, G.D.Magnuson: Phys. Rev. **122**, 1421 (1961)

4.378 R.Behrisch, G.Maderlecher, B.M.U.Scherzer, M.T.Robinson: Appl. Phys. **18**, 391 (1979)

4.379 L.G.Haggmark, W.D.Wilson: J. Nucl. Mater. **76**, 149 (1978)

4.380 G.K.Wehner: J. Appl. Phys. **25**, 270 (1954)

4.381 G.K.Wehner: J. Appl. Phys. **30**, 1762 (1959)

4.382 V.A.Molchanov, V.G.Telkovskii: Dokl. Akad. Nauk SSSR **136**, 801 (1961); Engl. transl.: Sov. Phys.-Doklady **6**, 137 (1961)

4.383 I.I.Dushikov, V.A.Molchanov, V.G.Tel'kovskii, V.M.Chicherov: Zh. Tekh. Fiz. **31**, 1012 (1961); Engl. transl.: Sov. Phys.-Tech. Phys. **6**, 735 (1962)

4.384 E.S.Mashkova, V.A.Molchanov: Zh. Tekh. Fiz. **9**, 2081 (1964); Engl. transl.: Sov. Phys.-Tech. Phys. **9**, 1601 (1965)

4.385 D.T.Goldman, A.Simon: Phys. Rev. **111**, 383 (1958)

4.386 R.S.Pease: In *Proc. 13th Course Int. Summer School "Enrico Fermi"* (Academic, London 1959) p. 157

4.387 G.Dupp, A.Scharmann: Z. Phys. **194**, 448 (1966)

4.388  C.E.Ramer, M.A.Narasinham, H.K.Reynolds, J.C.Alldred: J. Appl. Phys. **36**, 1673 (1964)
4.389  I.N.Evdokimov, V.A.Molchanov: Fiz. Tverd. Tela **9**, 2503 (1967); Engl. transl.: Sov. Phys.-Solid State **9**, 1967 (1968)
4.390  I.N.Evdokimov, V.A.Molchanov: Can. J. Phys. **46**, 779 (1968)
4.391  H.L.Bay, J.Bohdansky: Appl. Phys. **19**, 421 (1979)
4.392  T.Hoffmann, H.L.Dodds, M.T.Robinson, D.K.Holmes: Nucl. Sci. Eng. **68**, 204 (1978)
4.393  A.D.Marwick: Nucl. Instrum. Methods **132**, 313 (1976)
4.394  K.T.Rie: Jül-304-RW Kernforschungsanlage Jülich (1965)
4.395  M.W.Thompson: Philos. Mag. **4**, 139 (1959)
4.396  G.Ayrault, R.S.Averback, D.N.Seidman: Scr. Metall. **12**, 119 (1978)
4.397  G.Ayrault, D.N.Seidman: Preprint No. 2944. Materials Science Center, Cornell University (1978)
4.398  J.P.Biersack, W.Kaczerowski, J.Ney, B.K.H.Rahim, A.Riccato, G.R.Thacker, H.Uecker: J. Nucl. Mater. **76, 77**, 640 (1978)
4.399  J.P.Biersack, E.Santner, R.Neubert, J.Ney: J. Nucl. Mater **63**, 443 (1976)
4.400  H.H.Andersen: J. Nucl. Mater. **76**, 190 (1978)
4.401  P.Sigmund: Can. J. Phys. **46**, 731 (1968)
4.402  J.Schou, H.Sørensen, U.Littmark: J. Nucl. Mater. **76**, 359 (1978)
4.403  H.H.Andersen, T.Lenskjær, G.Sidenius, H.Sørensen: J. Appl. Phys. **47**, 13 (1976)
4.404  H.Sørensen: In *Proc. Int. Symp. on Plasma Wall Interactions* (Pergamon, London 1977) p. 437
4.405  O.Almén, G.Bruce: *Conf. on the Physics of Electromagnetic Separation Methods*, Orsay (1962) (unpublished)
4.406  H.H.Andersen: Radiat. Eff. **3**, 51 (1970)
4.407  H.H.Andersen: Radiat. Eff. **7**, 179 (1971)
4.408  D.Hildebrandt, R.Manns: Phys. Status Solidi (a)**38**, K155 (1976)
4.409  D.Hildebrandt, R.Manns: Radiat. Eff. **31**, 153 (1977)
4.410  H.Düsterhöft, R.Manns, D.Hildebrandt: Phys. Status Solidi (a)**36**, K93 (1976)
4.411  D.Hildebrandt, R.Manns, R.Rogaschewski: Radiat. Eff. **33**, 251 (1977)

# 5. Sputtering Yields of Single Crystalline Targets

Hans E. Roosendaal

With 27 Figures

The sputtering yield of a single crystal depends on the crystallographic orientation of the target relative to the incident beam direction. The measured energy dependence of the yield for selected incidence directions and the angular dependence for different ion energies are outlined and discussed in terms of the channeling model. The influence of the crystal temperature on the yield depends on the nature of the target: for semiconductors annealing of defects is the predominant effect, whereas for metals indications of an influence of the monocrystalline lattice on the emission process are found. The relation between sputtering yield and surface structure is discussed for special incidence directions.

## 5.1 First Observations of Single Crystal Effects in Sputtering

The sputtering process is strongly influenced by the crystalline state of the target. If a well-focused beam of keV ions is directed on a single crystal parallel to a low index crystallographic direction or plane, the observed sputtering yield is generally lower than for polycrystalline material. A larger yield than for a polycrystalline target may be observed for incidence of the ion beam along high index crystallographic directions.

This was observed in 1959 by *Rol* et al. [5.1]. Similar results were reported soon after by *Molchanov* et al. [5.2] and *Almén* et al. [5.3]. Figure 5.1 shows the results of *Rol* et al. for incidence of 20 keV $Ar^+$ on a (100) Cu crystal, rotated around an axis in the crystal surface, 7° off a [011] axis. Two broad minima at angles of incidence $\vartheta \cong 0°$ and 35°, separated by a high maximum at $\vartheta \cong 21°$ are found. A wider angular range $(-10° < \vartheta < 75°)$ was covered by *Molchanov* et al. in their measurements of 27 keV $Ar^+$ incidence onto (100) Cu, rotated around the [01$\bar{1}$] axis (Fig. 5.1), revealing minima at $\vartheta \cong 0°$, 20°, 35°, and 55°. The sputtering yield is reduced for incidence of the ion beam along the low index crystallographic directions of the target: [100], [411], [211], and [111] resp. (see insert Fig. 5.1). Between these relative minima the yields are higher than the corresponding yields for polycrystalline material. The yield of a monocrystalline target is seen to follow the overall angular dependence of the yield of a polycrystalline target. On this rather smooth angular variation, an angular structure is superimposed, which is characteristic of the monocrystalline surface and the rotation axis.

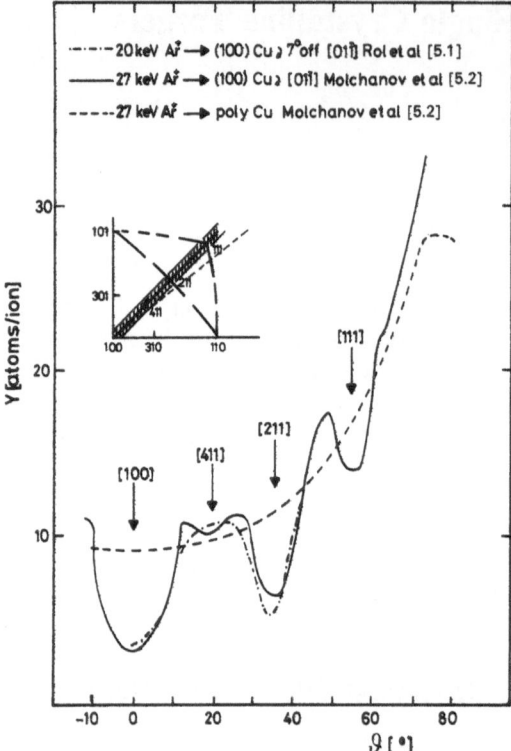

**Fig. 5.1.** The sputtering yield as a function of the angle of incidence of $Ar^+$ on a (100) Cu crystal [5.1, 2]

The first observation of an influence of the monocrystalline substrate structure on the sputtering yield dates back to 1912. Using the weight-loss method *Stark* and *Wendt* [5.4] determined that the sputtering yield of a Bismuth crystal, cut along the basal plane, and bombarded for one hour with ions accelerated through 3000V, was only one-third of the sputtering yield of a Bi crystal, polished to provide a surface almost perpendicular to the basal plane. In order to explain this effect the possibility of directed momentum transfer along a close-packed row of atoms was considered [5.5]. The authors, however, failed to find experimental verification for such a mechanism and rejected the idea. Another proposal was made by *Stark* [5.6] in his formulation of a concept of channeling: as a consequence of a series of correlated collisions an energetic particle could be constrained to a trajectory between lattice rows and therefore penetrate more deeply into the single crystal than in amorphous species, resulting in a lower sputtering yield.

This directional dependence of the single crystal sputtering yield will generally also affect the sputtering yield of a polycrystalline target containing any texture, since the yield represents an average over the yields of the differently oriented small crystals. The wide scatter in the experimental results [5.7], observed at different laboratories (see Chap. 4 by *Andersen* and *Bay*), can be attributed to a large extent to a different structure of the polycrystalline targets.

Monocrystalline effects are also revealed in the properties of the sputtered material. The angular distribution of particles sputtered from single crystals shows distinct peaks for ejection directions that can be crystallographically characterized by low Miller indices, as first observed in 1955 by *Wehner* [5.8]. Also in the energy spectra of sputtered particles, there are indications of a significant influence of the monocrystalline structure of the target on the energy distribution of the ejected particles [Ref. 5.9, Chap. 2].

As an explanation for the reduction in yield for ion incidence along a low index direction, the concept of transparency was introduced by various authors [5.7, 5.10–17]. The arrangement of atoms in a regular array brings atoms in subsurface layers into the shadow of surface layer atoms. This reduces the collision probability in these subsurface layers of the crystal. Although no stability of the particle trajectory between low index lattice rows is assumed, the sputtering yield for ion incidence along such a transparent direction will be reduced by virtue of the fact that sputtering is assumed to be predominantly determined by ion-atom collisions, which take place at only a shallow depth beneath the surface. In 1961, the channeling effect was discovered by *Robinson* et al. in computer calculations [5.18, 19] and confirmed by measurements [5.20–22] of the range of energetic ions in solids. It implies that part of the particle trajectories have a certain stability to remain near the center of the channel formed by low index lattice rows. This reduces the collision probability for small impact parameter collisions by a considerable amount.

Introducing the idea of channeling in sputtering theory, *Onderdelinden* [5.23, 24] assumed that channeled particles do not contribute to sputtering, whereas the nonchanneled part of the beam will give a contribution, equivalent to a beam in a structureless medium. *Onderdelinden* used the theoretical scheme of *Lindhard* [5.25] to describe the channeling phenomenon.

The existence of preferential ejection of particles sputtered from monocrystals led to the concept of focusing collision sequences, as formulated by *Silsbee* [5.26]. In consequence, theoretical models [5.16, 17, 27–29] have been proposed, where it was assumed that the main mechanism of the ejection of surface atoms is due to momentum transport towards the surface along close-packed rows of atoms.

At present, no comprehensive theory on the sputtering of single crystals exists, but it is endeavoured to formulate such a theory through computer simulation models [5.30–33]. The results are still far from being complete, (Chap. 3) and allow only a limited comparison with experimental results.

In the following, the experimental data on sputtering yields of single crystals and their dependence on primary energy and crystallographic orientation of the surface, the angle of incidence of the ion beam, substrate temperature and surface faceting will be examined and compared to corresponding data for polycrystalline material to pinpoint specific monocrystalline dependencies. The results will be discussed in terms of the channeling model of *Onderdelinden*

[5.23, 24]. This model provides a description of the simultaneous dependence of the sputtering yield on primary energy and crystallographic orientation. We therefore start our discussion of the data with the energy dependence of the sputtering yield for normal incidence of the ions onto low index surface planes, and postpone a discussion of the angular dependence till after.

## 5.2 Experimental Methods

The experimental methods used in single crystal sputtering experiments have to fulfil the same requirements on ion beam, vacuum and dose collection as discussed for polycrystalline materials in Chap. 4.

For absolute measurements of the total sputtering yield of single crystals, only the weight-loss method has been used up to now. Relative measurements of the sputtering yield have been performed by collection of sputtered material. The angular distribution as well as the charge state of the sputtered particles is a function of the angle of incidence. To obtain information on the dependence of the total sputtering yield on the angle of incidence, single crystalline effects in the emission distribution are integrated by a large acceptance angle of the detection system and by choosing a fixed detection angle with respect to the surface normal [5.17, 34, 35].

A critical parameter in the determination of the single crystal sputtering yield is the angular divergence of the ion beam, since variations in yield occur for only a small change in angle of incidence (Fig. 5.1). As the angular widths of the dips become narrower for higher primary energies, the condition on beam collimation is dependent on the primary energy of the ion beam. Typical values for the angular spread are of the order of 2°–3° in experiments, using beams of 1–30 keV [5.1, 14, 36]. The angular spread then is about 10% of the total angular width. *Elich* et al. used a much smaller angular spread of 0.5° [5.37, 38] and 0.3° [5.39].

It is extremely important to measure single crystal sputtering yields on freshly electropolished targets. Although the reproducibility in most measurements is about 5% or better, discrepancies of 10% or more were reported by *Southern* et al. [5.7] between sputtering yields of freshly polished and unpolished (111) Cu crystals under normal incidence. *Onderdelinden* [5.40] even found deviations of 25% between his measurements on a freshly polished (100) Cu crystal with the results of *Fluit* et al. [5.11], obtained with the same apparatus for crystals not polished between the measurements. In the latter case, the differences must be due to the build up of surface faceting during bombardment (see also Sect. 5.6).

In measurements on the temperature dependence of the sputtering yield performed at temperatures higher than room temperature [5.41–43], the target was heated and the temperature controlled by a thermocouple. Care must be taken to correct the weight loss of the target after sputtering at high temperatures for the weight loss due to evaporation [5.42]. Temperature dependent measurements at low temperatures ($\gtrsim 50$ K) have been described by *Elich* et al. [5.38, 39]. By simultaneous cooling and heating the target temperature is kept constant within 1 K during the ion beam irradiation. Measurements of the temperature dependence of the angular dependence of the sputtering yield [5.39] require a very small angular spread of the ion beam, since the temperature effects may be overshadowed by the beam spread [5.17].

Radiation damage influences the penetration of the ions into the crystalline solid and thus the sputtering yield. Measurements of the sputtering yield and of the angular distribution of the sputtered particles for ion incidence onto metal targets indicate that the crystalline structure is largely preserved during ion bombardment. Semiconductor targets at substrate temperatures below $\sim 500$–$600$ K turn amorphous under ion bombardment (see [5.44]). This is also reflected in the angular distribution of the sputtered particles and the sputtering yields. At higher substrate temperatures the damage is apparently annealed as in the case for metals at room temperature (see also Sect. 5.5).

## 5.3 On the Energy Dependence of the Sputtering Yield for Normal Incidence of Ions onto Low Index Crystallographic Planes

In this section we will discuss the dependence of the sputtering yield on projectile primary energy for normal incidence on low index planes and its dependence on the crystallographic properties of the surface under bombardment.

The most complete set of experimental data has been obtained for the projectile-target combination $Ar^+ \rightarrow Cu$. In the first part of this section we will therefore restrict the discussion to this particular set of data, since it illustrates most of the general features of the energy dependence.
The data for other projectile-target combinations will then be discussed in a later part of this section on the basis of the knowledge that is gathered from the considerations on the data for Argon bombardment onto monocrystalline Cu.

### 5.3.1 FCC Crystals

#### a) Experimental Data for $Ar^+ \rightarrow Cu$

The general features of the energy dependence of the sputtering yields $Y_{(uvw)}$ are most suitably demonstrated by the experimental data for the projectile-target combination $Ar^+ \rightarrow Cu$ for incidence onto the three main low index crystallographic planes (100), (110), and (111). A compilation of the available data for these planes is shown in Fig. 5.2, together with the corresponding data for the bombardment of a polycrystalline Cu target [5.2, 7, 11, 13, 14, 23, 24, 45].

For primary energies above about 10 keV, the single crystal sputtering yields for low-index single crystal planes $Y_{(uvw)}$ are all smaller than $Y_{poly}$, where

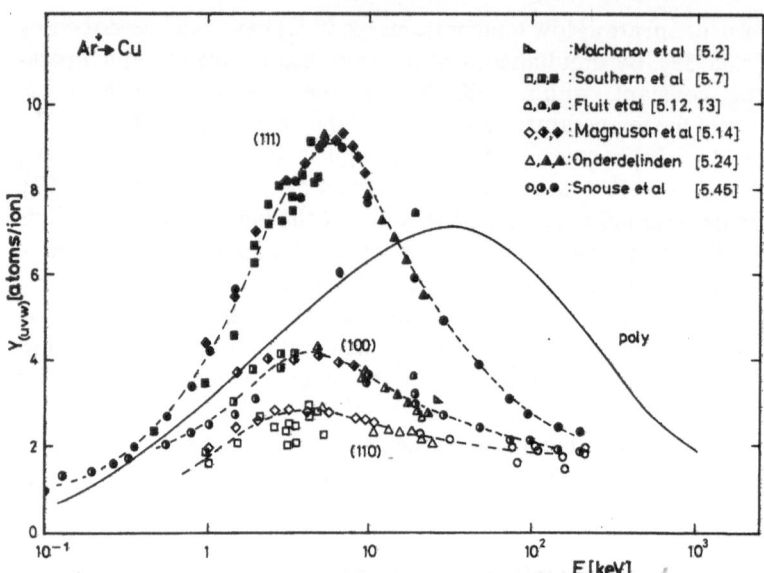

**Fig. 5.2.** The energy dependence of the sputtering yields for incidence of Ar⁺ on the (110), (100), and (111) plane of Cu. Also indicated is the energy dependence of the sputtering yield for polycrystalline Cu

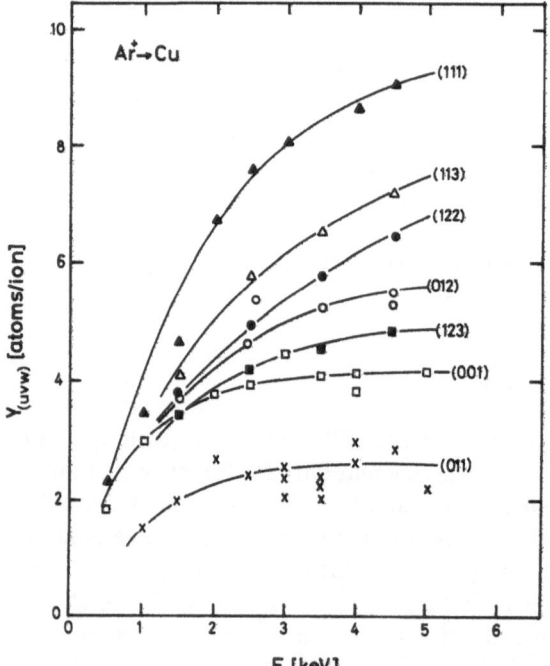

**Fig. 5.3.** The energy dependence of the sputtering yields of higher index Cu monocrystals for normal incidence of Ar⁺ [5.7]

$Y_{(111)} > Y_{(100)} > Y_{(110)}$. Below 10 keV, $Y_{poly}$ drops under the value for the (111) surface. Furthermore, $Y_{(100)}$ is slightly higher than $Y_{(111)}$ for primary energies smaller than 0.5 keV.

The single crystal yield curves show relatively sharp maxima at an energy $E_{max}(uvw)$. This energy depends on the crystallographic orientation of the surface plane in the order $E_{max}(111) > E_{max}(100) > E_{max}(110)$. The values for $E_{max}(uvw)$ for the projectile-target combination $Ar^{+} \to Cu$ are about 7, 4, 3 keV [5.40], respectively. These energies are much smaller than the energy, for which $Y_{poly}$ shows a maximum.

Measurements of the sputtering yield of higher index planes at normal incidence have been performed by *Southern* et al. [5.7] and are shown in Fig. 5.3. It is seen that the sputtering yields of the higher index planes constitute a series following the magnitude of the interatomic distance of these directions, as do the yields of low index planes. This series starts for the (123) plane at a lower yield value than is observed for the (111) plane.

Further information on the energy dependence of the sputtering yield is obtained from threshold measurements for single-crystalline targets [5.46], which indicate that the threshold energy for sputtering of the (111) surface plane is larger than for the (100) plane which is consistent with the observation that below 0.5 keV, $Y_{(100)} > Y_{(111)}$.

## b) Theoretical Considerations

The sputtering yield of polycrystalline material is proportional to the energy dissipated in elastic collisions in the surface layer and inversely proportional to the surface binding energy $U_0$ (Chap. 2). Range measurements [5.20–22] have indicated that the stopping power of keV ions, incident along a low index crystallographic direction, is reduced since part of the ion beam will be channeled.

*Onderdelinden* [5.23] therefore suggested that at energies above a few keV, channeling will also influence the sputtering process in monocrystalline targets. In his model, *Onderdelinden* relates the sputtering yield $Y_{(uvw)}$ to the sputtering yield $\hat{Y}$ of a structureless medium through [(3.3.15)]

$$Y_{(uvw)}(E) = \chi_{uvw}(E)\,\eta_{(uvw)}\,\hat{Y}(E), \tag{5.3.1}$$

with $E$ the primary energy of the projectile, $\chi_{uvw}(E)$ the nonchanneled fraction of the ion beam after entering the crystal surface and $\eta_{(uvw)}$ a fitting parameter. Thus it is assumed that only the nonchanneled part of the beam contributes to sputtering and that this contribution is equivalent to that of a beam in a structureless medium. The nonchanneled fraction of the beam $\chi_{uvw}(E)$, is calculated following the theoretical treatment on channeling by *Lindhard* [5.25] using his low-energy approximation to the Thomas-Fermi ion-atom interaction potential. If we neglect thermal vibrations, $\chi_{uvw}(E)$ is then given by [see also (3.3.17)]

$$\chi_{uvw}(E) = \pi N t_{uvw} p_m^2 \tag{5.3.2a}$$

$$= \pi N t_{uvw}^{3/2} \cdot (3a^2 Z_1 Z_2 e^2 / E)^{1/2} \tag{5.3.2b}$$

$$= \pi N t_{uvw}^3 \psi_{2uvw}^2 \tag{5.3.2c}$$

$$= (E_{uvw}^c / E)^{1/2} . \tag{5.3.2d}$$

$N$ is the atomic density of the target, $p_m$ is the minimum impact parameter for channeling $t_{uvw}$ the distance between the atoms measured along the $[uvw]$ direction, $Z_1$ and $Z_2$ the atomic number of the incident ion and the target atom, respectively, $a = 0.8853 a_0 (Z_1^{2/3} + Z_2^{2/3})^{-1/2}$ and $a_0$ the Bohr radius, $e$ is the elementary charge and $\psi_{2uvw}$ is the critical angle for channeling for incidence along the $[uvw]$ direction [5.25]:

$$\dot{\psi}_{2uvw} = (3a^2 Z_1 Z_2 e^2 / E t_{uvw}^3)^{1/4} . \tag{5.3.3}$$

$E_{uvw}^c$ is defined through

$$E_{uvw}^c = \pi^2 N^2 t_{uvw}^3 (3a^2 Z_1 Z_2 e^2). \tag{5.3.4}$$

According to *Onderdelinden*'s model, channeling will effectively reduce the sputtering yield for primary energies $E$ larger than $E_{uvw}^c$, bringing about a successive interference of channeling with sputtering for the different surface planes (110), (100), and (111), respectively.

For normal incidence on a low index plane, channeling will only reduce the sputtering yield, but it is found that for intermediate energies (0.5–10 keV) the measured sputtering yield $Y_{(uvw)}$ may be higher than $Y_{poly}$. The values obtained for $Y_{(uvw)}(E)/\chi_{uvw}(E) = \eta_{(uvw)} \hat{Y}(E)$ [see (5.3.1)] are generally larger than $Y_{poly}$. As the surface binding energy of the atoms in a monocrystalline surface plane will be higher than the surface binding energy of atoms situated on the surface of a polycrystal or high index planes, this would give an opposite effect.

Originally *Onderdelinden* [5.23] suggested $Y_{poly}(E)$ as a value for $\hat{Y}(E)$ and argued that $\eta_{(uvw)}$ must be larger than unity for the following reasons:

(i) the random part of the ion beam always collides in the top layers of the crystal;

(ii) part of the ions impinging on a polycrystalline target will channel as well;

(iii) the factor accounts for any orientation dependence of the ejection mechanism, such as, e.g., focusing collision sequences along close-packed directions.

For a comparison of (5.3.1) with the measured yields $Y_{(uvw)}(E)$ for $Ar^+$ on Cu, the values of *Yonts* et al. [5.47] for $Y_{poly}(E)$ were taken for $\hat{Y}(E)$ to calculate the curves in Fig. 5.4. A good fit was obtained for the following $\eta$-values:

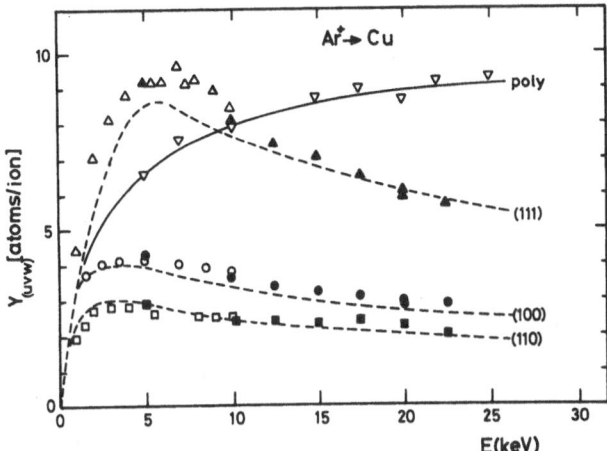

**Fig. 5.4.** Comparison of the theoretical predictions of the channeling model (dashed curves) with experiment for normally incident Ar⁺ onto low index planes of Cu [5.23]

$\eta_{(100)} = \eta_{(111)} = 1.3$ and $\eta_{(110)} = 1.6$. With these numbers, the shift of the maxima with orientation is also well predicted by the model.

However, the reasonable fit can only be obtained for the energy interval shown in Fig. 5.4. Inspecting (5.3, 2b, 2c), it is seen that for $E > E^c_{uvw} > E^c_{u'v'w'}$, the ratio of the nonchanneled fractions is given by

$$\chi_{uvw}(E)/\chi_{u'v'w'}(E) = (E^c_{uvw}/E^c_{u'v'w'})^{1/2}$$
$$= (t_{uvw}/t_{u'v'w'})^{3/2}$$
$$= \text{constant} \qquad (5.3.5)$$

and thus for the sputtering yields

$$Y_{(uvw)}(E)/Y_{(u'v'w')}(E) = (t_{uvw}/t_{u'v'w'})^{3/2} \cdot \eta_{(uvw)}/\eta_{(u'v'w')}. \qquad (5.3.6)$$

The ratio of the sputtering yields of particular surface planes should tend to a constant for $E > E^c_{uvw} > E^c_{u'v'w'}$, or at least to a slowly varying function with energy, for energies well below $\sim 1$ MeV. Such ratios were plotted by *Snouse* et al. [5.45] for the energy range from 0.1 to 200 keV (Fig. 5.5). But instead of a constant ratio, a strongly energy dependent function is observed.

Even more conclusive are the values of $\eta_{(uvw)} \hat{Y}(E)$, obtained by dividing the experimentally measured yields $Y_{(uvw)}(E)$ by the calculated channeled fraction $\chi_{uvw}$ of the ion beam [see (5.3.1, 2)]. The results for the same set of data as used in Fig. 5.2 are plotted in Fig. 5.6 and compared with the corresponding data for a polycrystalline target. As mentioned already, the data for $\eta_{(uvw)} \hat{Y}$ are higher than the data for $Y_{poly}$, and it is seen that the differences gradually increase with energy. Most interesting, however, is the behaviour at energies roughly above

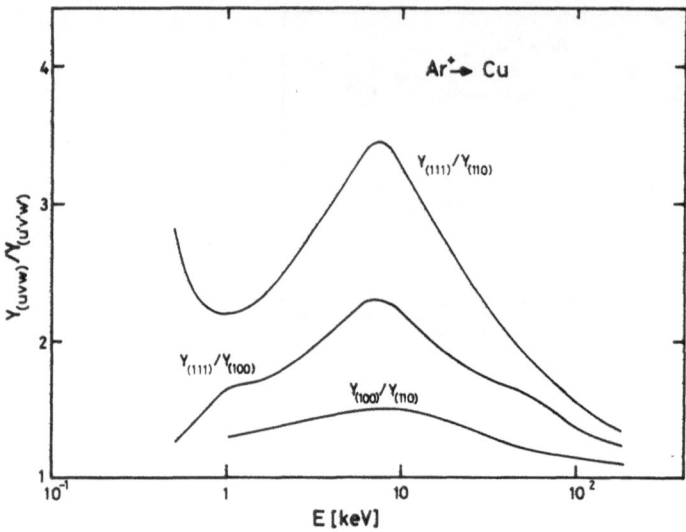

**Fig. 5.5.** The yield ratio as a function of the energy of normally incident Ar$^+$ [5.45]

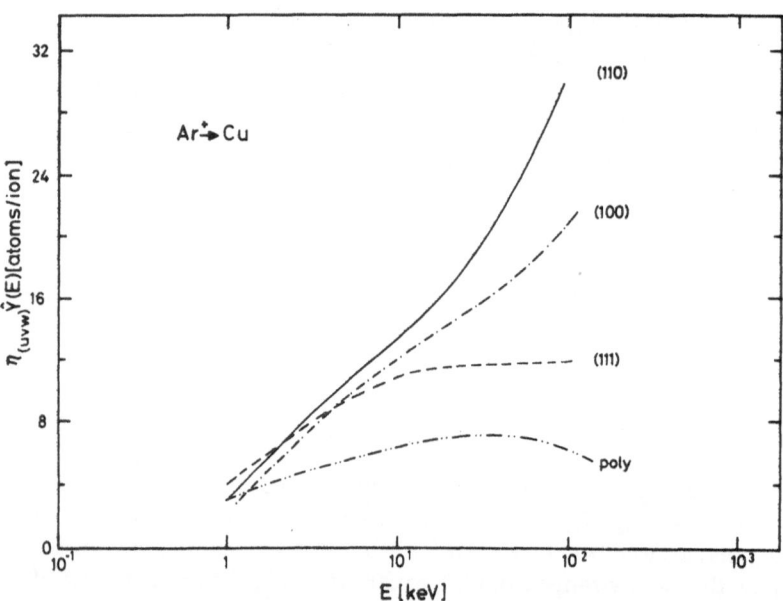

**Fig. 5.6.** $\eta_{(uvw)} \hat{Y}(E)$ as a function of the Ar$^+$-ion energy in comparison with $Y_{poly}$

10 keV. The $\eta_{(111)} \hat{Y}$ values bend over as do the values for $Y_{poly}$, but for the two other planes the values still increase, for the (110) plane faster than for the (100) plane.

At energies $\lesssim 5$–10 keV, the discrepancy between $\eta_{(uvw)} \hat{Y}(E)$ and $Y_{poly}$ cannot be attributed to a breakdown of the channeling model. Since $\chi_{uvw}(E)$ is large, small uncertainties of $\chi_{uvw}(E)$ will not influence $\eta_{(uvw)}\hat{Y}$ significantly. For higher energies, however, this model apparently tends to overestimate the reduction in yield. This effect is the larger the higher the primary energies and the smaller $t_{uvw}$, i.e., for smaller nonchanneled fractions $\chi_{uvw}(E)$ of the ion beam [see (5.3.2)]. Reasons for such an overrating of the channeled fraction of the ion beam are to be found in the neglect of dechanneling in the model and the very crude description of the flux profile of the ion beam as a function of depth, expecially for shallow depths near the surface.

At energies only somewhat above the threshold energy for sputtering, channeling is no longer important and the yields are governed by the surface binding energy $U_0$, to which the sputtering yield at keV energies is inversely proportional [5.48]. For a variety of fcc metals, *Jackson* [5.49] has calculated the dependence of the surface binding energy on the crystallographic orientation of the surface plane and found the following order: $U_{(111)} > U_{(100)} > U_{(110)}$. This gives an ordering of the sputtering yields in agreement with the results of Fig. 5.2.

In a theoretical treatment *Harrison* et al. [5.50] expected, that the threshold energy would be lowest for the (110) plane and highest for the (111) plane, so at very low energies one has $Y_{(110)} > Y_{(100)} > Y_{(111)}$, which is also in qualitative agreement with the above mentioned results of *Zdanuk* et al. [5.46].

Thus for energies above a few keV, the channeling model provides a fair qualitative explanation for the dependence of the sputtering yield on the lattice parameter $t_{uvw}$ and the implicitly connected dependence on primary energy. However, this direct dependence seems to be violated for planes with higher indices, as is demonstrated in Fig. 5.3. Although the [111] direction is more open than the [123], [012], [122], and [113] directions, still the highest yield is observed for the (111) plane. For 20 keV Ar$^+$ bombardment onto a (111) plane, but into the [132] and [021] direction, respectively, the latter directions were indeed found to be more opaque than the [111] direction (see also Sect. 5.4.1). For the energies, for which the data presented in Fig. 5.3 are measured, the application of the channeling model is certainly doubtful, but even then only a decisive dependence on the crystallographic orientation of the surface plane can explain the results of Fig. 5.3.

At energies below $\sim 0.5$ keV, the magnitude of the sputtering yield is predominantly determined by the surface binding energy $U_{(uvw)}$.

### c) Other Projectile-Target Combinations

Similar qualitative agreement between experiment and theoretical predictions can be established for other projectile-target systems.

For Kr$^+$ bombardment of Cu, *Onderdelinden* [5.24, 40] measured that $Y_{(111)}$ is larger than $Y_{poly}$ up to an energy of 40–50 keV, whereas $Y_{poly} > Y_{(100)} > Y_{(110)}$ over the measured energy span of 6–40 keV (Fig. 5.7). A comparison of Cu$^+ \to$ Cu single crystal yields [5.24, 40] with data for polycrystalline targets gives similar results.

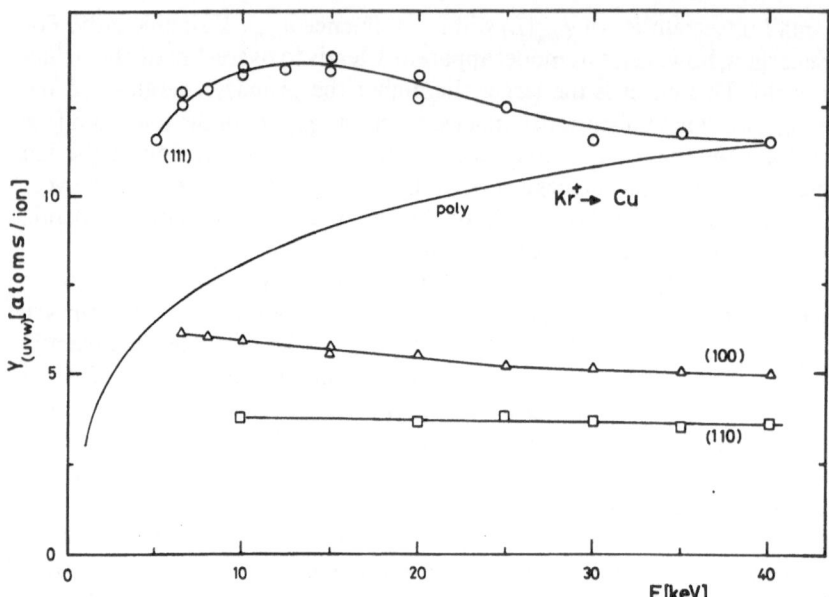

**Fig. 5.7.** The sputtering yields for $Kr^+$ bombardment of monocrystalline and polycrystalline Cu as a function of the primary energy [5.24]. Also indicated is the energy dependence of the sputtering yield for polycrystalline Cu

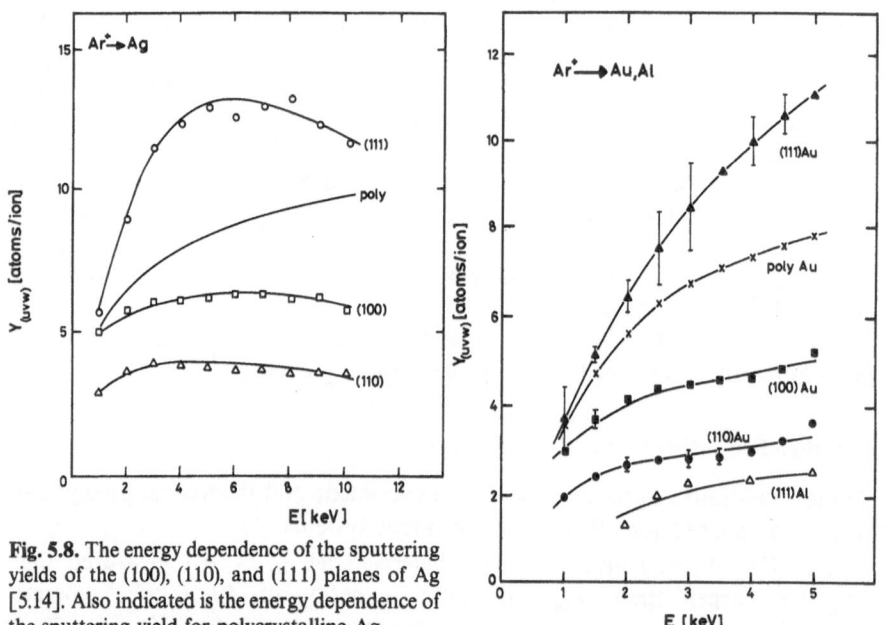

**Fig. 5.8.** The energy dependence of the sputtering yields of the (100), (110), and (111) planes of Ag [5.14]. Also indicated is the energy dependence of the sputtering yield for polycrystalline Ag

**Fig. 5.9.** The energy dependence of the sputtering yields of polycrystalline Au and monocrystalline Au and Al for normal incidence of $Ar^+$ [5.51]

Measurements of $Ar^+$ incidence onto the three main low index planes of Ag [5.14] are shown in Fig. 5.8, and compared with data for polycrystalline Ag. For incidence of 1–5 keV, $Ar^+$ sputtering yields have been determined [5.51] for bombardment of Au and Al (Fig. 5.9).

All these measurements confirm the general features already discussed for Cu as a target, i.e., $Y_{(111)} > Y_{poly} > Y_{(100)} > Y_{(110)}$ for bombarding energies of a few keV.

Low energy ($< 100$ eV) $Ar^+$ bombardment of Au [5.52] shows that $Y_{(100)}$ is larger than $Y_{(111)}$, and threshold measurements of the differential sputtering yields $(\Delta Y)_{[110]}$, ejected from the (110), (100), and (111) surface planes of Ag [5.53] imply that $Y_{(110)} > Y_{(100)}$ for primary energies just above threshold. The latter measurements again show quantitative agreement with the theory of *Harrison* et al. [5.50] for values of the surface binding energy, as calculated by *Jackson* [5.49].

### 5.3.2 HCP Crystals

Sputtering yields for a few keV $Ar^+$ bombardment of hcp crystals [5.54] are listed in Table 5.1. It is found that $Y_{(0001)} \gtrsim Y_{(10\bar{1}0)} > Y_{(11\bar{2}0)}$ for Mg,

**Table 5.1.** Sputtering yields of hcp monocrystals for normal incidence of $Ar^+$ [5.54]

| Metal | Surface plane | Ion energy $E$ [keV] | Observed yield $Y$ [atoms/ion] | $\chi$ | $\varepsilon \hat{Y}$ [atoms/ion] |
|-------|---------------|---------------------|-------------------------------|--------|-----------------------------------|
| Mg | (0001) | 5 | $3.13 \pm 0.15$ | 0.296 | $10.59 \pm 0.51$ |
| | (10$\bar{1}$0) | 5 | $2.70 \pm 0.12$ | 0.326 | $8.29 \pm 0.37$ |
| | (11$\bar{2}$0) | 5 | $1.64 \pm 0.12$ | 0.143 | $11.47 \pm 0.84$ |
| Zn | (0001) | 3 | $14.79 \pm 0.46$ | 0.730 | $20.26 \pm 0.63$ |
| | | 4 | $16.70 \pm 0.53$ | 0.632 | $26.41 \pm 0.84$ |
| | | 5 | $15.80 \pm 0.36$ | 0.565 | $27.93 \pm 0.64$ |
| | (10$\bar{1}$0) | 3 | $11.22 \pm 0.25$ | 0.658 | $17.83 \pm 0.40$ |
| | | 4 | $11.43 \pm 0.15$ | 0.570 | $20.98 \pm 0.28$ |
| | | 5 | $12.65 \pm 0.25$ | 0.510 | $25.96 \pm 0.51$ |
| | (11$\bar{2}$0) | 3 | $9.19 \pm 0.21$ | 0.289 | $31.83 \pm 0.73$ |
| | | 4 | $10.07 \pm 0.12$ | 0.250 | $40.27 \pm 0.48$ |
| | | 5 | $9.35 \pm 0.31$ | 0.224 | $41.28 \pm 1.39$ |
| Zr | (0001) | 5 | $1.12 \pm 0.02$ | 0.428 | $2.63 \pm 0.05$ |
| | (10$\bar{1}$0) | 5 | $1.56 \pm 0.03$ | 0.485 | $3.21 \pm 0.05$ |
| | (11$\bar{2}$0) | 4 | $0.74 \pm 0.07$ | 0.238 | $3.12 \pm 0.29$ |
| | | 5 | $0.64 \pm 0.03$ | 0.213 | $3.03 \pm 0.15$ |
| | | 10 | $0.65 \pm 0.01$ | 0.150 | $4.35 \pm 0.07$ |
| Cd | (0001) | 4 | $19.01 \pm 0.26$ | 0.619 | $30.72 \pm 0.42$ |
| | | 5 | $21.25 \pm 0.46$ | 0.554 | $38.39 \pm 0.83$ |
| | (10$\bar{1}$0) | 4 | $15.86 \pm 0.64$ | 0.545 | $29.10 \pm 1.17$ |
| | | 5 | $17.86 \pm 0.40$ | 0.487 | $36.65 \pm 0.82$ |
| | (11$\bar{2}$0) | 5 | $13.58 \pm 0.28$ | 0.214 | $63.51 \pm 1.31$ |

$Y_{(10\bar{1}0)} > Y_{(0001)} > Y_{(11\bar{2}0)}$  for  Zr,  whereas  for  Zn  and  Cd $Y_{(0001)} > Y_{(10\bar{1}0)} > Y_{(11\bar{2}0)}$. Nicely reflected in these results is the change in order in magnitude of the interatomic distance $t$ for the [0001] and [10$\bar{1}$0] direction with the value for $c/a$. For Cd and Zn ($c/a > 3^{1/2}$) we have $t_{11\bar{2}0} < t_{10\bar{1}0} < t_{0001}$; thus we can expect that $Y_{(0001)} > Y_{(10\bar{1}0)} > Y_{(11\bar{2}0)}$. For Mg and Zr ($c/a < 3^{1/2}$), $t_{0001} < t_{10\bar{1}0}$ and thus $Y_{(10\bar{1}0)} \gtrsim Y_{(0001)} > Y_{(11\bar{2}0)}$.

In analogy to the results for fcc metals we see that the $\eta \hat{Y}$ values are highest for the closest packed direction, i.e., for the (11$\bar{2}$0) plane and thus for the smallest fraction $\chi$ (Table 5.1). The values for $\chi$ for the (11$\bar{2}$0) plane are comparable with the $\chi$-values for 6–20 keV Ar$^+$ onto (110) Cu. So the high $\eta \hat{Y}$ values can be due either to an overestimate of the channeled fraction of the beam or to a possible significant influence of monocrystalline effects, e.g., focusing collisions on the ejection mechanism (Sect. 5.3.1b).

### 5.3.3 Other Crystals

For a bcc crystal, some data [5.14] are known for 5 keV Ar$^+$ bombardment of Mo, for normal incidence onto the (110) and (100) planes and for normal incidence onto polycrystalline Mo. It is found that $Y_{(110)} > Y_{poly} > Y_{(100)}$, thus essentially the same result is found as for bombardment of fcc metals: the sputtering yield for the direction with the largest atomic distance (for bcc the [110] and for fcc the [111] direction), is larger than the sputtering yield for polycrystalline material.

In the case of semiconductor targets such as Si and Ge, i.e., for crystals with a diamond type lattice, at a substrate temperature well above the transition temperature (see also Sect. 5.5) it is generally found that $Y_{(100)} > Y_{(111)} > Y_{(110)}$, as one would expect on the basis of the previous considerations. At low primary energies of the ion beam ($\lesssim E^c_{(uvw)}$), the sputtering yield of the amorphous target is about equal to the yield for the (111) plane [5.55]. For higher primary energies, where the sputtering yield is effectively reduced by channeling, $Y_{amorph}$ is always larger than $Y_{(uvw)}$ [5.56, 57].

## 5.4 On the Angular Dependence of the Sputtering Yield

From inspection of the results of *Rol* et al. [5.1] and *Molchanov* et al. [5.2] (Fig. 5.1), it is seen that the sputtering yield for monocrystalline targets $Y_{(uvw)}(\vartheta, \psi)$ is both a function of the angle $\vartheta$ with respect to the surface normal [uvw], as well as of the angle $\psi$ with respect to the nearest low index direction [u'v'w']. In measurements on the angular dependence of the sputtering yield, the ion beam is aligned along the [u'v'w'] direction by rotating the crystal around an axis [u''v''w''] in the surface. For incidence onto (100) Cu, but rotated around the

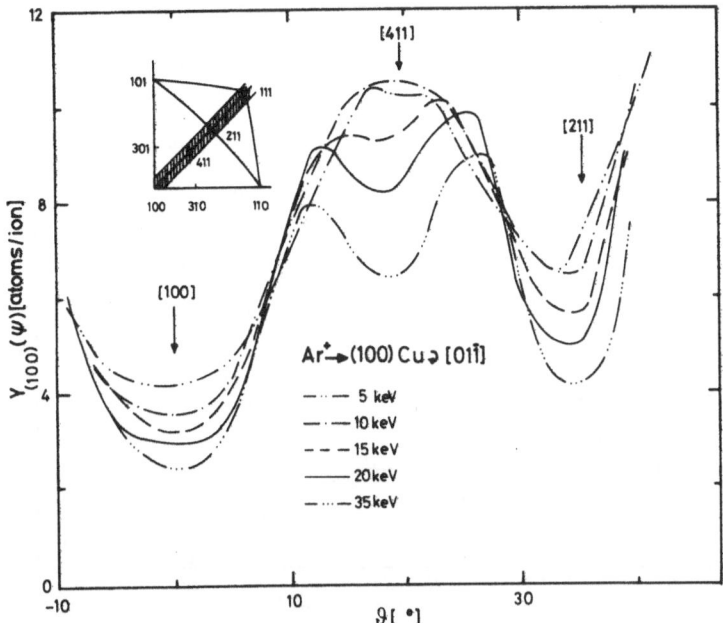

**Fig. 5.10.** The energy dependence of the angular variation of $Y_{(100)}(\vartheta, \psi)$ for incidence of $Ar^+$ onto (100) Cu, rotated around the [01$\bar{1}$] axis [5.24]

[001] axis, *Fluit* et al. [5.12] observed, e.g., minima for incidence along the [100], [310], and [110] directions.

*Mashkova* et al. [5.58] observed a reduction in yield along the [110], [130], and [010] directions for incidence of 30 keV $Ar^+$ onto (110) Cu, rotated around the [001] axis, and along the [110], [221], [111], and [112] directions for rotation around the [$\bar{1}$10] axis.

Measurements of the angular dependence were also performed for incidence on Ni crystals [5.2] and on Zn [5.59], i.e., a hcp crystal.

### 5.4.1 Absolute Measurements of the Angular Dependence of the Sputtering Yield

Systematic studies of the dependence of the angular variation of the total sputtering yield on primary energy and substrate mass were executed by *Onderdelinden* [5.24] for incidence of $Ar^+$ onto the (100) surface plane of fcc metals rotated about the [01$\bar{1}$] axis.

The energy dependence of the angular variation of $Y_{(100)}(\vartheta, \psi)$ in the energy range 5–35 keV is demonstrated in Fig. 5.10. For all primary energies, $Y_{(100)}(\vartheta, \psi)$ shows minima for incidence along the [100] axis ($\vartheta=0$) and along the [211] axis ($\vartheta=35°$). A third minimum at $\vartheta=19°$ (i.e., corresponding to the [411] axis) appears for primary energies above 10 keV. The sputtering yields in

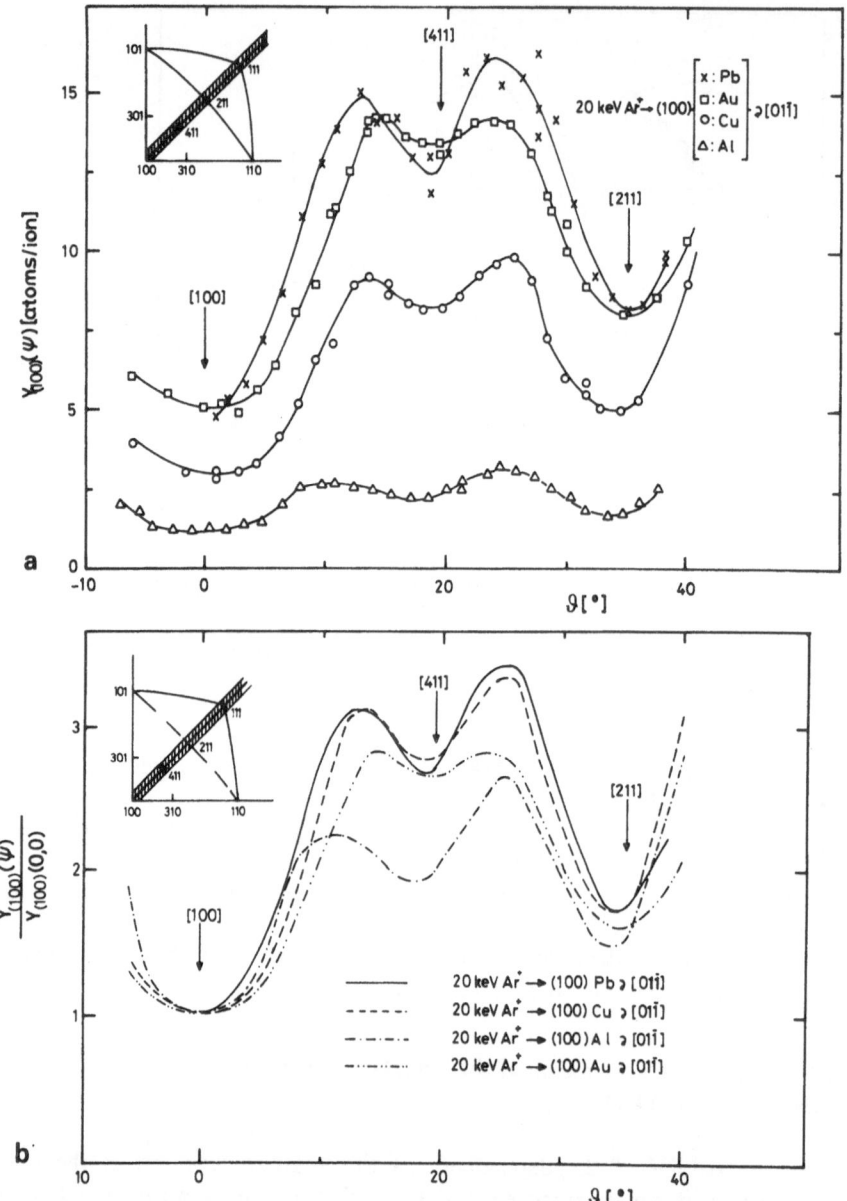

**Fig. 5.11a, b.** The dependence on substrate mass of the angular variation of $Y_{(100)}(\vartheta, \psi)$ for incidence of 20 keV Ar⁺. Rotation axis is the [01$\bar{1}$] axis (**a**) absolute yields, (**b**) yields normalized to $Y_{(100)}(0, 0)$

the minima are seen to decrease with energy both absolutely as well as relatively to the yield in the maxima. For a larger primary energy a smaller width of the angular dips is observed.

The dependence on the substrate mass is demonstrated in Fig. 5.11a, and in Fig. 5.11b after normalization on $Y_{(100)}(0, 0)$. The angle $\vartheta$, at which the first

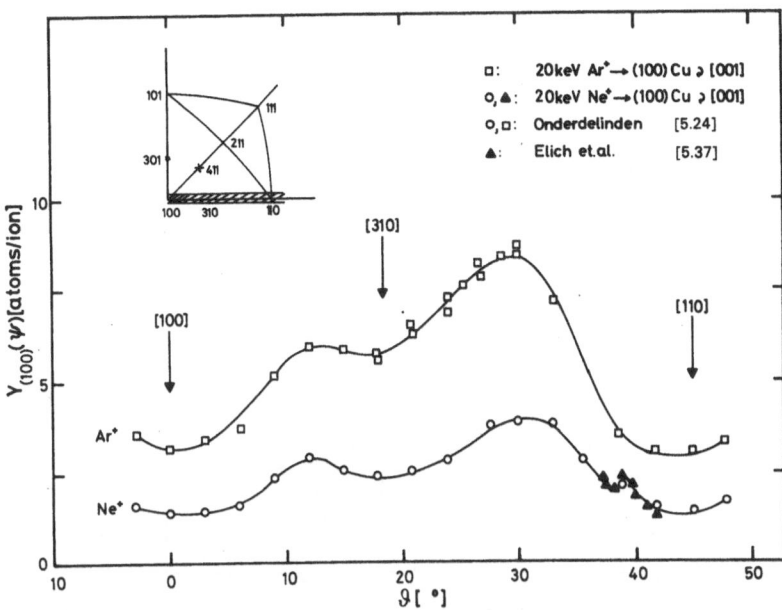

**Fig. 5.12.** The sputtering yield $Y_{(100)}(\vartheta, \psi)$ for incidence of 20 keV Ar$^+$ and 20 keV Ne$^+$. The crystal was turned around the [001] axis

maximum appears, is slightly dependent on substrate mass ($\vartheta = 11°$ for Al, 13° for Cu and Pb, and 15° for Au, see Fig. 5.11b). The relative change of the sputtering yield turns out to be identical for Cu and Pb, notwithstanding the quite different values for the absolute sputtering yields. This is due to the fact that $Z_2/t^3$ is about the same for the two targets.

A smaller projectile mass will reduce the widths of the angular dips and results in a larger relative reduction in yield for incidence along a low index direction, as can be seen from a comparison of the results for 20 keV Ne$^+$ bombardment with the results for 20 keV Ar$^+$ bombardment, both onto (100) Cu, rotated around the [001] axis [5.24, 37] (Fig. 5.12). The sputtering yield shows minima for incidence along the [100] axis ($\vartheta = 0°$), along the [310] axis ($\vartheta = 18°$) and along the [110] axis ($\vartheta = 45°$). For the Ne$^+$ measurements, an additional peak appears in the sputtering yield curve for incidence at $\vartheta = 39°$. We will come back to this interesting feature in Sect. 5.6.

The dependence on projectile mass is also shown in the results of *Elich* et al. [5.39] for bombardment of the (111) plane of a Cu crystal (Fig. 5.13) rotated around the [01$\bar{1}$] axis [with minima at $\vartheta = 0°$ ([111] axis) and $\vartheta = 19°$ ([211] axis)]. For rotation about the [11$\bar{2}$] axis [5.39], the sputtering yield is reduced when the beam is parallel to the [111] ($\vartheta = 0°$), the [132] ($\vartheta = 22°$), and the [021] axis ($\vartheta = 39°$) (Fig. 5.14). In the latter case no planar channeling is possible, as is the case for the other rotation axes [5.24]. Note also that the reduction in yield for the [132] and [021] axes is smaller than for the [111] axis, as one would expect, since the [111] direction is more open than the two other directions (see also Sect. 5.3.1b).

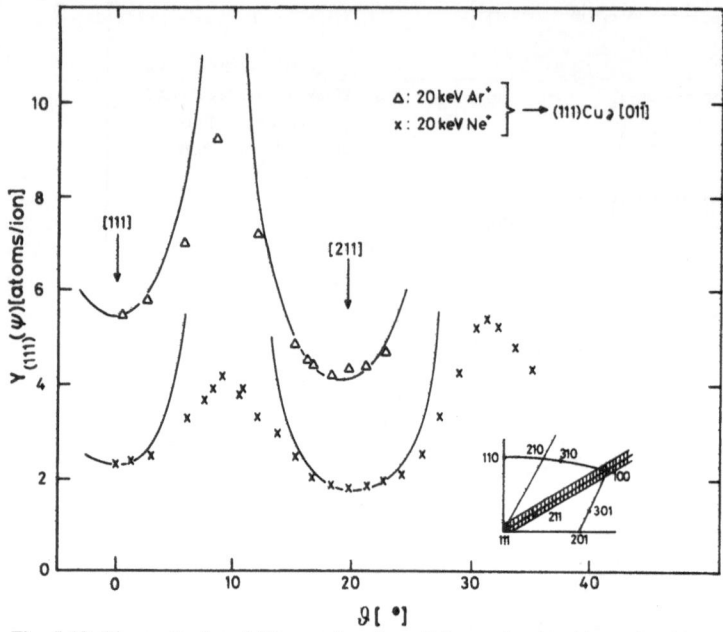

**Fig. 5.13.** The sputtering yields as a function of the angle of incidence for 20 keV Ar⁺ and 20 keV Ne⁺ ions on a (111) Cu crystal turned around the [01$\bar{1}$] axis. Substrate temperature 122 K. The points are experimental, the lines theoretical [5.39]

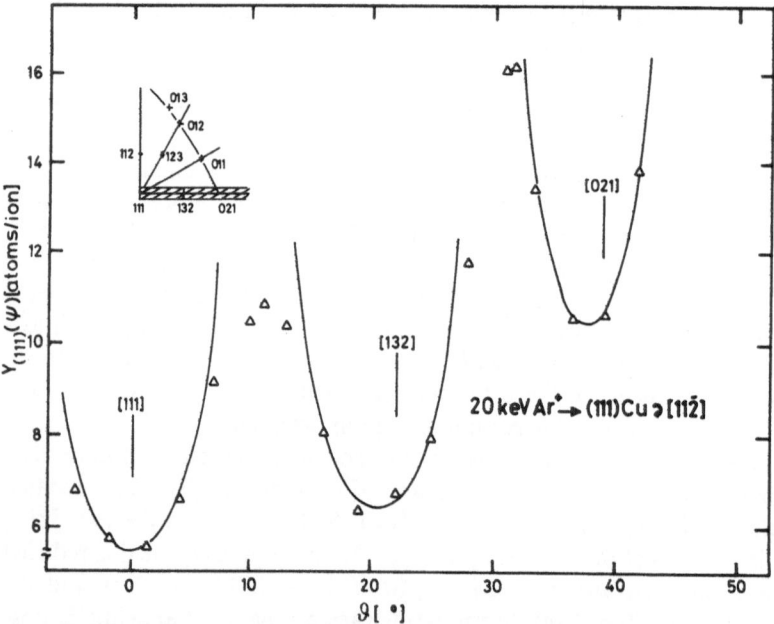

**Fig. 5.14.** The sputtering yield as a function of the angle of incidence for 20 keV Ar⁺ ions on a (111) Cu crystal turned around the [11$\bar{2}$] axis. Substrate temperature: 122 K. The points are experimental, the lines theoretical [5.39]

## 5.4.2 Theoretical Considerations

Several models have been developed throughout the years to explain the angular dependence of the sputtering yield of monocrystalline targets. These models are based on the idea of transparency, first introduced by *Fluit* [5.10–12] and extended by various authors [5.7, 14–17].

But, since these transparency models do not provide a consistent description of the energy dependence of the angular dependence of the sputtering yield, we will restrict our considerations to the channeling model of *Onderdelinden* [5.23, 24, 40] (also Sect. 3.3.4).

In order to account for the angular dependence, (5.3.1) has to be generalized [5.23, 24, 40] (3.3.15):

$$Y_{(uvw)}(\vartheta, E, \psi) = \eta_{(uvw)} \chi_{u'v'w'}(\psi, E) \, \hat{Y}(\vartheta, E), \tag{5.4.1}$$

with (3.3.18):

$$\chi_{u'v'w'}(\psi, E) = \chi_{u'v'w'}(0, E)$$
$$\cdot \{1 - [1 - \chi_{u'v'w'}(0, E)] (\psi/f\psi_2)^2\}^{-1}, \; \psi < f\psi_2, \tag{5.4.2}$$

where $\chi_{u'v'w'}(0, E)$ is defined through (5.3.2), $\psi_2$ through (5.3.3) and $f$ is a constant $(\sim \sqrt{3})$ [5.24].

The channeling model thus lends itself to a direct comparison with experimental quantities such as the halfwidth of the angular dips and the relative yield or minimum yield in the angular dip. In Fig. 5.15 a plot is given of the halfwidth at full maximum for the [100] minimum, obtained from the energy dependent and substrate mass dependent measurements of *Onderdelinden* [5.24, 40], versus a reduced primary energy $E/E_k$ (with $E_k = 3aZ_1Z_2e^2/t_{100}^3$) of the projectiles. For such a restricted energy range, the experimental data are seen to obey an energy dependence $E^{-1/4}$, as predicted in (5.3.3).

Using (5.4.1, 2), a description of the angular behaviour over the complete angular range can be given. This description is compared with experiment in Fig. 5.16 for the incidence of 20 keV Ar$^+$ on (100) Cu, rotated around the [01$\bar{1}$] axis. The fit in Fig. 5.16 is obtained with $f = 1.2$. Further fits of the model (for $f = 1.3$) to experimental results are presented in Figs. 5.13 and 5.14. Except for the values in the maxima, the agreement is seen to be good. The discrepancy in the maxima can be lifted by the inclusion of planar channeling in the calculation, as was done by *Francken* et al. [5.60].

The disappearance of the [411] minimum for primary energies below 10 keV (see Fig. 5.11) is consistently explained in terms of the channeling model, since according to (5.3.2), $E_{411}^c \sim 10$ keV. The channeling model thus provides a consistent theoretical framework for a description of the angular variation of the sputtering yield.

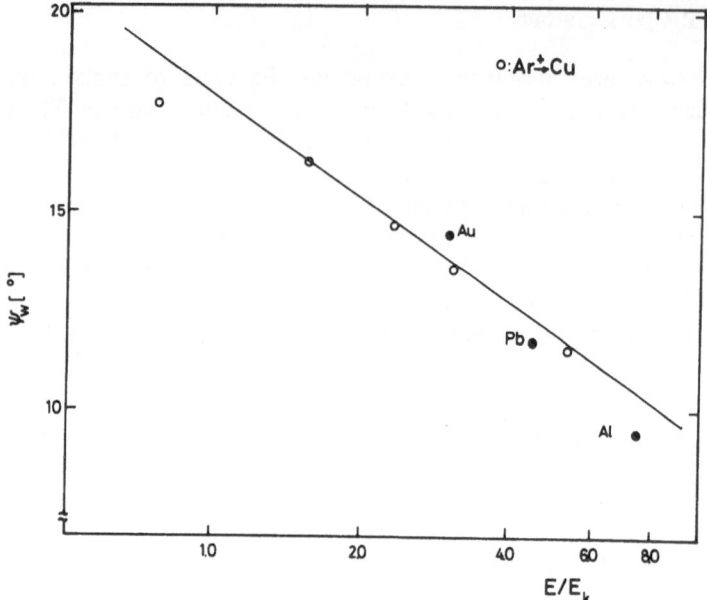

**Fig. 5.15.** The halfwidth at full maximum of the [100] minimum as a function of the reduced energy $E/E_k$ ($E_k = 3aZ_1Z_2e^2/4\pi\varepsilon_0 t_{100}^3$). The open circles are from 5, 10, 15, 20, and 35 keV ions on Cu, the black dots from 20 keV ions on Au, Pb, and Al crystals

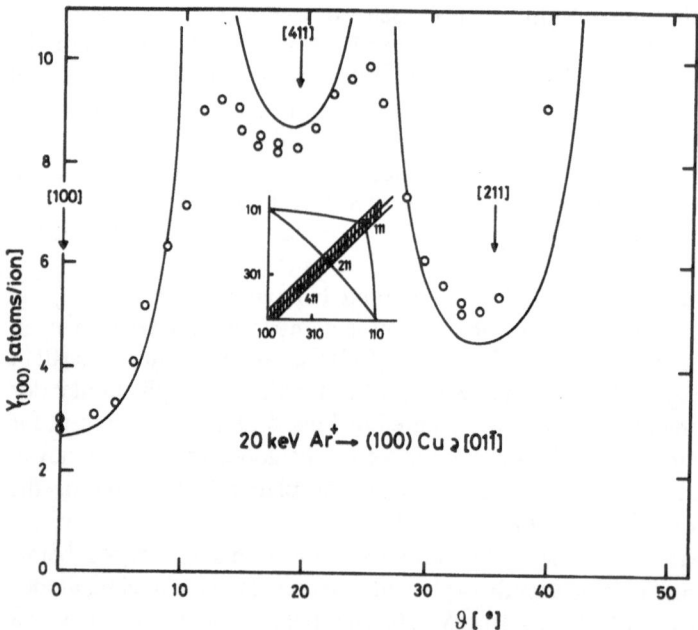

**Fig. 5.16.** The sputtering yield $Y_{(100)}(\vartheta, \psi)$ for incidence of 20 keV Ar$^+$. Rotation axis is the [01$\bar{1}$] axis. The open circles are experimental, the lines theoretical

From a comparison of their computer simulation data of the sputtering yield with the experimental data, *Harrison* et al. [5.33] reached contrary conclusions about the applicability of the channeling model to explain the angular dependence of the sputtering yield of monocrystalline target materials. Their model, however, tends to shift the sputtering minima towards smaller angles $\vartheta$. It is interesting to note that for incidence of $Ar^+$, a similar additional peak is predicted [5.33], as is experimentally observed for $Ne^+$ bombardment at $\vartheta = 39°$ [5.37]. The predicted angular location of this peak is shifted proportionally to the shifts of the sputtering minima. Experimentally, no such peak was found for $Ar^+$ bombardment (see also Sect. 5.6). The shift in angle sheds doubt on the potential function used in these computer calculations [5.31, 33] (Chap. 3).

### 5.4.3 Relative Measurements of the Angular Variation of the Sputtering Yield

In the preceding sections, we have discussed the angular variation of the sputtering yield, obtained from measurements of the total sputtering yield, determined by the weight-loss method (Chap. 4). Valuable information on the relative change of the sputtering yield with angle of incidence can also be obtained by the direct detection of sputtered particles, ions or neutrals.

By measuring the angular variation of the sputtering yield through the change of resistance of a collector, which was rigidly connected to the target manipulator, *Drentje* [5.17] found an energy dependence $E^{-0.30 \pm 0.02}$ for the halfwidths of the angular dips from measurements for the target-projectile combination $Ar^+ \rightarrow Cu$ at the energies 34, 53, and 80 keV.

The detection of positive ion emission [5.35, 61–65] for various projectile-target combinations in the energy range 10–200 keV seemed to suggest an $E^{-1/3}$ dependence rather than an $E^{-1/4}$ dependence, as one would expect on the basis of *Lindhard*'s [5.25] low energy approximation to the Thomas-Fermi ion-atom potential. Furthermore, the angular width of positive ion emission turned out to be identical to the widths of the angular dips observed in the intensity of backscattered particles, measured by means of the integral method [5.62]. By plotting a wider range of data, *Grahmann* et al. [5.66] demonstrated, however, that the observed values [5.62] fit an approximate $E^{-1/4}$ dependence.

Extensive intercomparative studies of the halfwidths of the angular dips and the relative reduction in yield have been performed by *Zwangobani* et al. [5.34] for 5–15 keV inert gas ($Ne^+$, $Ar^+$, $Kr^+$) ion bombardment, incident along the [100] and [110] directions of semiconductor targets (Ge, InSb). In order to avoid radiation damage build-up, the targets were held at a temperature well above the transition temperature during the experiment. The results are shown in Table 5.2. It is seen that the experimental halfwidths $0.5\psi_w$ deviate in absolute value from the theoretical prediction for $\psi_{2uvw}$, where the deviation is larger for the more open [110] direction. The directional dependence of $\psi_{2uvw}$ turns out to agree

**Table 5.2.** Comparisons of half-widths of yield minima [5.34]

(a) InSb

| Ion | $E$ [keV] | $\left(\dfrac{5\text{ keV}}{E\text{ [keV]}}\right)^{1/4}$ | $\dfrac{\psi\,[110]}{\psi\,[100]}$ | $(\psi\,5\text{ keV}/\psi_w)$ | | $(\psi_2/\tfrac{1}{2}\psi_w)$ | |
|---|---|---|---|---|---|---|---|
| | | | | [100] | [110] | [100] | [110] |
| Ne | 5 | 1.0 | 1.12 | 1.0 | 1.0 | 1.06 | 1.25 |
| Ne | 10 | 1.2 | 1.12 | 1.2 | 1.2 | 1.08 | 1.26 |
| Ne | 15 | 1.3 | 1.08 | 1.3 | 1.35 | 1.06 | 1.27 |
| Ar | 5 | 1.0 | 1.21 | 1.0 | 1.0 | 1.15 | 1.22 |
| Ar | 10 | 1.2 | 1.16 | 1.14 | 1.19 | 1.1 | 1.23 |
| Ar | 15 | 1.3 | 1.16 | 1.27 | 1.34 | 1.12 | 1.24 |
| Kr | 5 | 1.0 | 1.3 | 1.0 | 1.0 | 1.23 | 1.22 |
| Kr | 10 | 1.2 | 1.3 | 1.13 | 1.12 | 1.16 | 1.15 |
| Kr | 15 | 1.3 | 1.25 | 1.24 | 1.3 | 1.16 | 1.21 |

(b) Ge

| Ion | E [keV] | $\left(\dfrac{10\text{ keV}}{15\text{ keV}}\right)^{1/4}$ | $\dfrac{\psi\,[110]}{\psi\,[100]}$ | $(\psi\,10\text{ keV}/\psi_w)$ | | $(\psi_2/\tfrac{1}{2}\psi_w)$ | |
|---|---|---|---|---|---|---|---|
| | | | | [100] | [110] | [100] | [110] |
| Ne | 10 | 1.0 | 1.15 | 1.0 | 1.0 | 1.05 | 1.18 |
| Ne | 15 | 1.2 | 1.20 | 1.12 | 1.08 | 1.06 | 1.15 |
| Ar | 10 | 1.0 | 1.25 | 1.0 | 1.0 | 1.13 | 1.16 |
| Ar | 15 | 1.1 | 1.34 | 1.07 | 1.09 | 1.17 | 1.13 |

better with the theoretical predictions for heavier projectiles. In this restricted energy range no deviations of $\psi_{2uvw}$ from the predicted energy dependence are observed.

The experimental relative yield is always larger than the nonchanneled fraction $\chi_{uvw}$. The deviation is larger for smaller $\chi_{uvw}$. This may well result from incompletely annealed radiation damage and thermal dechanneling.

## 5.5 On the Temperature Dependence of the Single-Crystal Sputtering Yield

In the preceding sections not very much attention was paid to the substrate temperature, since in most experiments this temperature was not strictly controlled, but the targets were about or slightly above room temperature. The sputtering process and thus the sputtering yield will, however, be submitted to various temperature dependent processes.

Experiments have shown that the effect of target temperature on sputtering is, in general, small for temperatures not too close to the melting point. The sputtering yield rises sharply when the temperature approaches the melting point [5.42, 67, 68].

An interpretation of this effect is found in the thermal spike theory, which incorporates the target temperature as a temperature level relative to which a spike temperature can be defined [5.68, 69]. Thus a relationship between the sputtering yield and target temperature can be derived.

Since we are interested in specific monocrystalline thermal effects, we restrict ourselves in this section to substrate temperatures well below the melting point, i.e., in the temperature range from $\sim 50$–$1000$ K. In this temperature régime, the changes in the sputtering yield of monocrystalline targets will be mainly due to

(i) the influence of thermal vibrations on single-crystalline directional effects like

    (a) focusing and
    (b) channeling, and

(ii) annealing of radiation damage.

In experiments at temperatures lower than room temperature, condensation on the surfaces has to be prevented by suitable vacuum conditions.

### 5.5.1 Metal Targets

For metal targets, no drastic changes in the sputtering yield have been observed in experiments, but the effects are of a rather moderate nature.

In the temperature range 300–1000 K, *Carlston* et al. [5.41] observed a decrease of $Y_{(111)}$ for 2, 5, and 10 keV Ar$^+$ bombardment of Cu; this reduction also turned out to be energy dependent (24, 12, and 16% at 2, 5, and 10 keV, respectively). For 5 keV Ar$^+$ bombardment perpendicular to (110) Cu, no measurable change in the sputtering yield was observed, as was also for polycrystalline Cu.

Similar observations were made by *Evdokimov* et al. [5.42] at a higher primary energy (30 keV Ar$^+ \rightarrow$Cu) in the temperature range 400–1300 K. A slight decrease in yield for $Y_{(111)}$ and a slight increase for $Y_{(110)}$ was observed (Fig. 5.17) with increasing substrate temperature.

*Elich* et al. [5.38] found for 20 keV Ar$^+$ onto Cu, a more pronounced increase for $Y_{(110)}$ in the temperature range 100–600 K (Fig. 5.18). In this range, an increase of $Y_{(100)}$ and $Y_{(111)}$ was also found (Fig. 5.18), the latter in contrast to the result of *Evdokimov* et al. [5.42]. In this context it has to be remarked that the absolute value of $Y_{(111)}$, reported by *Evdokimov* et al. [5.42] at a temperature of 423 K, is considerably higher than one would expect from a comparison with the data, reported in Fig. 5.2, Sect. 5.3 [$Y_{(111)}$ (30 keV) $\sim 7$ [5.42], where from Fig. 5.2, $Y_{(111)} \sim 5$]. No explanation for this discrepancy can be found in the paper of *Evdokimov* et al. [5.42]. The reason for the difference between their results on the temperature dependence and the results of *Elich* et al. [5.38], however, may well be found in this discrepancy.

For fixed angles of incidence, *Evdokimov* et al. [5.42] also measured the change in yield for bombardment of the (111) Cu surface, rotated around the

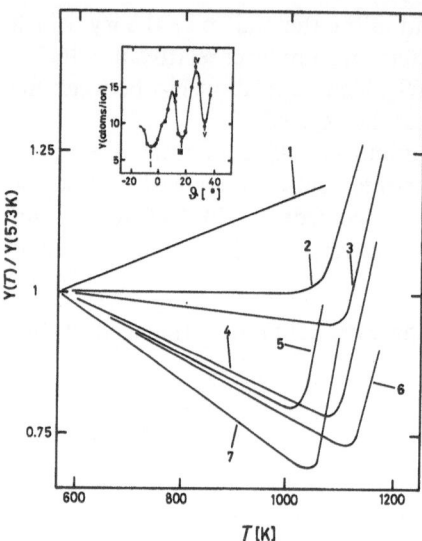

**Fig. 5.17.** The temperature dependence of the sputtering yield of Cu for the incidence of 30 keV Ar$^+$ in the temperature range 573–1200 K, normalized to the yield at $T = 573$ K [5.42]; *1* normal incidence onto the (110) surface; *2* polycrystalline Cu; *3* incidence on (111) surface, point I (see insert); *4* point III; *5* point II; *6* point V; *7* point IV

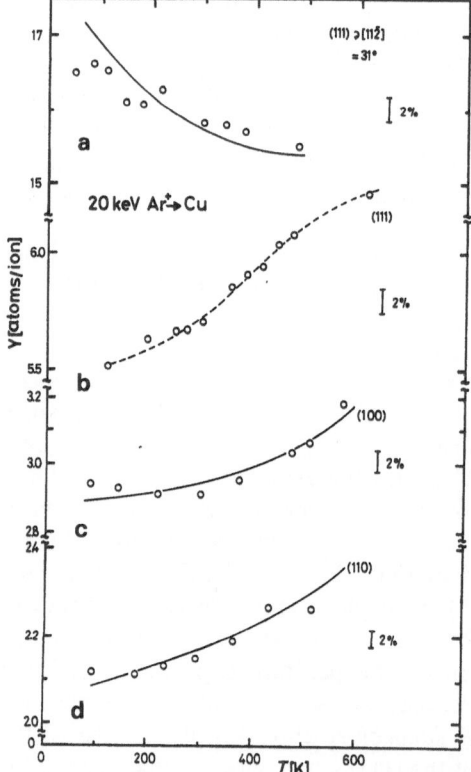

**Fig. 5.18.** The sputtering yield of monocrystalline Cu crystals for normal incidence of 20 keV Ar$^+$ as a function of the substrate temperature [5.38]. *a* (110) Cu, *b* (100) Cu, *c* (111) Cu, *d* $\vartheta = 31°$ for incidence on (111) Cu turned around the [11$\bar{2}$] axis. ○: experimental results; solid line: calculated values

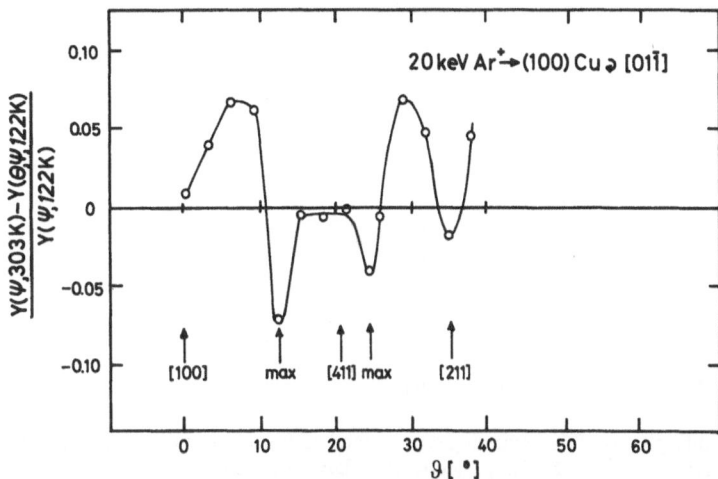

**Fig. 5.19.** The relative increase of the sputtering yield at 303 K with respect to 122 K as a function of the angle of incidence of 20 keV Ar$^+$ on (100) Cu turned around the [01$\bar{1}$] axis [5.39]

[11$\bar{2}$] axis (Fig. 5.17). In all cases a decrease in yield with temperature was observed, where the decrease is larger the more opaque the direction of incidence. For the same incidence conditions, but in the temperature range 50–600 K and at a primary energy of 20 keV, such a decrease was also observed by *Elich* et al. [5.38] at an angle of incidence $\vartheta = 31°$, and, but less pronounced, for incidence along the [021] direction.

To illustrate the variation of the sputtering yield with temperature as a function of the angle of incidence, we can define [5.38]

$$Q(\vartheta, \psi) = \frac{Y(\vartheta, \psi, T_2) - Y(\vartheta, \psi, T_1)}{Y(\vartheta, \psi, T_1)}, \ T_2 > T_1, \tag{5.5.1}$$

$Q(\vartheta, \psi)$ is thus also a function of both $\vartheta$ and $\psi$, just as $Y(\vartheta, \psi)$. A plot of $Q(\vartheta, \psi)$ is shown in Fig. 5.19 for incidence of 20 keV Ar$^+$ on (100) Cu, rotated around the [01$\bar{1}$] axis.

The following general tendency is noticed: $Q$ is a few percent for a low index direction (i.e., $\psi = 0$), increases with $\psi$ to a maximum at around the critical angle for channeling and then drops sharply to zero or even to negative values for incidence along directions for which the sputtering yield is at its maximum, i.e., along opaque directions.

One can also plot [5.70] $Q$ vs the nonchanneled fraction $\chi$ of the ion beam, as calculated from the *Onderdelinden* model [5.23, 24, 40]. This is done in Fig. 5.20 for directions of incidence which can be regarded as representing pure axial or planar channeling or genuine opaque situations.

With the exception of the [111] direction, a general behaviour is observed: $Q$ is positive for small values of $\chi$, and becomes negative for large values of $\chi$.

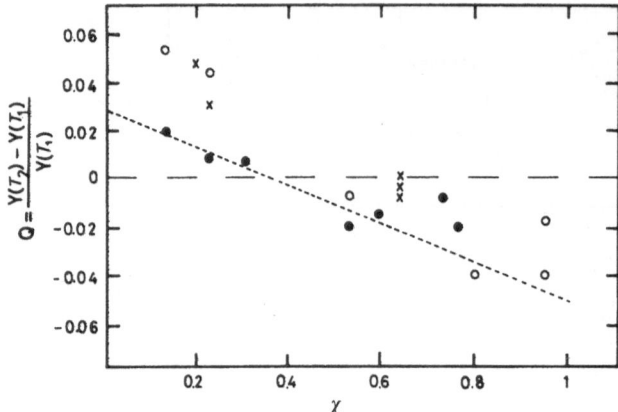

**Fig. 5.20.** Relative difference $Q$ of the sputtering ratio vs random fraction $\chi$ of the ion beam at the surface. $\bigcirc$: 20 keV Ar$^+$→Cu, $T_1 = 120$ K, $T_2 = 300$ K. ●: 20 keV Ar$^+$→Cu, $T_1 = 300$ K, $T_2 = 500$ K. ×: 20 keV Ne$^+$→Cu, $T_1 = 120$ K, $T_2 = 300$ K [5.70]

These $Q$ values [5.39, 70], both as a function of the angle of incidence and as a function of the nonchanneled fraction $\chi$, enable one to draw the following qualitative conclusion. At large values of $\chi$, the sputtering yield is largely determined by the sputtering ejection mechanism of the substrate atoms. Such a mechanism is apparently inhibited by a higher substrate temperature. A possible reason is the declining influence of focusing collision sequences, since the pathlength of focusing collison sequences will be reduced by thermal vibrations [5.71, 72]. From Fig. 5.20 it is seen that the $Q$ values are higher for the combination $T_1 = 300$ K, $T_2 = 500$ K than for the combination $T_1 = 120$ K, $T_2 = 300$ K. Such behaviour would be consistent with the reduction of focusing pathlengths in the different temperature régimes [5.71, 72].

For small values of $\chi$ it is observed that $Q > 0$, indicating a second temperature dependent effect which is apparently to do with a change in the collision probability of the impinging ions in the first layers of the crystal through a temperature dependent dechanneling mechanism. It is evident that dechanneling is more important for a larger channeled fraction of the ion beam, i.e., for smaller values of $\chi$.

The angular behaviour of the $Q$ values can then be understood through competition of the two mechanisms mentioned above. For small $\psi$ values dechanneling will increase, and thus $Q$ will increase; with increasing angle $\psi$ the channeled fraction of the beam continuously decreases, thus enhancing the relative importance of the ejection mechanism in the determination of the sputtering yield, as is experimentally observed (Fig. 5.19).

Both mechanisms were taken into account in the empirical model proposed by *Elich* et al. [5.38]. They calculated the variation of the sputtering yield for incidence along low index directions and an opaque direction. The non-channeled fraction of the ion beam was modified by allowing depth dependent dechanneling of the ions by thermally vibrating string atoms, whereas the influence of the ejection mechanism was taken into account in a rather schematic way through a temperature dependence of the "depth important for sputtering" [5.47]. The calculations are seen to give a fair agreement with the experimental results (see Fig. 5.18), despite the rather crude underlying assumptions.

For 10 keV Cs bombardment of (100) Cu at normal incidence, *Olson* et al. [5.73] report no change of the sputtering yield with temperature in the temperature range 77–473 K, as one would expect on the basis of the previous discussions, since $\chi \approx 0.5$. The same holds for the measurements of *Bhattacharya* et al. [5.43] for incidence of 6 keV Kr onto (110) Ag. For incidence of 6 keV Ne$^+$ onto (110) Ag, only a slight effect can be anticipated and that was beyond the detection limits of their experiment. The measurements of *Carlston* et al. [5.41] for the (111) plane of Cu are expected to show a decrease of the sputtering yield with temperature since $\chi$ is large. Such a conclusion is also valid for the measurements of *Hasseltine* et al. [5.74] on sapphire and of *Snouse* et al. on Cu [5.75].

Using Mo as a target, *Carlston* et al. [5.41] found an increase in yield with temperature for the incidence of 5 keV Ar$^+$ onto the (110) plane and no dependence for the (100) plane, but contrary to Cu an even more pronounced temperature effect was observed for bombardment of polycrystalline Mo. This temperature effect is reduced for incidence into a low index, but opaque direction [i.e., for incidence onto the (110) plane] and is even more reduced for incidence along a low index channeling direction [for incidence onto the (100) plane]. For the incidence of 1–7.5 keV Cs onto (100) Mo, i.e., when no channeling takes place, *Green* et al. [5.76] have observed a pronounced decrease of the sputtering yield with increasing temperature for substrate temperatures between 77 and 473 K; the effect is more pronounced for higher primary energies in this series.

### 5.5.2 Semiconductor Targets

A strong temperature effect in the sputtering yield is observed for sputtering of semiconductors. The sputtering process completely changes its character within a restricted temperature range. Only at temperatures elevated above a certain transition temperature $T_a$ is the influence of the monocrystalline substrate structure on the sputtering process observed [5.55–57, 77–84] (Fig. 5.21).

So for $T < T_a$ the angular dependence of the sputtering yield shows a monotonous increase with $\vartheta (\vartheta \leq 60°)$ as observed for structureless targets, whereas for $T > T_a$ it shows the behaviour characteristic for sputtering of crystalline targets [5.85–87] (Fig. 5.22). The transition temperature $T_a$ depends on the primary energy of the projectile [$T_a(E') > T_a(E)$ for $E' > E$, for energies of a few hundred eV], on the dose rate of the impinging ions [$T_a(I') > T_a(I)$ for $I' > I$] and on the mass of the projectiles [$T_a(M'_1) > T_a(M_1)$ for $M'_1 > M_1$] [5.55, 84]. These

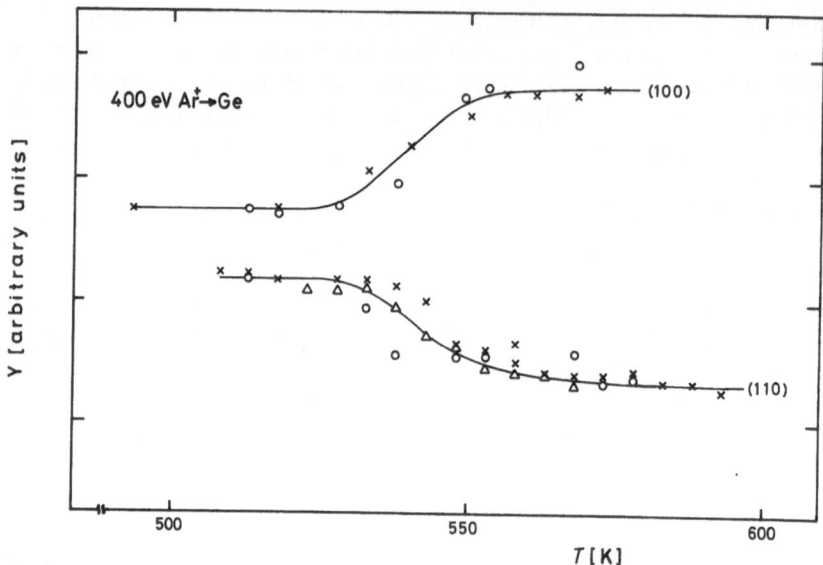

**Fig. 5.21.** Dependence of the relative sputtering yield of monocrystalline Ge on the surface plane and target temperature [5.55]

results suggest that an understanding of this behaviour has to be found in terms of a competition between damage production and damage annealing throughout the bombardment [5.55, 84].

For 40 keV $Ar^+$ bombardment of Si and Ge [5.56] and 50 keV $Ge^+$ bombardment of Ge [5.57] (Fig. 5.23), first an increase of the sputtering yield with increasing temperature was observed before the yield dropped due to channeling of the ion in the apparently largely crystalline lattice at higher temperatures. In this temperature interval, anisotropic ejection patterns were observed [5.56] for a rearranged surface with a lower surface binding energy than a polycrystal. The increase of the sputtering yield turned out to be dependent on the beam target orientation [5.57] (Fig. 5.23).

### 5.5.3 Ferromagnetic Materials

The temperature effect observed for ferromagnetic materials when the substrate temperature passes through the Curie point $T_c$ is of quite a different nature. The dependence of the sputtering yield on the magnetic state of the target for ferromagnetic materials has been studied by *Yurasova* et al. [5.88] for incidence of 15 keV $Ar^+$ onto (110) and (111) Ni. At the Curie point the sputtering yield is seen to rise sharply (Fig. 5.24), followed by a gradual decline to values

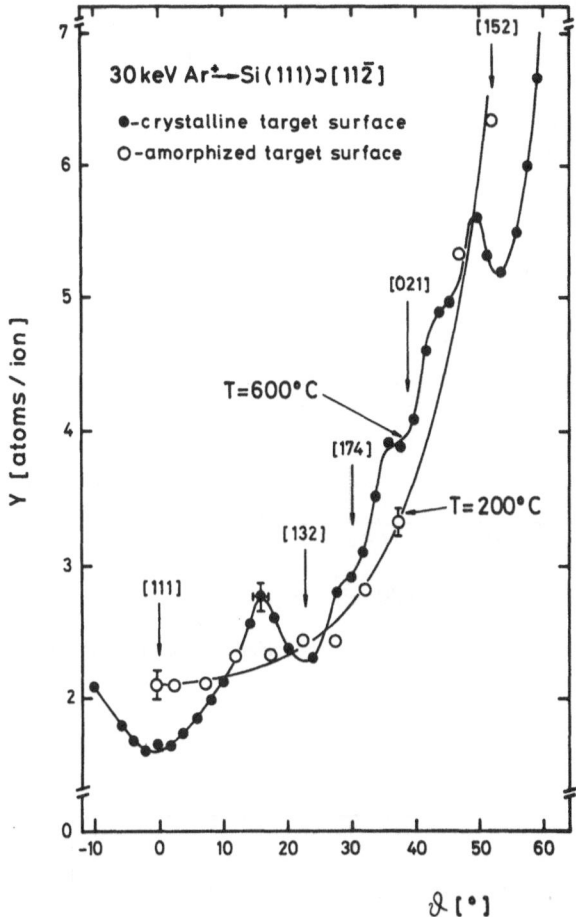

**Fig. 5.22.** The sputtering yield as a function of the angle of incidence for $30\,keV\,Ar^+$ on a (111)Si crystal rotated around the $[11\bar{2}]$ axis [5.87]

approximately equal to values at temperatures below $T_c$. The authors also studied the relative intensities of the [110] emission spots as compared to the [100] spots for both the ferromagnetic and the paramagnetic state of the target. In the ferromagnetic state the intensity in the [110] spots turned out to be more intense with respect to the intensity in the [100] spots than for the paramagnetic state [Ref. 5.9, Chap. 2]. This lead the authors to suggest that the origin of the observed phenomenon must be found in a change in the atom-atom interaction potential, accompanying the ferromagnetic-paramagnetic transition, but a change of the surface properties with temperature (e.g. blistering) cannot be ruled out.

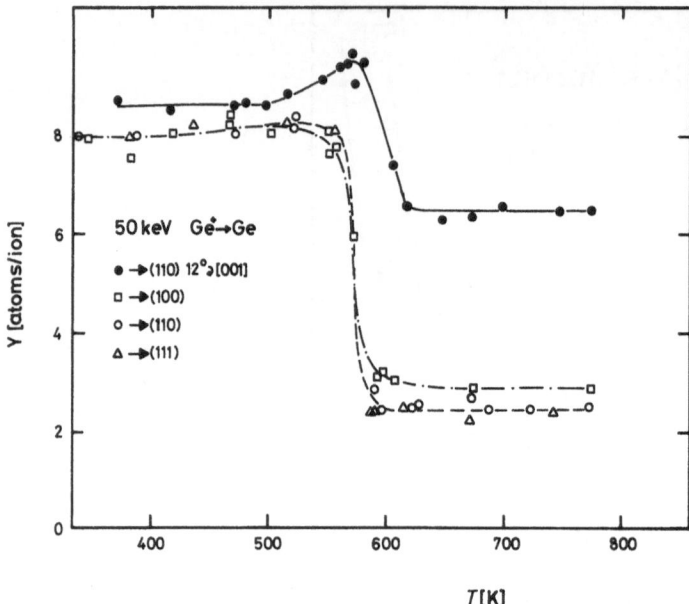

**Fig. 5.23.** Temperature dependence of the sputtering yield for incidence of 50 keV Ge onto Ge [5.57]

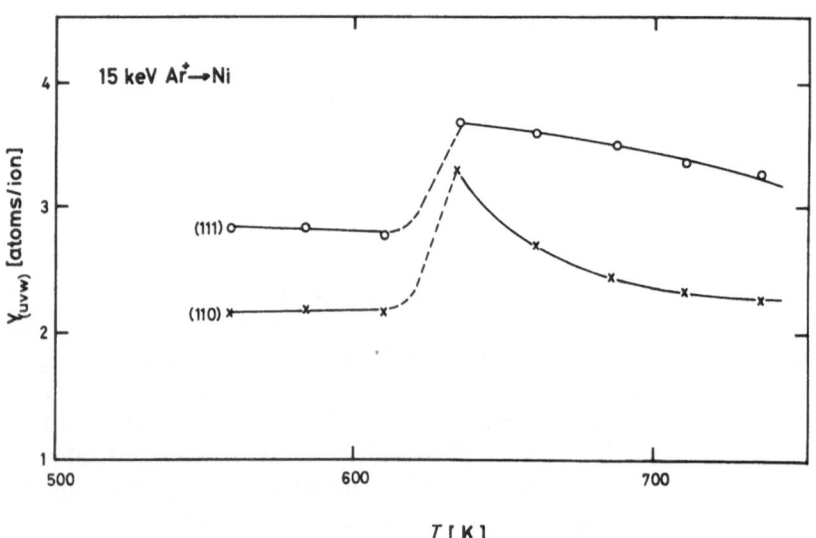

**Fig. 5.24.** The sputtering yields vs. target temperature for incidence of 15 keV Ar$^+$ onto Ni monocrystals [5.88]

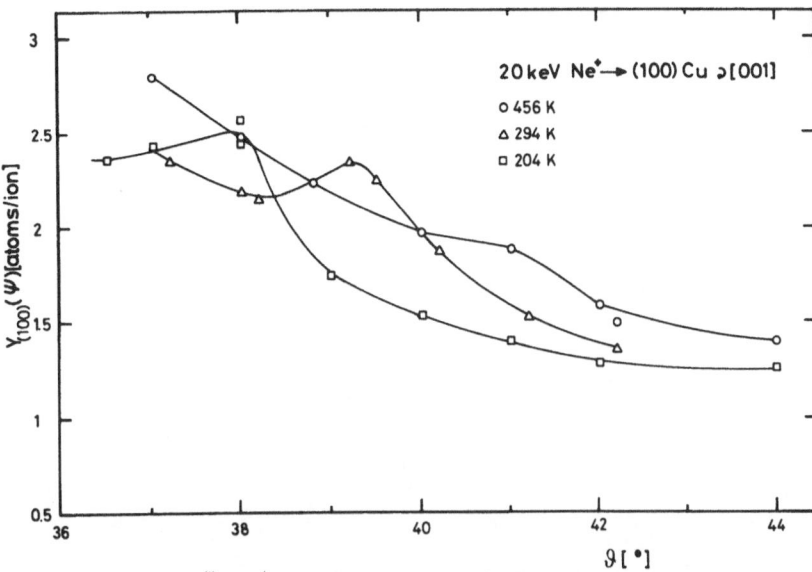

**Fig. 5.25.** The sputtering yield for 20 keV Ne⁺ ions on a (100) Cu crystal turned around a [001] axis in the surface. The target temperature was 204 K, 294 K, and 456 K [5.37]

## 5.6 Angular Fine Structure of the Sputtering Yield and Its Relation to Surface Structure

The angular dips of the sputtering yield have been seen to behave accordingly to the channeling model as described in Sect. 5.4. The sputtering yield increases smoothly from a minimum value at $\psi=0$ to a maximum at an angle approximately corresponding to the critical angle for channeling [5.23, 24, 37, 39, 40].

For incidence of 10 and 20 keV Ne⁺ onto (100) Cu rotated around a [001] axis, *Elich* et al. [5.37] have observed deviations from this general behaviour for incidence of the ion beam in a direction close to the [110] minimum. The observed angular fine structure is shown in Fig. 5.25 for three target temperatures 204, 294, and 456 K. The overall angular dependence is seen to follow the already observed general behaviour, where the angular dips become narrower for higher substrate temperatures, but there is also an additional peak in the sputtering yield at an angle $\vartheta=38°$ ($\psi=7°$) at 204 K, $\vartheta=39°$ ($\psi=6°$) at 294 K, and $\vartheta=41°$ ($\psi=4°$) at 456 K for the incidence of 20 keV Ne⁺. For 10 keV Ne⁺, similar features were observed though less pronounced, whereas for the incidence of 20 keV Ar⁺ only a faint indication of this anomaly is observed at the target temperature of 456 K.

The increase of the sputtering yield was accompanied by the occurrence of strong faceting of the crystal surface (Fig. 5.26). The surface structure was

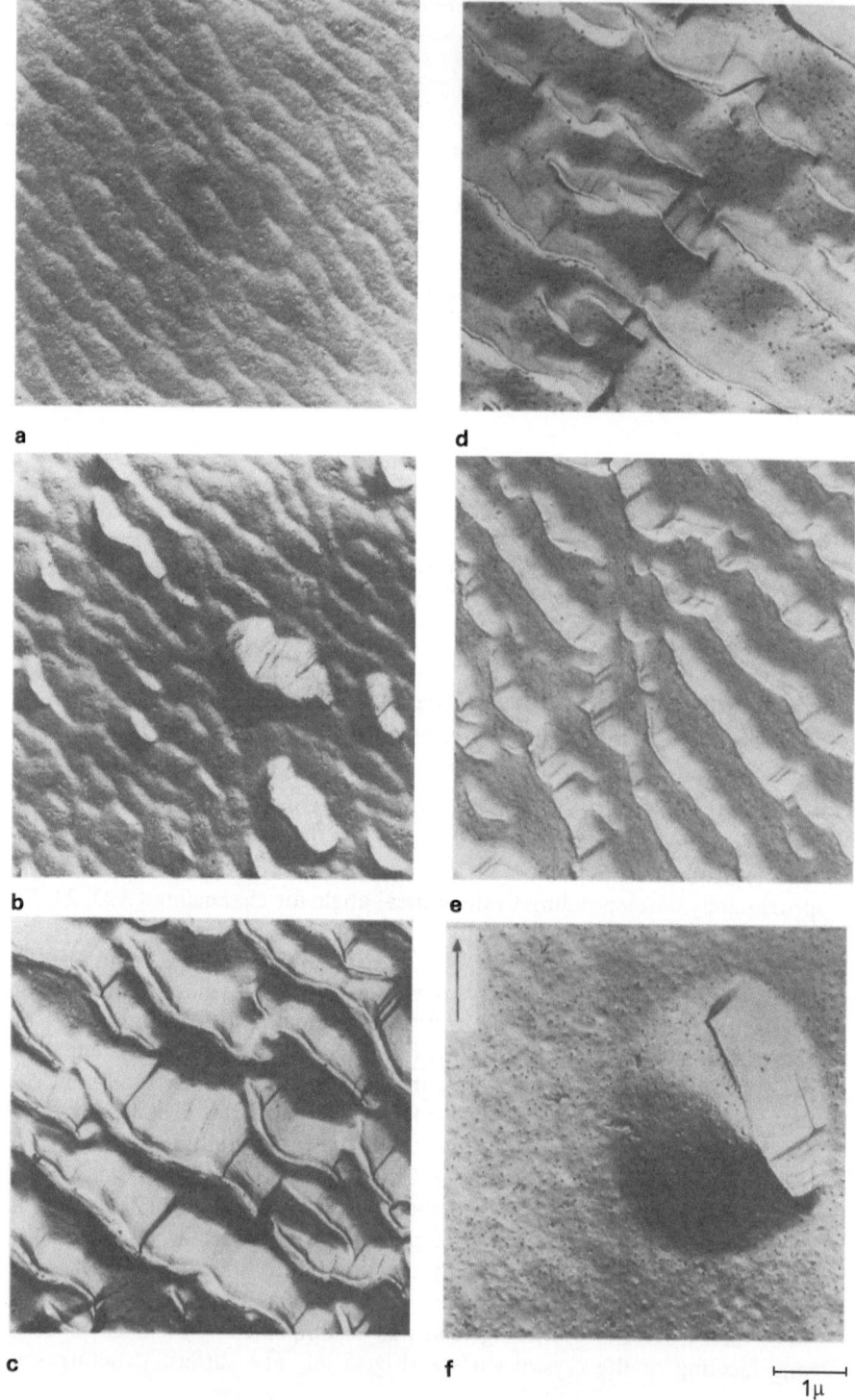

a

d

b

e

c

f

1μ

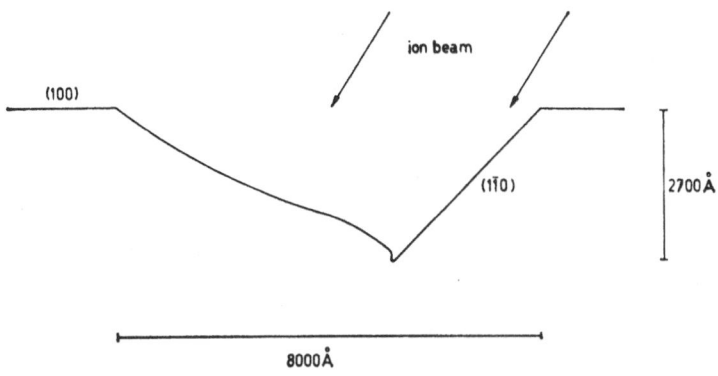

**Fig. 5.27.** Schematic profile of a pit [5.37]

measured by the replica technique. It is seen that the anomaly in the sputtering yield is apparently strongly coupled to the surface structure.

The sputtering yield is both a function of the angle $\psi$ with the low index direction and a function of the angle $\vartheta$ with respect to the surface normal (Sect. 5.4). For a faceted surface the sputtering yield will therefore be a function of the angles $\vartheta$ with the different facets, developed during ion bombardment. So the total sputtering yield will be the sum of all local sputtering yields weighed over their relative abundance [5.37, 89].

The facets that are found are schematically shown in Fig. 5.27. Part of the surface will be almost perpendicular to the beam direction ($\vartheta = 0$), whereas the beam hits the other part at almost glancing incidence ($\vartheta \sim 80°–85°$) where the sputtering yield is known to be very high [5.90, 91]. Following the calculation of *Elich* et al. [5.37], this results in an additional increase of the sputtering yield, which necessarily coincides with the occurrence of a fully developed faceted surface produced through continuous erosion of dislocations by the impinging ion beam [5.37, 89].

The authors suggest that the nucleation of the dislocations will be stimulated by a perpendicular focusing mechanism, producing radiation damage in deeper atomic layers. Under the right geometrical conditions, i.e., for incidence of the ions near the [110] direction, sufficient momentum will be transferred in the [Ī10] direction to start a focusing replacement sequence [5.37]. Evidence that such a mechanism exists has been obtained in experiments on the angular distribution of the sputtered particles ([Ref. 5.9, Chap. 2]

◄ **Fig. 5.26a–f.** The surface structure developed after 20 keV Ne$^+$ bombardment of a (100) Cu crystal, rotated around the [001] axis. The target temperature was 294 K. The angle of incidence $\vartheta$ was **(a)** 37.2°, **(b)** 38.2°, **(c)** 39.2°, **(d)** 40.2°, **(e)** 41.2°, and **(f)** 42.2°. The shadowing with platinum took place against the arrow direction under an angle of 30° with the surface [5.37] and [5.92]). At angles symmetric to the [110] direction (i.e., $\vartheta > 45°$), the perpendicular focusing mechanism will enhance the contribution of sputtered particles in the [1$\bar{1}$0] spot, opposite to the incoming beam. This enhancement reaches a maximum for $\vartheta = 51°$ or at an angle $\psi = 6°$ from the [110] direction (Fig. 5.28), thus under the same angle $\psi$, under which the angular fine structure in the sputtering yield is observed. Both effects also appear over a similar restricted angular interval around $\psi = 6°$ and have a similar temperature dependence, thus suggesting a strong correlation between the two phenomena.

Recently *Hauffe* [5.93] reported enhanced emission for incidence of 10 keV $Ne^+$ onto (100) Ag ($T = 90$ K), rotated around the [001] axis, for $\vartheta = 52°$ [i.e., $\psi = 7°$] and surface faceting, similar to that shown in Fig. 5.25, for $\vartheta = 37°$ [i.e., $\psi = 8°$]. He also reported enhanced emission and surface faceting for incidence directions ($\psi \approx 6°$) symmetric around the [111] direction for incidence of 10 keV $Ar^+$ onto (110) Fe, thus a bcc crystal. The crystal was rotated around the [1$\bar{1}$0] axis. The enhanced emission was observed around the [11$\bar{1}$] spot. In both cases a correlation between faceting and enhanced emission for directions of incidence symmetric around the low index close-packed direction was observed. *Hauffe* [5.93], however, ascribed the development of faceting to the large fraction of the ion beam reflected at close-packed crystal planes. In such a description, faceting will also occur in a restricted angular interval around $\psi \approx 6°-10°$.

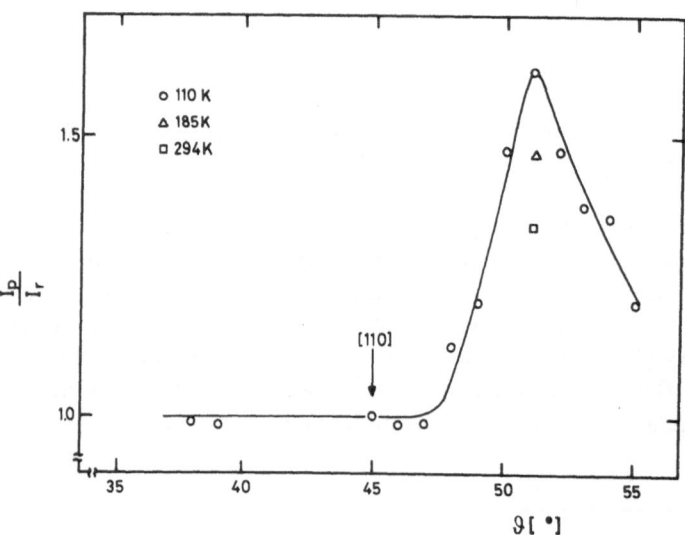

**Fig. 5.28.** The deposit thickness $I_p$ in the maximum of the [$\bar{1}$10] spot opposite the ion beam, normalized to the deposit thickness $I_r$ in the maximum of the other [110] spots as a function of the angle of incidence [5.92]

## 5.7 Conclusions

The sputtering yield of monocrystalline targets shows anisotropic features due to the regular arrangement of the atoms in rows and planes. For incidence of projectiles with energies of a few keV and higher along a low index crystallographic direction, the sputtering yield is reduced as a consequence of channeling. At these energies, the effect is reasonably well described by the *Onderdelinden* model [5.23, 24, 40], though the model seems to break down for improved channeling conditions. This is particularly apparent at higher energies and low $Z_1$ projectiles, or very transparent low index directions. The reason for this discrepancy must undoubtedly be found in the crude assumption of the depth dependence of the flux of channeled particles near the target surface. The sputtering yield critically depends on the flux of channeled particles in the surface region, since the initiating ion-atom collisions leading to sputtering will take place at a shallow depth beneath the surface of the target. The larger the fraction of channeled particles, the more critical the collisions at the surface of the crystal will be compared to the collisions in the subsequent atom layers. That is to say, the sputtering yield at higher energy will be predominantly determined by the transparency of the interface.

Such conclusions were also drawn by *Onderdelinden* [5.40], although based on different arguments. In order to account for these effects, *Onderdelinden* [5.24, 40] formulated a preliminary model in which he combined the transparency in the first layers with channeling in the subsequent layers.

A shortcoming of *Onderdelinden*'s channeling model is certainly the omission of dechanneling, a conclusion that is supported by the temperature dependence of the sputtering yield for incidence along low index directions at intermediate energies and well defined beam conditions. An attempt to account for dechanneling due to thermal vibrations was made by *Elich* et al. [5.38].

If the primary energy of the ion beam is reduced, channeling will become gradually less influential, but the dependence on the monocrystalline structure of the sputtering yield does not disappear. This dependence cannot be understood in terms of the anisotropy of the surface binding energy $U_{(uvw)}$ or in terms of channeling, and it might be questionable if such an effect is due to single crystalline effects in the ion penetration.

However, the sputtering yield $Y_{(uvw)}$ is seen to be roughly proportional to the interatomic spacing $t_{uvw}$ indicating that the above-mentioned effects are due to transparency or collision probability effects, or momentum reversibility of the lattice. But on the other hand, it cannot be ruled out that the discrepancy between the sputtering yield for monocrystalline and polycrystalline material is due to monocrystalline effects in the ejection mechanism of the target atoms themselves. An example of such a monocrystalline influence on the ejection mechanism would be momentum transport through focusing collision sequences, which would enlarge the effectivity of the ejection mechanism.

An increase of the target temperature will reduce the effectivity of such an ejection mechanism based on the regularity of the monocrystalline lattice. This is not at variance with the observations made in Sect. 5.5 on the temperature dependence, where we also had to include two competing mechanisms, i.e., the collision probability of the incoming ion and the effectivity of the ejection mechanism.

Definite conclusions can, however, only be reached from differential measurements of the absolute yield of sputtered material and of the energy distribution of sputtered particles.

*Acknowledgement.* It is a pleasure to acknowledge the invaluable discussions with D. Onderdelinden and J. J. Ph. Elich during the preparation and completion of this manuscript.

# References

5.1   P.K.Rol, J.M.Fluit, F.P.Viehböck, M.deJong: In *Proc. 4th Int. Conf. Phen. Ion. Gases*, Uppsala, ed. by N.R.Nilsson (North-Holland, Amsterdam 1959) p. 257
5.2   V.A.Molchanov, V.G.Tel'kovskii, V.M.Chickerov: Sov. Phys.-Doklady 6, 3, 222 (1961)
5.3   O.Almén, G.Bruce: Nucl. Instrum. Methods 11, 257 (1961)
5.4   J.Stark, G.Wendt: Ann. Phys. (Leipzig) 38, 921 (1912)
5.5   J.Stark, G.Wendt: Ann. Phys. (Leipzig) 38, 941 (1912)
5.6   J.Stark: Phys. Z. 13, 973 (1913)
5.7   A.L.Southern, W.R.Willis, M.T.Robinson: J. Appl. Phys. 34, 1, 153 (1963)
5.8   G.K.Wehner: Phys. Rev. 102, 690 (1956)
5.9   R.Behrisch (ed.): *Sputtering by Particle Bombardment*, III, Topics in Applied Physics (Springer, Berlin, Heidelberg, New York 1981) to be published
5.10  J.M.Fluit: Thesis, University of Leiden (1963)
5.11  J.M.Fluit, P.K.Rol, J.Kistemaker: J. Appl. Phys. 34, 3, 690 (1963)
5.12  J.M.Fluit, P.K.Rol: Physica 30, 857 (1964)
5.13  J.M.Fluit, C.Snoek, J.Kistemaker: Physica 30, 144 (1964)
5.14  G.D.Magnuson, C.E.Carlston: J. Appl. Phys. 34, 11, 3267 (1963)
5.15  D.D.Odintsov: Sov. Phys.-Solid State 5, 4, 813 (1963)
5.16  Y.V.Martynenko: Sov. Phys.-Solid State 6, 1581 (1965)
5.17  S.A.Drentje: Thesis, University of Groningen (1968)
5.18  M.T.Robinson, O.S.Oen: Appl. Phys. Lett. 2, 30 (1963)
5.19  M.T.Robinson, O.S.Oen: Phys. Rev. 132, 2385 (1963)
5.20  H.O.Lutz, R.Sizmann: Phys. Lett. 5, 113 (1963)
5.21  G.R.Piercy, F.Brown, J.A.Davies, M.McCargo: Phys. Rev. Lett. 10, 399 (1963)
5.22  R.S.Nelson, M.W.Thompson: Philos. Mag. 8, 1677 (1963)
5.23  D.Onderdelinden: Appl. Phys. Lett. 8, 8, 189 (1966)
5.24  D.Onderdelinden: Can. J. Phys. 46, 739 (1968)
5.25  J.Lindhard: K. Dan. Vidensk. Selsk. Mat. Fys. Medd. 34, 14 (1965)
5.26  R.H.Silsbee: J. Appl. Phys. 28, 1246 (1957)
5.27  M.W.Thompson: In *Proc. 5th Int. Conf. Phen. Ion. Gases*, Munich, 1961 (North-Holland, Amsterdam 1962)
5.28  C.Lehmann: Nucl. Instrum. Methods 38, 263 (1965)
5.29  J.B.Sanders, D.Onderdelinden: In *Proc. 7th Int. Conf. Phen. Ion. Gases*, Beograd, 1966, ed. by B.Perović, D.Tošć (Gradevinska, Knjiga 1966) p. 149
5.30  D.E.Harrison, Jr., N.S.Levy, J.P.Johnson III, H.M.Effron: J. Appl. Phys. 39, 8, 3742 (1968)
5.31  M.T.Robinson: J. Appl. Phys. 40, 2670 (1969)
5.32  D.E.Harrison, Jr.: J. Appl. Phys. 40, 9, 3871 (1969)
5.33  D.E.Harrison, Jr., W.L.Moore, Jr., H.T.Holcombie: Radiat. Eff. 17, 167 (1973)
5.34  E.Zwangobani, R.J.MacDonald: Radiat. Eff. 20, 81 (1973)

5.35  A. van Wijngaarden, E. Reuther, J. N. Bradford: Can. J. Phys. **47**, 411 (1969)

5.36  D. Onderdelinden, F. W. Saris, P. K. Rol: In *Proc. 7th Int. Conf. Phen. Ion. Gases*, Beograd, 1966, ed. by B. Perović, D. Tosić (Gradevinska, Knijiga 1966) p. 157

5.37  J. J. Ph. Elich, H. E. Roosendaal, H. H. Kersten, D. Onderdelinden, J. Kistemaker, J. D. Elen: Radiat. Eff. **8**, 1 (1971)

5.38  J. J. Ph. Elich, H. E. Roosendaal, D. Onderdelinden: Radiat. Eff. **10**, 175 (1971)

5.39  J. J. Ph. Elich, H. E. Roosendaal, D. Onderdelinden: Radiat. Eff. **14**, 93 (1972)

5.40  D. Onderdelinden: Thesis, University of Leiden (1968)

5.41  C. E. Carlston, G. D. Magnuson, A. Comeaux, P. Mahadevan: Phys. Rev. **138**, 3A, A759 (1965)

5.42  I. N. Evdokimov, V. A. Molchanov, D. D. Odintsov, V. M. Chicherov: Sov. Phys.-Doklady **12**, 11, 1050 (1968)

5.43  R. S. Bhattacharya, D. K. Mukherjee, S. B. Karmohapatro: Nucl. Instrum. Methods **99**, 509 (1972)

5.44  J. W. Mayer, L. Eriksson, J. A. Davies: *Ion Implantation in Semiconductors* (Academic Press, New York, London 1970)

5.45  T. W. Snouse, L. C. Haughney: J. Appl. Phys. **37**, 2, 700 (1966)

5.46  E. J. Zdanuk, S. P. Wolsky: J. Appl. Phys. **36**, 5, 1683 (1965)

5.47  O. C. Yonts, C. E. Normand, D. E. Harrison: J. Appl. Phys. **31**, 447 (1960)

5.48  P. Sigmund: Phys. Rev. **184**, 383 (1969)

5.49  D. P. Jackson: Radiat. Eff. **18**, 185 (1973)

5.50  D. E. Harrison, Jr., G. D. Magnuson: Phys. Rev. **122**, 5, 1421 (1961)

5.51  M. T. Robinson, A. L. Southern: J. Appl. Phys. **38**, 7, 2969 (1967)

5.52  G. J. Ogilvie, J. V. Sanders, A. A. Thomson: J. Phys. Chem. Solids **24**, 247 (1963)

5.53  L. L. Tongson, C. Burleigh Cooper: Radiat. Eff. **24**, 187 (1975)

5.54  M. T. Robinson, A. L. Southern: J. Appl. Phys. **39**, 7, 3463 (1968)

5.55  G. S. Anderson: J. Appl. Phys. **38**, 4, 1607 (1967)

5.56  J. Nizam, N. Benazeth-Colombie: Rev. Phys. Appl. **10**, 183 (1975)

5.57  G. Holmén: Radiat. Eff. **24**, 7 (1975)

5.58  E. S. Mashkova, V. A. Molchanov, D. D. Odintsov: Sov. Phys.-Solid State **5**, 2516 (1964)

5.59  M. Balarin, V. A. Molchanov, V. G. Tel'kovskii: Sov. Phys.-Doklady **7**, 11, 1005 (1963)

5.60  L. Francken, D. Onderdelinden: In *Proc. Int. Conf. Atomic Coll. Phen. in Solids*, Brighton (1969), ed. by D. W. Palmer, M. W. Thompson, P. D. Townsend (North-Holland, Amsterdam, London, New York 1970) p. 266

5.61  A. van Wijngaarden, B. Miremadi, N. M. A. Finney, J. N. Bradford: Phys. Rev. **185**, 2, 490 (1969)

5.62  E. Reuther, J. N. Bradford: Phys. Rev. **188**, 654 (1969)

5.63  E. Reuther, J. N. Bradford, A. van Wijngaarden: In *Proc. Int. Conf. Atomic Coll. Phen. in Solids*, Brighton (1969), ed. by D. W. Palmer, M. W. Thompson, P. D. Townsend (North-Holland, Amsterdam, London, New York 1970) p. 278

5.64  R. S. Bhattacharya, S. B. Karmohapatro: Nucl. Instrum. Methods **109**, 191 (1973)

5.65  R. S. Bhattacharya, D. Basu, S. B. Karmohapatro: Radiat. Eff. **21**, 275 (1974)

5.66  H. Grahmann, A. Feuerstein, S. Kalbitzer: Radiat. Eff. **29**, 117 (1969)

5.67  R. S. Nelson: Philos. Mag. **11**, 291 (1965)

5.68  R. S. Nelson: *The Observation of Atomic Collisions in Crystalline Solids* (North-Holland, Amsterdam 1968)

5.69  M. W. Thompson, R. S. Nelson: Philos. Mag. **7**, 2015 (1962)

5.70  H. E. Roosendaal, J. J. Ph. Elich: Radiat. Eff. **19**, 127 (1973)

5.71  R. S. Nelson, M. W. Thompson, H. Montgomery: Philos. Mag. **7**, 1385 (1962)

5.72  J. B. Sanders, J. M. Fluit: Physica **30**, 129 (1964)

5.73  N. T. Olson, H. P. Smith: Phys. Rev. **157**, 241 (1967)

5.74  E. H. Hasseltine, F. C. Hurlbut, N. T. Olson, H. P. Smith: J. Appl. Phys. **38**, 11, 4313 (1967)

5.75  T. W. Snouse, M. Bader: In *Trans. 8th National Vacuum Symp. Washington*, Vol. 1 (Pergamon Press, New York 1962) p. 271

5.76  J. B. Green, N. T. Olson, H. P. Smith: J. Appl. Phys. **37**, 4699 (1966)

5.77  G.S.Anderson, G.K.Wehner, H.J.Olin: J. Appl. Phys. **34**, 12, 3492 (1963)
5.78  G.S.Anderson: J. Appl. Phys. **37**, 7, 2838 (1966)
5.79  G.S.Anderson: J. Appl. Phys. **37**, 9, 3455 (1966)
5.80  R.J.MacDonald: Phys. Lett. **29**A, 256 (1969)
5.81  R.J.MacDonald: Philos. Mag. **21**, 171, 519 (1970)
5.82  R.J.MacDonald: Radiat. Eff. **3**, 131 (1970)
5.83  E.Zwangobani, R.J.MacDonald: Phys. Lett. **32**A, 308 (1970)
5.84  E.D.Zwangobani, R.J.MacDonald: Radiat. Eff. **11**, 215 (1971)
5.85  Kh.Zommerfel'dt, E.S.Mashkova, V.A.Molchanov: Sov. Phys.-Doklady **14**, 9, 901 (1970)
5.86  H.Sommerfeldt, E.S.Mashkova, V.A.Molchanov: Radiat. Eff. **9**, 267 (1971)
5.87  H.Sommerfeldt, E.S.Mashkova, V.A.Molchanov: Phys. Lett. **38**A, 237 (1972)
5.88  V.E.Yurasova, V.S.Chernysh, M.V.Kuvakin, L.B.Shelyakin: JETP Lett. **21**, 3, 88 (1975)
5.89  J.Kistemaker, H.E.Roosendaal: Jpn. J. Appl. Phys., Suppl. **2**, Pt. 2 (1974)
5.90  P.K.Rol: Thesis, University of Amsterdam (1960)
5.91  V.A.Molchanov, V.G.Tel'kovskii, Sov. Phys.-Doklady **6**, 137 (1961)
5.92  J.J.Ph.Elich, H.E.Roosendaal: Phys. Lett. **33**A, 235 (1970)
5.93  W.Hauffe: Thesis B, University of Dresden (1978)

# Additional References with Titles

## Chapter 1

1. E. Taglauer, W. Heiland: *Inelastic Particle-Surface Collisions*, Springer Series in Chemical Physics, Vol. 17, (Springer, Berlin, Heidelberg, New York 1981)

## Chapter 3

1. A. van Veen: "Sputtering and scattering by interaction of low energy noble gas ions with monocrystalline metal surface", Thesis, Utrecht, 1979
2. J. Giber, J. Kazsoki, L. Koblinger: Collision cascades and the disturbed zone during sputtering processes (model computation). Acta Phys. Acad. Sci. Hung. **45**, 275–279 (1978
3. B. Poelsema, A. L. Boers: A new method for the calculation of ion scattering from single crystals; method and potentialities. Radiat. Eff. **41**, 229–237 (1979)
4. J. E. Adams, J. E. Doll: Dynamics of ion channeling at low energies: preliminary trajectory studies. J. Chem. Phys. **73**, 2137–2144 (1980)
5. D. J. Martin: The influence of correlated atomic vibrations on low energy ion scattering. I. Quasi-triple scattering from a (100) row of copper atoms. Surf. Sci. **97**, 586–594 (1980)
6. N. Winograd, B. J. Garrison, D. E. Harrison, Jr.: Mechanisms of CO ejection from ion bombarded single crystal surfaces. J. Chem. Phys. **73**, 3473–3479 (1980)
7. D. P. Jackson: The influence of surface trapping on low energy light ion reflection. Radiat. Eff. **49**, 233–237 (1980)
8. Y. Yamamura, W. Takeuchi: Ion focusing effects on total reflection coefficient near the semichannel direction, Radiat. Eff. **49**, 251–254 (1980)
9. A. S. Mosunov, L. B. Shelyakin, V. E. Yurasova, D. Čirič, B. Perovič, I. Terzič: Computer simulation of ion scattering by polycrystals. Radiat. Eff. **52**, 85–89 (1980)
10. D. P. Jackson: "The Theory of Sputtering", in *Symposium on Sputtering*, ed. by P. Varga, G. Betz, F. P. Viehböck (Institut für allgemeine Physik, Technische Universität Wien, 1980) pp. 2–35
11. D. E. Harrison, Jr.: "Full Lattice Simulation of Atomic Ejection Mechanisms and Sputtering", reference 10, pp. 36–61
12. M. Hou: "The Influence of Induced Surface Point Defects on Sputtering from Metals", reference 10, pp. 101–111
13. V. E. Yurasova, V. A. Eltekov: "Models of Single Crystal Sputtering", reference 10, pp. 134–202
14. V. A. Eltekov, L. P. Razvina, O. A. Popova, V. E. Yurasova: "Computer Simulation of Single Crystal Sputtering by Dynamic Models of Small Atom Blocks", reference 10, pp. 203–215
15. A. van Veen, A. G. J. de Wit, J. M. Fluit: "Surface Ejection of Random and Focussed Metal Recoils from Ion Bombarded Single Crystals", reference 10, pp. 226–235
16. J. J. Szafarz, M. V. Kuvakin, A. V. Lusnikov: "The Shape of Surface Potential Barrier for Cu and Ni", reference 10, pp. 327–336

# Chapter 4

1. C. Abatino, G. Luzzi, L. Pagagno: Sputtering yield coefficients of thin copper films bombarded by 3–15 keV Ar⁺ at an angle of incidence of 30°. Thin Solid Films **56**, 291 (1979)
2. H.H. Andersen, H.L. Bay: A survey of sputtering-yield data for plasma-wall-interaction calculations. J. Nucl. Mater. **93+94**, 625 (1980)
3. O. Auciello, R. Kelly, R. Iricibar: New insight into the development of pyrimidal structures on bombarded copper surfaces. Radiat. Eff. **46**, 105 (1980)
4. R. Behrisch, J. Roth, J. Bohdansky, A.P. Martinelli, B. Schweer, D. Rusbüldt, E. Hintz: Dependence of light-ion sputtering yields of iron on ion fluence and oxygen partial pressure. J. Nucl. Mater. **93+94**, 645 (1980)
5. J. Bohdansky: Important Sputtering Yield Data for Tokamaks: a Comparison of Measurements and Estimates. J. Nucl. Mater. **93+94**, 44 (1980)
6. J. Bohdansky, J. Roth: Formation of various coatings and their behaviour under particle bombardment. J. Nucl. Mater. **85+86**, 1145 (1979)
7. J. Børgesen, J. Schou, H. Sørensen: Erosion of Thin Films of D, by keV Light Ions", in *Symposium on Sputtering*, ed. by P. Varga, G. Betz, F.P. Viehböck (Institut für allgemeine Physik, Technische Universität Wien, Wien 1980) p. 822
8. M.I. Current, D.N. Seidmann: Sputtering of tungsten: An atomic view of a near surface depleted zone created by a single 30 keV ⁶³Cu⁺ projectile. Nucl. Instrum. Methods **170**, 377 (1980)
9. A. Elbern, P. Mioduszewski: Measurements of the sputtering yield of Fe from an oxidized stainless steel target using a pulsed dye laser. J. Vac. Sci. Technol. **16**, 2090 (1979)
10. D. Ghose: On the angular dependence of sputtering yields. Jn. J. Appl. Physics **18**, 1849 (1979)
11. W.H. Gries: "An Implantation Collector for Accurate Sputtering Measurements of Monolayer Resolution", in *Symposium on Sputtering*, ed. by P. Varga, G. Betz, F.P. Viehböck (Institut für allgemeine Physik, Technische Universität Wien, Wien 1980) p. 842
12. L.G. Haggmark, J.P. Biersack: Monte Carlo calculations of light-ion-sputtering as a function of the angle of incidence. J. Nucl. Mater. **93+94**, 664 (1980)
13. E. Hechtl, J. Bohdansky, J. Roth: "Sputtering Yield Dependence on Ion Mass at Low Energy for Ta and W", in *Symposium on Sputtering*, ed. by P. Varga, G. Betz, F.P. Viehböck (Institut für allgemeine Physik, Technische Universität Wien, Wien 1980) p. 834
14. S.T. Kang, R. Shimizu, T. Okutani: Sputtering of Si with keV Ar⁺-ions. I. Measurement and Monte Carlo calculations of sputtering yield. Jn. J. Appl. Phys. **18**, 1717 (1979)
15. G. Kiriakides, C.E. Christodoulides, G. Carter, J.S. Colligon: An RBS technique for measurement of the erosion rate of ion implanted films. Appl. Phys. **19**, 191 (1979)
16. G. Kiriakides, J.S. Colligon, S.P. Chenakin: Secondary ion mass spectrometric study of self-sputtered copper. Radiat. Eff. **41**, 119 (1979)
17. R. Kirscher, S. Hofman: Coverage of foreign atoms on surfaces as a function of adsorption, sputtering and diffusion rates. Surf. Sci. **83**, 296 (1979)
18. S.K. Lam, M. Kaminsky: Sputtering of annealed aluminium and sintered aluminium powder (SAP 985) under D⁺ and ⁴He⁺ bombardment: Study of microstructural effects. J. Nucl. Mater. **89**, 205 (1980)
19. K. Libbrecht, J.E. Griffith, R.A. Weller, T.A. Tombrello: Energy dependence of the trapping of uranium atoms by aluminium oxide surfaces. Radiat. Eff. **49**, 195 (1980)
20. N. Matsunami, Y. Yamamura, Y. Itikawa, N. Itoh, Y. Kazumata, S. Miyagawa, K. Morita, R. Shimizu: A semiempirical formula for the energy dependence of sputtering yield. Radiat. Eff. Lett. **57**, 15 (1980)
21. K.L. Merkle, W. Jäger: Direct observation of spike effects in heavy-ion sputtering. Philos. Mag., in press (1981)
22. Y. Okayima: Estimation of sputtering rate by bombardment with argon gas ions. J. Appl. Phys. **51**, 715 (1980)
23. S. Okuda, H. Akimure: Surface erosion of metal molybdenum bombarded with energetic hydrogen and helium atoms. Jn. J. Appl. Phys. **18**, 1335 (1979)

24. A.R.Oliva-Florio, E.V.Alonso, R.A.Boragiola, J.Ferron, M.M.Jakos: Energy dependence of the molecular effect in sputtering. Radiat. Eff. Lett. **50**, 3 (1979)
25. L.Pagagno, G.Luzzi, M.Meuti: Sputtering yields of thin overlayer films by attennation of the substrate auger signal. Thin Solid Films **60**, 307 (1979)
26. J.B.Roberto, R.A.Zuhr, J.L.Moore, G.D.Alton: Low energy hydrogen sputtering of Au, Ni and stainless steel. J. Nucl. Mater. **85 + 86**, 1073 (1979)
27. J.Roth: "Sputtering with Light Ions", in *Symposium on Sputtering*, ed. by P.Varga, G.Betz, F.P.Viehböck (Institut für allgemeine Physik, Technische Universität Wien, Wien 1980) p. 773
28. J.Roth, J.Bohdansky, R.S.Blewer, W.Ottenberger: Sputtering of Be and BeO by light ions. J. Nucl. Mater. **85 + 86**, 1077 (1979)
29. J.Roth, J.Bohdansky, A.P.Martinelli: Low energy light ion sputtering of metals and carbides. Radiat. Eff. **48**, 213 (1980)
30. R.Sartwell: Thin film sputtering yields for Fe, Cr and an Fe-Cr alloy measured by proton-induced x-rays. J. Appl. Phys. **50**, 7887 (1979)
31. D.J.Sharp, J.K.G.Panitz, D.M.Mattox: Applications of a Kaufmann ion source to low energy ion erosion studies. J. Vac. Sci. Technol. **16**, 1897 (1979)
32. P.Sigmund, C.Claussen: "Sputtering from Elastic-Collision Spikes", in *Symposium on Sputtering*, ed. by P.Varga, G.Betz, F.P.Viehböck (Institut für allgemeine Physik, Technische Universität Wien, Wien 1980) p. 113
33. J.N.Smith, Jr.: Surface cleaning and sputtered ion production. J. Nucl. Mater. **82**, 179 (1979)
34. R.Smith, J.M.Walls: The development of surface topography during depth profiling in auger electron spectroscopy. Surf. Sci. **80**, 557 (1979)
35. Ch.Steinbrüchel, D.M.Gruen: Absolute measurement of sputtered ion fractions using matrix isolation spectroscopy. Surf. Sci. **93**, 299 (1980)
36. Ch.Steinbrüchel, D.M.Gruen, J.Dawson: Application of matrix irradiation spectroscopy to the measurements of sputtering yields. J. Vac. Sci. Technol. **16**, 251 (1979)
37. M.Szymonoski: Sputtering of Cu and Zn atoms from elemental and alloy targets. Appl. Phys. **23**, 89 (1980)
38. E.Taglauer, W.Heiland: Mass and energy dependence of the equilibrium surface composition of sputtered tantalum oxide. Appl. Phys. Lett. **33**, 950 (1978)
39. E.Taglauer, W.Heiland: "Changes of the Surface Composition of Compounds due to Light Ion Bombardment", in *Symposium on Sputtering*, ed. by P.Varga, G.Betz, F.P.Viehböck (Institut für allgemeine Physik, Technische Universität Wien, Wien 1980) p. 423
40. E.Taglauer, W.Heiland, U.Beitat: The influence of adsorption energies on ion impact desorption of surface layers. Surf. Sci. **89**, 710 (1979)
41. E.Taglauer, W.Heiland, R.J.MacDonald: The study of sputtering effects in oxides and metal-adsorbed-gas systems using combined analytical techniques. Surf. Sci. **90**, 661 (1979)
42. D.A.Thompson: "Non-Linear Effects in Sputtering", in *Symposium on Sputtering*, ed. by P.Varga, G.Betz, F.P.Viehböck (Institut für allgemeine Physik, Technische Universität Wien, Wien 1980) p. 62
43. D.A.Thompson: Sputtering of Ag, An and Pt by heavy atomic and molecular ion bombardment. J. Appl. Phys. **52**, (1981) (in press)
44. P.F.Tortorelli, C.J.Altstetter: The sputtering yield of polycrystalline materials. Radiat. Eff. **51**, 241 (1980)
45. P.F.Tortorelli, C.D.Altstetter: Sputtering of two-Phase polycrystalline metals. J. Vac. Sci. Technol. **16**, 804 (1979)
46. H.Uecker, A.Riccator, G.R.Thacker, J.Ney, J.P.Biersack: Experiments on sputtering of niobium by 14–16 MeV protons and Monte Carlo calculations for proton and neutron sputtering. J. Nucl. Mater. **93 + 94**, 670 (1980)
47. P.Williams: Anomalous sputter yields due to cascade mixing. Appl. Phys. Lett. **36**, 758 (1980)
48. K.Wittmaack: On the mechanism of cluster emission in sputtering. Phys. Lett. **69A**, 322 (1979)
49. K.Wittmaack, P.Blank, W.Wach: High fluence retention of noble gases implanted in silicon. Radiat. Eff. **39**, 81 (1978)
50. G. van Wyk: The dependence of ion bombardment induced preferential orientation on the direction of the ion beam. Radiat. Eff. Lett. **57**, 45 (1980)

51. G. van Wyk: The influence of ion species on ion bombardment induced preferential orientation. Radiat. Eff. Lett. **57**, 165 (1981)
52. G.N. von Wyk, H.J.Smith: Crystalline reorientation due to ion bombardment. Nucl. Instrum. Methods **170**, 433 (1980)
53. R. Yamada, K. Nakamura, K.Sone, M.Saidoh: Measurement of chemical sputtering yields of various types of carbon. J. Nucl. Mat. **95**, 278 (1980)
54. R. Yamada, M.Saidoh, K.Sone, H.Ohtsuka: Dose and microstructural effects on surface topography change and sputtering yield in polycrystalline molybdenum bombardment with 2keV Ne$^+$-ions. J. Nucl. Mater. **82**, 155 (1979)
55. R. Yamada, K.Sone, M.Saidoh: Surface microstructural effects on angular distribution of molybdenum particles sputtered with low energy Ne$^+$-ions. J. Nucl. Mater. **84**, 101 (1979)
56. A.Žabkar, P.Panjau, B.Navinšek: "Sputtering Yield Measurements of Nickel Using a Quartz Oscillator Microbalance", in *Contributed Papers of SPIG* 80, ed. by B.Čobič (Boris Kidrič Institute of Nuclear Sciences, Beograd 1980) p. 132

# Chapter 5

1. V.S.Chernysh, A.Johannsen, L.Sarholt-Kristensen: "Sputtering yield measurements on hcp and fcc cobalt" Rad. effects Letters **57**, 119 (1980)

# List of Symbols

| | | | |
|---|---|---|---|
| $A$ | Mass ratio target atom to projectile; $A = M_2/M_1$ | $d_{hkl}$ | Unit translation perpendicular to planes |
| $a, c$ | Cell edges in hexagonal crystals | $d\sigma(E, T)$ | Differential cross section for transferring an energy $(T, dT)$ to a target atom |
| $a_{BM}, C_{BM}$ | Born-Mayer potential parameters | | |
| $a_{MO}, C_{MO}$ | Morse Potential Parameters | $E$ | Energy of a projectile |
| $a_i$ | Area per adsorbed gas species $i$ | $E_b, W$ | Binding energy of an atom to its lattice site |
| $a_0$ | Bohr radius; $a_0 = 0.529$ Å | $E_d$ | Threshold energy for producing stable atomic displacements |
| $a_{12}, a$ | Screening length for atomic potentials; $a_{12} = 0.885$ $a_0(Z_1^{2/3} + Z_2^{2/3})^{-1/2}$ (Thomas-Fermi, Lindhard) $a_{12} = 0.885$ $a_0(Z_1^{1/2} + Z_2^{1/2})^{-2/3}$ (Firsov) | $E_F$ | Thomas-Fermi energy unit $E/\varepsilon$ |
| | | $E_f$ | Focusing energy |
| | | $E_r$ | Relative energy; $E_r = A E/(1 + A)$ |
| | | $E_{th}$ | Threshold energy for sputtering |
| | | $E_0$ | Energy of a moving target atom |
| $\alpha$ | Dimensionless function of the mass ratio $M_2/M_1$, the angle of incidence $\theta$ and the ion energy $E$ in the sputtering yield | $E_1$ | For $E \geq E_1$ Coulomb scattering dominates |
| | | $e$ | Elementary charge; $e^2 = 14.39$ eV Å |
| $c_i$ | Atomic concentration of atoms $i$ | $\varepsilon$ | Lindhard's reduced energy, $\varepsilon = EaM_2/ Z_1Z_2e^2(M_1 + M_2)$ |
| $C_m$ | Cross section constant for power potentials | $\varepsilon_\alpha$ | Energy of an excited state |
| $\chi_m$ | Minimum yield in a channeling experiment | $\varepsilon_F$ | Fermi level |
| $\chi_{uvw}, \chi(\psi, E)$ | Relative dechanneling yield, or non channeled fraction of an ion beam in a crystal | $\eta_{(uvw)}$ | Fitting parameter in single crystal sputtering yields |
| | | $\eta(E)$ | Average amount of energy ending up in |

$F_D(E, \boldsymbol{\Omega}, \boldsymbol{r})$ Spatial distribution of the density of deposited energy in $(\boldsymbol{r}, d^3r)$ for an ion starting at $r=0$ with energy $E$ and direction $\boldsymbol{\Omega}$

$F_D(E, \theta, x)$ depth distribution of deposited energy

$F(E, E_0)dE_0$ Recoil density = mean number of atoms recoiling into an energy interval $(E_0, dE_0)$; $F(E, E_0)$ $=d[n(E, E_0)]/dE_0$

$F_p(E, \boldsymbol{\Omega}, \boldsymbol{r})$ Mean momentum deposited in $(\boldsymbol{r}, d^3r)$ by an incoming ion of initial energy $E$ and direction $\boldsymbol{\Omega}$

$f(\boldsymbol{r},\boldsymbol{v},t)d^3rd^3v$ Distribution function, statistical average over the number of atoms in a volume element $(rd^3r)$ with velocities $(\boldsymbol{v},d^3v)$ at time $t$

$F_R(x,E,0)$ Penetration profile

$G(E,E_0)dE_0$ Mean number of atoms moving at any time with energy $(E_0,dE_0)$ if a source supplies $\psi$ particles per unit time; $G(E,E_0)dE_0 = \psi\, n(E,E_0)dt_0$

$\Gamma$ Flux of residual gas atoms to a surface

$\Gamma_m, \lambda_m$ Dimensionless functions of $m$ for power potentials.

$\gamma$ Energy transfer factor; $\gamma=4M_1 M_2(M_1 + M_2)^{-2}$.

Also used for Axial ratio $\gamma=c/a$ in hcp crystals

$\gamma_i$ Trapping coefficient

$\gamma_{i,s}$ Sticking probability of a species $i$ on a surface $s$

$\gamma_N^{-1}$ Distance from a surface over which the level width decreases to $1/2.781$ of its bulk value

$\gamma(\vartheta)$ Energy reflection coefficient, sputtering efficiency

$(hkl)$ Normal to a crystal plane

$\hbar$ Planck constant; $\hbar=1.05459\cdot10^{-34}$ Js

$I$ Projectile flux

$k$ Parameter in electronic stopping function

$K(\boldsymbol{v},\boldsymbol{v}_1 ; \boldsymbol{v}',\boldsymbol{v}'')d^3v'd^3v''$ Differential cross section for scattering a projectile from $\boldsymbol{v}$ to $\boldsymbol{v}'d^3v'$ and the target atom from $\boldsymbol{v}_1$ to $\boldsymbol{v}''d^3v''$

$\kappa$ Number of atoms in the elementary interval along a row

$\Lambda$ Material constant entering into the sputtering yield

$\Lambda(\vartheta)$ Focusing parameter in collision sequences

$M_1$ Incident ion mass

$M_2$ Target mass

$m$ Atom potential parameter for power potentials.

$\mu$ $\cos\vartheta$

$N$ Density of atoms

$N_0$ Avogadros number; $N_0=6.022\ 10^{23}$ atoms per mol

$n(E,E_0)$ Mean number of atoms set in motion with an initial energy greater than $E_0$ in a cascade initiated by a primary atom with energy $E$

$n_i$ — Number of bombarding projectile atoms

$v(E)$ — Average amount of energy transferred to atomic nuclei during the entire slowing down process of energetic particles starting with energy $E$ in a solid; $\eta(E) + \gamma(E) = E$

$\Omega$ — Directional vector

$\Omega(E)$ — Cascade volume

$P(E_0, \theta_0)$ — Probability for an atom with energy $E_0$ moving at an angle $\theta_0$ to escape from the surface

$P_a$ — Probability for an excited state $a$

$p$ — Impact parameter

$p_c$ — Critical approach distance at which individual atomic rows may be isolated

$p_m$ — Minimum impact parameter for ions in respect to a lattice row in channeling

$\Phi$ — Evaporation rate

$\phi'$ — Laboratory scattering angle for an ion in a binary collision with a target atom in Chap. 2 ($\theta$ in Chap. 3)

$\phi, \phi''$ — Laboratory scattering angle of recoil target atom

$\psi$ — Angle of an ion trajectory relative to a close packed lattice row

$\psi_1, \psi_2$ — Critical angle for channeling at high and low energies

$\psi_w$ — Angular half width for channeling

$\psi_\perp$ — Limiting angle for channeling in the transparency model

$R$ — Apsis of a collision (in Chap. 2 also used for distance of colliding atoms)

$R_f$ — Range of a focusing collision sequence

$R_0$ — Ion reflection coefficient (Also used for Apsis of a head on collision in Chap. 3)

$R(E)$ — Total path length for energetic ions in a solid

$R_p(E)$ — Projected range for energetic ions in a solid

$r$ — Distance of colliding atoms

$\mathbf{r}$ — Vector distance of an ion in a solid from the entering point

$\varrho$ — Spike radius

$\varrho_{uvw}$ — Density of atomic rows in a plane perpendicular to the rows

$S_e$ — Electronic stopping cross section

$S_n$ — Nuclear stopping cross section

$s_n(\varepsilon)$ — Reduced nuclear stopping cross section

$\sigma$ — Cross section

$\mathcal{T}$ — Temperature

$T$ — Transferred energy

$t_{uvw}$ — Unit translation along an axis

$\tau$ — Time integral for a binary collision. (also used as time constant for a spike in Chap. 2)

| | | | |
|---|---|---|---|
| $\vartheta, \theta$ | Angle of incidence relative to the surface normal ($\vartheta$ is also used for the laboratory scattering angle of an ion in a binary collision in Chap. 3) | $u_1^2$ | Mean square amplitude of one dimensional thermal vibrations |
| | | $v$ | Velocity of a moving atom |
| $\Theta$ | Barycentric scattering angle | $V(R), V(r)$ | Two body interaction potential |
| $\Theta$ | Average energy per atom in a spike | $x$ | Length, thickness, distance from a row or from a plane |
| $U_0$ | Cohensive energy ($\simeq$ latent heat of sublimation corrected to absolute zero) mostly taken as the average binding energy of surface atoms for structureless matter | $Y$ | Sputtering yield for amorphous and polycrystalline material |
| | | $Y_{(uvw)}$ | Sputtering yield for single crystalline material |
| | | $Y_i$ | Partial sputtering yield for multicomponent materials |
| $U_s$ | Surface binding energy | $\partial Y/\partial E_1, \partial^2 Y/\partial^2 \Omega_1$ | Differential sputtering yields |
| $U_{uvw}(x)$ | Average potential of a row or plane | | |
| $[uvw]$ | Direction of an axis | $Z_1, Z_2$ | Charge numbers of incident and target atoms |

# Author Index

In this index the numbers in brackets refer to the relevant references

Abe, T. 152, 181 [4.66]
Adams, C.T. 158 [4.195]
Adams, D. 76 [3.25]
Agranovich, V.M. 105, 109, 115, 116, 118, 128, 161 [3.120, 3.150, 4.245]
Aizentson, A.I. 152, 160 [4.64]
Akaishi, K. 152, 176, 180, 201 [4.63, 4.349]
Akimune, H. 155 [4.113]
Allan, G. 80 [3.52]
Alldred, J.C. 200 [4.388]
Almén, O. 2, 4, 13, 15, 34, 74, 75, 91, 92, 147, 149, 152, 164–168, 170–173, 175, 176, 179–186, 188, 189, 196, 200, 201, 206, 219 [1.18, 2.26, 3.8, 4.16, 4.33, 4.310, 4.405, 5.3]
Anderman, A. 120 [3.192]
Andersen, C.A. 58 [2.131]
Andersen, H.H. 7, 15, 16, 34, 35, 47, 49, 50, 54, 55, 57, 60, 63, 98, 105, 121, 128, 146, 147, 149–151, 153, 155–158, 160, 161, 163–165, 169, 176, 182, 187, 188, 191–193, 203, 204, 206–208 [1.33, 2.52, 2.73, 2.96, 2.101, 2.143, 2.150, 2.164, 3.108, 3.109, 3.122, 4.7, 4.12, 4.34, 4.35, 4.70, 4.71, 4.72, 4.73, 4.74, 4.79, 4.83, 4.88, 4.161, 4.193, 4.227, 4.270, 4.296, 4.400, 4.403, 4.406, 4.407]
Andersen, N. 20, 39, 60–63 [2.77, 2.148]
Anderson, G.S. 78, 112, 151, 154, 156, 163, 232, 245, 246 [3.45, 3.46, 4.41, 4.93, 4.273, 4.274, 5.55, 5.77, 5.78, 5.79]
Andersson, S. 76 [3.27]
Appelbaum, J.A. 80 [3.51]

Appleton, B.R. 76 [3.29]
Arifov, U.A. 185 [4.362]
Ariyasu, R.G. 124, 125, 132 [3.199]
Arminen, E. 152, 174, 175, 183 [4.46, 4.47, 4.48]
Ashton, D.H. 116, 117, 125, 132 [3.165]
Askerov, G.Sh. 175, 176, 180, 184, 185, 196 [4.340]
Ato, Y. 164 [4.302]
Augustyniak, W.M. 18, 153, 169, 175, 186, 188 [2.55, 4.86]
Austin, L.W. 12 [2.12]
Averback, R.S. 49, 96, 205 [2.100, 3.103, 4.396]
Ayrault, G. 49, 96, 205 [2.100, 3.103, 4.396, 4.397]
Ayukhanov, A.Kh. 185 [4.362]

Bach, H. 155 [4.107, 4.108, 4.109, 4.110]
Bader, M. 245 [5.75]
Baglin, J.E.E. 154 [4.102]
Balarin, M. 94, 233 [3.92, 5.59]
Ball, G.C. 14 [2.38]
Barrett, J.H. 76, 92, 95, 127, 129 [3.29, 3.90, 3.96, 3.229]
Baruah, J.N. 152 [4.54]
Bastings, L.C. 161 [4.240]
Basu, D. 239 [5.65]
Bates, J. 161, 187, 188 [4.239]
Bay, H.L. 15, 16, 35, 49–51, 54, 55, 57, 146, 147, 149–153, 155, 158–160, 164–166, 169–171, 173, 175, 176, 181, 182, 185, 187, 188, 191–194, 196–199, 202–204 [2.52, 2.101, 2.105, 2.164, 4.7, 4.12, 4.13, 4.34, 4.35, 4.62,

4.70, 4.71, 4.72, 4.73, 4.187, 4.206, 4.220, 4.305, 4.325, 4.337, 4.339, 4.361, 4.374, 4.391]
Beeler, J.R., Jr. 115, 116, 119, 126–128 [3.146, 3.160, 3.161, 3.207, 3.208, 3.209, 3.210, 3.211, 3.212]
Beeler, M.F. 116, 119, 126, 128 [3.160, 3.161]
Behrisch, R. 1–7, 10–12, 18, 37, 43, 49, 51, 52, 57, 58, 65, 74, 127, 145–147, 149, 150, 152, 153, 156, 160, 165, 176, 180, 182, 194, 196, 205, 221, 247, 251 [1.5, 1.6, 1.19, 1.24, 1.30, 2.4, 2.7, 2.102, 2.104, 2.109, 3.1a, 3.221, 4.1a, 4.3, 4.29, 4.36, 4.37, 4.50, 4.57a, 4.58, 4.61, 4.84, 4.133, 4.221, 4.378, 5.9]
Beitat, U. 148 [4.19]
Bellina, J.J. 77 [3.39]
Benazeth-Colombie, N. 169, 232, 245, 246 [4.320, 5.56]
Benes, E. 158 [4.183]
Benninghoven, A. 7, 156, 157, 167, 182, 183, 188 [1.35, 4.130, 4.155]
Berg, J. 165 [4.291]
Berliner, A. 12 [2.16]
Bernas, H. 55 [2.115]
Bernhardt, F. 158 [4.175]
Besco, D.G. 127 [3.207, 3.208, 3.209, 3.210]
Besenbacher, F. 160 [4.227]
Bespalova, N.S. 119 [3.186]
Bethe, H.A. 24 [2.61]
Betz, G. 7, 45, 63, 65 [1.36, 2.88, 2.152]
Bhattacharya, R.S. 63, 146, 182, 188, 222, 239, 245 [2.153, 4.11, 4.357, 5.43, 5.64, 5.65]

# Subject Index

# Inelastic Particle–Surface Collisions

Proceedings of the Third International Workshop on Inelastic Ion-Surface Collisions Feldkirchen-Westerham, Fed. Rep. of Germany, September 17–19, 1980
Editors: E. Taglauer, W. Heiland
1981. 194 figures. VIII, 329 pages
(Springer Series in Chemical Physics, Volume 17)
ISBN 3-540-10898-X

**Contents:** Electron Emission. – Electron and Photon Impact. – Electron Transfer. – Polarized Light Emission. – Excited Particle Emission. – Index of Contributors.

# Secondary Ion Mass Spectrometry SIMS-II

Proceedings of the Second International Conference on Secondary Ion Mass Spectrometry (SIMS II) Stanford University, Stanford, California, USA, August 27–31, 1979
Editors: A. Benninghoven, C. A. Evans, Jr., R. A. Powell, R. Shimizu, H. A. Storms
1979. 234 figures, 21 tables. XIII, 298 pages
(Springer Series in Chemical Physics, Volume 9)
ISBN 3-540-09843-7

**Contents:** Fundamentals. – Quantitation. – Semiconductors. – Static SIMS. – Metallurgy. – Instrumentation. – Geology. – Panel Discussion. – Biology. – Combined Techniques. – Postdeadline Papers.

Springer-Verlag
Berlin
Heidelberg
New York

M. A. Van Hove, S. Y. Tong

# Surface Crystallography by LEED

**Theory, Computation and Structural Results**

1979. 19 figures, 2 tables. IX, 286 pages
(Springer Series in Chemical Physics, Volume 2)
ISBN 3-540-09194-7

**Contents:** Introduction. – The Physics of LEED. – Basic Aspects of the Programs. – Symmetry and Its Use. – Calculation of Diffraction Matrices for Single Bravais-Lattice Layers. – The Combined Space Method for Composite Layers: by Matrix Inversion. – The Combined Space Method for Composite Layers: by Reverse Scattering Perturbation. Stacking Layers by Layer Doubling. – Stacking Layers by Renormalized Forward Scattering (RFS) Perturbation. – Assembling a Program: The Main Program and the Input. – Subroutine Listings. – Structural Results of LEED Crystallography. – Appendices. – References. – Subject Index.

# Vibrational Spectroscopy of Adsorbates

Editor: R. F. Willis
1980. 97 figures, 8 tables. XII, 184 pages
(Springer Series in Chemical Physics, Volume 15)
ISBN 3-540-10429-1

**Contents:** Introduction. – Theory of Dipole Electron Scattering from Adsorbates. – Angle and Energy Dependent Electron Impact Vibrational Excitation of Adsorbates. – Adsorbate-Induced Optical Phonons. – Inelastic Electron Tunnelling Spectroscopy. – Inelastic Molecular Beam Scattering from Surfaces. – Neutron Scattering Studies. – Reflection Absorption Infrared Spectroscopy: Application to Carbon Monoxide on Copper. – Raman Spectroscopy of Adsorbates at Metal Surfaces. – Vibrations of Monatomic and Diatomic Ligands in Metal Clusters and Complexes. – Analogies with Vibrations of Adsorbed Species on Metals. – Coupling Induced Vibrational Frequency Shifts and Island Size Determination: CO on Pt $\{001\}$ and Pt $\{111\}$.

# Electron Spectroscopy for Surface Analysis

Editor: H. Ibach
1977. 123 figures, 5 tables. XI, 255 pages
(Topics in Current Physics, Volume 4)
ISBN 3-540-08078-3

**Contents:** *H. Ibach:* Introduction. – *D. Roy, J. D. Carette:* Design of Electron Spectrometers for Surface Analysis. – *J. Kirschner:* Electron-Excited Core Level Spectroscopies. – *M. Henzler:* Electron Diffraction and Surface Defect Structure. – *B. Feuerbacher, B. Fitton:* Photoemission Spectroscopy. – *H. Froitzheim:* Electron Energy Loss Spectroscopy.

# Inelastic Electron Tunneling Spectroscopy

Proceedings of the International Conference, and Symposium on Electron Tunneling, University of Missouri-Columbia, USA, May 25–27, 1977
Editor: T. Wolfram
1978. 126 figures, 7 tables. VIII, 242 pages
(Springer Series in Solid-State Sciences, Volume 4)
ISBN 3-540-08691-9

**Contents:** Review of Inelastic Electron Tunneling. – Applications of Inelastic Electron Tunneling. – Theoretical Aspects of Electron Tunneling. – Discussions and Comments. – Molecular Adsorption on Non-Metallic Surfaces. – New Applications of IETS. – Elastic Tunneling.

H. Raether

# Excitation of Plasmons and Interband Transitions by Electrons

1980. 121 figures, 17 tables. VIII, 196 pages
(Springer Tracts in Modern Physics, Volume 88)
ISBN 3-540-09677-9

**Contents:** Introduction. – Volume Plasmons. – The Dielectric Function and the Loss Function of Bound Electrons. – Excitation of Volume Plasmons. – The Energy Loss Spectrum of Electrons and the Loss Function. – Experimental Results. – The Loss Width. – The Wave Vector Dependency of the Energy of the Volume Plasmon. – Core Excitations Application to Microanalysis. – Energy Losses by Excitation of Cerenkov Radiation and Guided Light Modes. – Surface Excitations. – Different Electron Energy Loss Spectrometers. – Notes Added in Proof. – References. – Subject Index.

# Theory of Chemisorption

Editor: J. R. Smith
1980. 116 figures, 8 tables. XI, 240 pages
(Topics in Current Physics, Volume 19)
ISBN 3-540-09891-7

**Contents:** *J. R. Smith:* Introduction. – *S. C. Ying:* Density Functional Theory of Chemisorption of Simple Metals. – *J. A. Appelbaum, D. R. Hamann:* Chemisorption on Semiconductor Surfaces. – *F. J. Arlinghaus, J. G. Gay, J. R. Smith:* Chemisorption on d-Band Metals. – *B. Kunz:* Cluster Chemisorption. – *T. Wolfram, S. Ellialtioğly:* Concepts of Surface States and Chemisorption on d-Band Perovskites. – *T. L. Einstein, J. A. Hertz, J. R. Schrieffer:* Theoretical Issues in Chemisorption.

# Springer-Verlag Berlin Heidelberg New York